THE REQUIREMENT FOR AN INVENTION IN PATENT LAW

The Requirement for an Invention in Patent Law

JUSTINE PILA

OXFORD
UNIVERSITY PRESS

OXFORD
UNIVERSITY PRESS

Great Clarendon Street, Oxford, OX2 6DP,
United Kingdom

Oxford University Press is a department of the University of Oxford.
It furthers the University's objective of excellence in research, scholarship,
and education by publishing worldwide. Oxford is a registered trade mark of
Oxford University Press in the UK and in certain other countries

First published 2010

Published in the United States of America by Oxford University Press
198 Madison Avenue, New York, NY 10016, United States of America

British Library Cataloguing in Publication Data
Data available

Library of Congress Cataloging in Publication Data
Data available

ISBN 978–0–19–929694–1

For Jonathan

Preface

Properly construed, the requirement for an invention in patent law sets the boundaries of the patent system in two ways. The first is by defining the categories of subject matter capable of supporting a patent, and the second is by restricting the protection conferred by a patent to subject matter conceived *qua* invention. In serving these functions, the requirement for an invention helps to fulfil the public benefit objectives of the patent system by mediating the balance struck by patents between individual patentees and the public.

The purpose of this book is to study the requirement for an invention in United Kingdom patent law in its historical and international statutory context. My hope is that those who read it will emerge with a better understanding of legal conceptions of the invention, the issues which the law's construction of the requirement for an invention raise, and the historical development of the UK patent system from before the first English patent legislation of 1623 through to the system's recent phase of Europeanization.

A central theme of the book is that the requirement for an invention cannot be separated out from the other requirements of patentability; and indeed, it has often been when determining novelty, inventive step, and industrial applicability that the courts have engaged implicitly or otherwise with the conception of subject matter *qua* invention. Of particular importance in this regard has been the requirement that patent claims be sufficiently supported in the specification, and it is for this reason that my study takes in that requirement.

This book is an adaptation of my University of Melbourne PhD thesis, parts of which were adapted for publication as stand-alone articles either during or shortly after completing my doctorate.[1] In revising my thesis for publication as a book I have therefore both drawn on and revised substantially those articles, in reflection of the further work and thinking I have done since writing them, and as part of the

[1] See J Pila, 'The Common Law Invention in its Original Form' [2001] IPQ 209–243; J Pila, 'Methods of Medical Treatment Within Australian and United Kingdom Patents Law' (2001) 24 UNSWLJ 421–462; J Pila, 'Inherent Patentability in Anglo-Australian Law: A History' (2003) 14 AIPJ 109–166; J Pila, 'Article 52(2) of the Convention on the Grant of European Patents: What Did the Framers Intend? A Study of the *Travaux Préparatoires*' (2005) 36 IIC 755–778; J Pila, 'Dispute Over the Meaning of "Invention" in Article 52(2) EPC: The Patentability of Computer-Implemented Inventions in Europe' (2005) 36 IIC 173–191. The book also includes material adapted from J Pila, 'Chemical Products and Proportionate Patents Before and After *Generics v Lundbeck*' (2009) 20 KLJ 489–526, and (with the permission of Hart Publishing) J Pila, 'Chemical Product Patents and *Biogen* Insufficiency Before the House of Lords' (2009) 125 LQR 573–578. J Pila, 'Inherent Patentability in Anglo-Australian Law: A History' (2003) 14 AIPJ 109–166 is published by, and available for separate purchase from, Thomson Reuters. For all subscription inquiries please phone, from Australia: 1300 304 195, from Overseas: +61 2 8587 7980, or go online at <http://www.thomsonreuters.com.au/catalogue>.

process of realizing my original study in the form in which it was intended to be realized, namely, as a history of the law of inherent patentability. At the time of going to press, that law is under review at the highest levels of three jurisdictions: by the US Supreme Court, the EPO's Enlarged Board of Appeal, and the Australian Government.[2]

A lot of the doctoral research from which this book is derived was undertaken in the idyllic setting of the Institute for Advanced Study, in Princeton, during a period when my husband was a visitor there. Its publication is a welcome reminder of that time of my life, which fell soon after the death of my father, the birth of my eldest daughter, and my husband's decision to return to academia. Without the luxury and pleasure of that time with my family, and the substantial support of the IAS, I cannot imagine my doctorate (nor consequently this book) having been written.

There are many others to whom I owe a substantial debt, and I am grateful for the opportunity to acknowledge some of them here. My PhD supervisor, Sam Ricketson, provided essential support before, during, and after my graduate studies, while Jane Ginsburg made it possible for me to finish writing my thesis in the United States. The Australian Government's Department of Foreign Affairs and Trade, and the University of Melbourne, provided the generous scholarships that enabled me to undertake a doctorate. The DG 5 Law Library of the European Patent Office offered critical assistance with my research; I am especially grateful to Gabriele Schaefer.

Some time has passed since I finished my thesis. In accounting for my delay in completing this book it is tempting to blame the interventions of life, which for me have included two international moves, the birth of my second daughter, and taking up my first academic post. There is also, however, another reason, which is that until now I have not been ready to complete it. For its preparedness to wait I am grateful to OUP, including especially my two editors, Chris Rycroft and Luke Adams.

Oxford has been an enormously stimulating place for me to think and write about the things I regard as important, and I am indebted to the students and colleagues who have made it thus. From my time at Oxford, I am particularly indebted to my former Catz and IP Law colleague, Michael Spence. I am also grateful to the staff in the Law Bodleian Library for research assistance, to the Oxford Law Faculty for agreeing to my eleventh hour request for support in the last two weeks of work

[2] With respect to the US Supreme Court's review see <http://www.supremecourtus.gov/orders/courtorders/060109zor.pdf> (Order of the US Supreme Court granting certiorari in *Bilski v Kappos* (Case No 08-964) following the decision in *Re Bilski* 545 F 3d 943 (Fed Cir, 2008); <http://www.supremecourtus.gov/oral_arguments/argument_transcripts/08-964.pdf> (transcript of the oral arguments presented to the Court on 9 November 2009). With respect to the EBA's review see G_3/08 (*PRESIDENT'S REFERENCE/Computer Program Exclusion*) [2009] EPOR 9. With respect to the Australian Government's review see <http://www.acip.gov.au/reviews.html#%23subject> ('Review of Patentable Subject Matter') and further links.

on my manuscript, and to Graduate Research Assistant, Michael Bolding, for his careful reading of the manuscript during that period. For their own care in the production of this book I thank the OUP production team, including especially my Production Editor, Lucy Page.

This leaves my greatest debts, which are to my family. My mother, Faith, was a formidable early role model, and I am grateful for her unstinting encouragement and support over the years. I am also grateful to my father, John, for his close friendship in the period before his death, after many years of us living in different cities and countries. While my academic leanings were not something he particularly embraced, he was a person of great intellectual and personal strength, and has long been a source of inspiration for me.

Three people above all have sustained me, and to them I give special thanks. They are my husband, Jonathan, and daughters, Dominique and Davita. Despite their young ages, Dominique and Davita have made an impressive attempt to engage with my project. Jonathan has engaged with and supported it from its inception, and for that and many other reasons I dedicate this book to him.

Justine Pila
Oxford
August 2009

Contents

Table of Cases xvii
Table of Statutes and Official Papers xxvii
List of Abbreviations xxxix

Introduction 1

A. The Requirement for an Invention and the Concept of Inherent Patentability 1
B. The Term 'Invention' 2
C. The Origins of the Contemporary (EPC) Definition of the 'Invention' 5
D. The Uncertainty and Controversy Surrounding the Requirement for an Invention 5
E. The Need for a Conception and Ontology of the Invention . . . 7
F. . . . Formulated in Light of the Purpose of the System, the Meaning of Europeanization, and Insights from the History and Philosophy of Science and Technology 8

I THE REQUIREMENT FOR AN INVENTION IN HISTORICAL PERSPECTIVE

1. The Origins of Patents for Inventions, and Early Conceptions of the Invention: Pre-1623 13

A. Introduction 13
B. The Royal Custom of Encouraging the Establishment of New Industries 14
C. The Elizabethan Policy of Granting Monopolies to Encourage 'Others to Study and Seke for the Knowledge of . . . Good Engines and Devyses' and Other 'Things Useful for the Public' 16
D. The Abuse of the Elizabethan Policy and Devolution of Patent Jurisdiction to the Common Law Courts 18
E. Section 6 of the Statute of Monopolies: The Common Law Invention in its Original Form 21
F. Conclusion 24

2. The Requirement for an Invention in the Pre-Modern Era of Mechanical and Chemical Manufacture: 1623 to 1882 25

A. Introduction 25

B. 1623 to 1794: The Post-Specification Invention as an Idea or Discovery 26

C. 1795 to 1829: The Invention in and after *Boulton v Bull* 29

1. The Invention as a Manufacture 29
2. The Requirement for an Accurate and Complete Representation of the Invention 41

D. 1830 to 1851: The (Patentable) Invention as a Mechanically Inventive Subject Matter 43

1. The Invention as a Principle 'Turned to Account' 43
2. The Exclusion of Manifestly Non-Inventive (Non-Ingenious) Subject Matter: Analogous Uses and *Losh v Hague* 45

E. 1852 to 1882: The Basis of *Losh v Hague* in the Requirement for an 'Invention' the Patenting of which was not 'Generally Inconvenient' 47

F. Conclusion 68

3. The Requirement for an Invention in the Modern Era of 'New' Technologies: 1883 to 1976 71

A. Introduction 71

B. 1883 to 1959: The Invention as a Manner of *New* Manufacture, Including an *Alleged* Invention 72

1. Literal Interpretations of the Statutory Definition of the 'Invention' 72
2. The Requirement for 'Invention with Reference to a Manufacture': *Cooper's Application* 73
   (a) The Exclusion of Subject Matter not Resulting in a (Vendible) Product 76
      (i) Methods of Medical Treatment 78
      (ii) Ephemeral and Biological Subject Matter 79
3. The Uncertain Position of (Chemical) Products 82
   (a) 1883 to 1918: Products as Manners of New Manufacture 82
   (b) 1919 to 1949: The Statutory Exclusion of Chemical Products and Fair Basis/*Mullard Radio* Solutions to the Problem of Obstructive Claims 84
   (c) 1949 to 1959: The (Inherent) Patentability of Chemical Products 87
4. The Exclusion of Immoral Subject Matter 88
5. Conclusion 90

C. Post-1959: The Invention as a 'Proper Subject of Letters Patent According to the Principles . . . Developed for the Application of s 6 of the Statute of Monopolies' 90

1. The Invention as a Human Action on the Physical World Producing an Artificial End of Practical Utility: *National Research Development Co v Commissioner of Patents* 90

2. Biotechnological and Chemical Subject Matter, Including Products 95
3. Revisiting the Exclusion of Surgical and Therapeutic Methods 98
4. Revisiting the Exclusion of (Other) Non-Manufacturing Methods 101
(a) Information-Related Subject Matter, Including Computer Programs 101
(b) Methods of Presentation 103
(c) 'Schemes', Including Business Methods 104
D. Conclusion 104

II THE REQUIREMENT FOR AN INVENTION IN THE ERA OF THE EUROPEAN PATENT CONVENTION

4. The EPC Requirement for an Invention in its Interpretive Context 109
A. Introduction 109
B. The EPC Requirement for an Invention in its International and European Statutory Context 112
1. International and European Patent Law-Making: An Overview 112
2. The Paris Convention 1883 (WIPO): A Mechanism for Creating 'Special Agreements for the Protection of Industrial Property' 115
3. The Strasbourg Convention 1963 (CofE): A Convention to Facilitate the Harmonization of Substantive Patent Law 115
4. The Patent Cooperation Treaty 1970 (WIPO): A Record of Patentability Exclusions 117
5. The Community Patent Convention 1975 (EEC): An Attempt to Create a Unitary Patent 121
6. The European Patent Convention 1973 (Munich Conference): A System of Law for the Grant of European Patents 122
7. The TRIPS Agreement 1994 (WTO): The Principle of Patents for Inventions in All Fields of Technology 124
8. Ongoing Attempts to Create a Community Patent 124
(a) EEC Initiatives 124
(b) EU (and European Patent Organisation) Initiatives 125
9. Conclusion 126
C. The European Drafters' Intent Regarding the Requirement for an Invention 126
1. The European Committee of Experts on Patents, 1951 to 1963 129
2. The EEC Patents Working Party, 1961 to 1965 141
3. The Luxembourg and Munich Conferences, 1969 to 1973 145
(a) Article 52(2) and (3) in Draft: Article 9(2) 147
(b) The Contents of and Relationship Between Articles 9(2) and 10(b) 148
(c) The Arrangement of the Exclusions 149
(d) Scientific and Mathematical Theories 149

(e) Natural Products and Information 150
(f) The Restriction of the Exclusions by Use of the Phrase 'As Such' 151
(g) Therapeutic Methods 152
(h) Computer Programs 153
(i) Substances for Use in Therapeutic Methods 155
(j) The Application of the Therapeutic Methods Exclusion to Animals 155
4. Conclusion 159
(a) The Intent Behind Article 52(2) and (3) of the EPC 159
(b) The Meaning of Inherent Patentability at the Time of the EPC's Conclusion 162
D. The European Drafters' Intent Regarding Chemical Product Patents and the Requirement for Equiparating Claims 164
1. The Inherent Patentability of Chemical Products 164
(a) The European Committee of Experts on Patents, 1951 to 1963 164
(b) The Luxembourg and Munich Conferences, 1969 to 1973 168
(c) Conclusion 171
2. The Requirement for Proportionate Claims 172
3. Conclusion 174
E. Other European Perspectives on Chemical Product Patents: A Question of Non-Discrimination 174
F. The Intention Behind Section 1(2) of the UK Patents Act 179
1. The Patents Act in its Legislative Context 179
2. The PLG Report, 1965 180
3. The Banks Report, 1970 182
4. The White Paper, 1975 186
5. Conclusion 189
G. Conclusion 191

5. Article 52(2) and (3) of the EPC According to the European Patent Office 195
A. Introduction 195
B. Different Approaches for Different Categories of Subject Matter: 1977 to 1987 196
1. Inherent Patentability in Light of the Business Methods Exclusion 197
(a) The Exclusion of Abstract Methods Producing an Immaterial Good: *CHRISTIAN FRANCERIES/Traffic Regulation* 197
(b) The Exclusion of Subject Matter not Described as Requiring Technical Means: *STOCKBURGER/Coded Distinctive Mark* 198
2. Inherent Patentability in Light of the Computer Programs Exclusion 199
(a) The Patentability of Computer Programs as Technical Methods for Producing Physical Entities: *VICOM/Computer-Related Invention* 199

(b) The Patentability of Computer Programs as Part of a Technical Apparatus: *KOCH & STERZEL/X-Ray Apparatus* 203
(c) The Patentability of Computer Programs as Solutions to a Technical Problem: *IBM/Computer-Related Invention* 205
3. Inherent Patentability in Light of the Aesthetic Works Exclusion 206
4. Conclusion 207
C. Article 52(2) and (3) of the EPC as a Requirement for Technical Character: 1988 to 1999 210
1. The Whole Contents and Contribution Theories of Article 52(2) and (3) 211
2. Inconsistent Conceptions of 'Technical Character' in the Context of Automated Transactional Schemes 215
3. Biotechnology as a Threat to Legalistic Conceptions of the Requirement for an Invention 220
D. Inherent Patentability in an Era of Harmonization: 1999 to Present 224
1. The EC Biotech Directive 1998 224
2. The TRIPS Agreement 1994 228
(a) Revisiting the Computer Programs Exclusion in Light of Harmonization: *IBM/Computer Programs* 228
(b) Revising Article 52 in Light of the TRIPS Agreement 231
3. Technical Information and Automated Transactional Schemes after *IBM/Computer Programs* 233
(a) Presentations of Information 234
(b) Automated Business Methods 235
(i) *PBS PARTNERSHIP/Controlling Pension Benefits System* 235
(ii) *PITNEY BOWES/A System for Processing Mail and HITACHI/Auction Method* 236
(iii) *DUNS LICENSING ASSOCIATES/Estimating Sales Activity* 238
a. EC Software Patenting Directive Proposal 241
b. 'Classical Notions of Invention' 244
E. Conclusion 246

6. Inherent Patentability in UK Law Since 1977 251
A. Introduction 251
B. Different Approaches for Different Categories of Subject Matter: 1977 to 1989 253
1. Inherent Patentability in Light of the Business Methods Exclusion 254
2. Inherent Patentability in Light of the Discoveries Exclusion 258
3. Reconciling *Genentech Inc's Patent* with Section 1(2) of the Patents Act 261
4. Conclusion 264

C. Inherent Patentability as a Fact-Specific Enquiry: Computer-Related Subject Matter after *Merrill Lynch Ltd's Application* 265
1. Conflicting Approaches to Computer Products: *Gale's Application* in the Court of Appeal 266
2. Fact-Specificity as a Mask for Legal Uncertainty 271
(a) Inherent Patentability as a Fact-Specific Enquiry 272
(b) Fact-Specificity as a Mask for Legal Uncertainty 275
(i) *Lux Traffic Controls Ltd v Pike Signals Ltd* 275
(ii) *Raytheon Co's Application* 277
(iii) *Fujitsu Ltd's Application* 280
(iv) Conclusion 287
D. Inherent Patentability as a Requirement for Subject Matter Sufficiently Supported *Qua* Invention: Biotechnology after *Genentech Inc's Patent* 294
1. Conflicting Approaches to Biotechnology: *Biogen* and *Chiron* in the Court of Appeal 294
2. Restricting Protection to Subject Matter Sufficiently Supported *Qua* Invention: *Biogen v Medeva* and *Kirin-Amgen v Hoechst Marion* in the House of Lords 298
3. Chemical Products *Qua* Invention: *Generics v Lundbeck* in the House of Lords 303
4. Conclusion 308
E. Revisiting the Computer Programs Exclusion in Light of Harmonization: 2005 to Present 308
1. Inherent Patentability as a Requirement for a (Potentially) Patentable *Invention*: *CFPH* and *Halliburton v Smith* 312
2. Requesting an Enlarged Board Decision on the EPC Requirement for an Invention: *Aerotel v Telco* in the Court of Appeal 314
3. Retreating from *Aerotel v Telco* (and *CFPH* and *Halliburton v Smith*) in the Interests of Harmonization: *Symbian v Comptroller-General* in the Court of Appeal 318
4. Conclusion 321

Conclusion 323
A. The Requirement for an Invention in Historical Context 323
B. Legal Conceptions of the Invention Prior to the EPC 1973 325
1. The Invention at the Time of the Statute of Monopolies 1623 325
2. The Invention in the Pre-Modern Era of Chemical and Mechanical Manufacture 326
3. The Invention in the Modern Era of 'New' Technologies 328
C. Concluding Reflections on the Requirement for an Invention in Patent Law 331

Index 339

Table of Cases

A & H's Application (1927) 44 RPC 298 (SG) 89, 90, 336
A D Goldhaft's and Others' Application [1957] RPC 276 (SE) 81
A E W's Application (1924) 41 RPC 529 (SG) 74, 76, 103
A F's Application (1913) 31 RPC 58 (SG) 76
A L's Application (1924) 41 RPC 615 (SG) 76
A W Loth's Application (1924) 41 RPC 273 (SG) 75
Acetylene Illuminating Co v United Alkali Co Ltd (1904) 22 RPC 145 (HL) 72, 82, 105
Actavis UK Ltd v Merck & Co Inc [2008] EWCA Civ 444 .... 251, 319
Adhesive Dry Mounting Co Ltd v Trapp & Co (1910) 27 RPC 341 (Ch) 82
Aerotel Ltd v Telco Holdings [2006] EWCA Civ 1371 .... 5, 6, 110, 111, 211, 251, 287, 306, 310, 314, 315, 316, 317, 318, 319, 320, 322, 334
Agrevo/Triazole Herbicides T0939/92 [1996] EPOR 171 240
Akzo/Automatic Revocation T0026/88 [1990] OJ EPO 4 111
Alderton's Application (1941) 59 RPC 56 (PAT) 75
Allen v Rawson (1845) 1 CB 551, 135 ER 656 (CP) 38
American Chemical Paint Company's Application [1958] RPC 47 (PAT) 77, 80
American Cyanamid Company (Dann's) Patent [1971] RPC 425 (HL) 95, 99, 105, 222, 259, 260, 330, 331
American Cyanamid Company v Berk Pharmaceuticals Limited [1976] RPC 231 (HC) 259
American Optical Company's Application [1958] RPC 40 (PAT) 74, 75
Arkwright v Nightingale (1785) G 14, Web Pat Cas 60, 1 Carp Pat Cas 38, Dav Pat Cas 37 (CP) 27, 67
Arnold v Bradbury (1871) LR 6 Ch 706 (LC) 59
Asahi Kasei Kogyo KK's Application [1991] RPC 485 (HL) 251, 295
AT&T/Computer System T0204/93 [2001] EPOR 300 .... 211, 212, 217
AT&T Knowledge Ventures LP [2009] EWHC 343 .... 7, 9, 321
Autodesk Inc v Dyason (No 2) (1993) 176 CLR 300 (HCA) 332
Axien's Application (1935) 53 RPC 127 (PAT) 76

B A's Application (1915) 32 RPC 348 (SG) 75, 76, 91
Badger Company Incorporated's Application [1970] RPC 36 (PAT) 101, 102
Baigent & Leigh v Random House Group [2007] EWCA Civ 24 333
Bailey v Robertson (1878) G 26, LR 3 App Cas 1055 (HL) 63
Bainbridge v Wigley (1810) G 30, 1 Carp Pat Cas 270 (NP) 41, 45
Bamlett v Picksley (1875) 1 Gr 40 (ExCh) 49, 62, 67
BBC/Colour Television Signal T0163/85 [1990] EPOR 599 .... 212, 213, 214, 234
Beattie/Marker T0603/89 [1992] EPOR 221 .... 212, 214, 215, 219
Bewley v Hancock (1856) 43 ER 1285 (LC) 48
Bilski v Kappos (Case No 08-964) 249
Binney v Feldtman (1871–5) 1 Gr 49 (HL) 60
Bio-Digital Sciences Incorporated's Application [1973] RPC 668 (PAT) 99
Biogen Inc v Medeva Plc [1995] RPC 25 (CA); [1997] RPC 1 (HL) 1, 6, 240, 291, 294, 295, 297, 298, 299, 300, 301, 302, 303, 304, 305, 306, 307, 308, 311, 333
Birch v Harrap (1896) 13 RPC 615 (VC) 72

Bircot's Case (1573) Co 3rd Inst 181 (ExCh) 23, 325
Blanchard v Hill (1742) 2 Atk 484, 26 ER 692 (LC) 28
Booth v Kennard (1856) 156 ER 1311 (ExCh) 49
Bosch/Electronic Computer Components T0164/92 [1995] EPOR 585 .... 201, 212
Boulton v Bull (1795) G 70, 2 HB1 463, 126 ER 651, Dav Pat Cas 162, 1 Carp Pat Cas 117 (CP); (1796) 3 Ves Jr 140, 30 ER 937, 1 Ves Sup 345, 34 ER 820, 1 Carp Pat Cas 155 (LC) 3, 25, 28, 29, 31, 40, 42, 43, 45, 77, 78, 88, 92, 326, 331
Bovill v Keyworth (1857) 7 E & B 725, 119 ER 1415 (QB) 49
Bovill v Moore (1815) G 74, 2 Coop G 56, 47 ER 1048 (LC); (1816) 1 Carp Pat Cas 320 (NP) 41
Bovill v Pimm (1856) 156 ER 1019 (Ex) 49
Bovingdon's Application (1946) 64 RPC 20 (PAT) 77
Bramah v Hardcastle (1789) 1 Carp Pat Cas 168, Web Pat Cas 44n, 76n, 194n (NP) 28, 44
British Petroleum Coy Ld's Application for a Patent [1958] RPC 253 (PAT) 77, 79
British Thomson-Houston Company Ld v Corona Lamp Works Ld (1921) 39 RPC 49 (HL) 89, 99
British Thomson-Houston v Charlesworth (1925) 42 RPC 180 (HL) 85, 105
Brook v Aston (1857) 120 ER 178 (QB); 28 LJ QB 175 (ExCh) 47, 49, 67
Broselow/Measuring Tape for Directly Determining Physical Treatment and Physiological Values T0077/92 [1998] EPOR 266 .... 212, 213, 214, 235
Brunton v Hawkes (1820) G 97, 1 Carp Pat Cas 405 (NP); (1821) 4 B & Ald 541, 106 ER 1034, 1 Carp Pat Cas 410 (KB) 33, 42, 46, 66, 326
Burroughs Corporation (Perkins') Application [1974] RPC 147 (PAT) 101, 102, 331
Bush v Fox (1852) Macr 140 (NP); (1856) Macr 179, 5 HL Cas 707, 10 ER 1080 (HL) 39, 47, 48

C's Application (1919) 37 RPC 247, SG 74, 76
C & W's Application (1914) 31 RPC 235 (SG) 78, 81, 89, 90, 336
C G R's Application (1924) 42 RPC 321 (SG) 76
C M's Application (1944) 61 RPC 63 (PAT) 79
Calmic Engineering Company Limited's Application [1973] RPC 684 (PAT) 99
Calvert v Ashburn (1862) JPM 84 (ExCh) 53, 67
Canada – Patent Protection of Pharmaceutical Products, Complaint by the European Communities and their Member States; Report of the Panel (17 March 2000) WT/DS114/R 177
Canada – Patent Protection of Pharmaceutical Products, Request for the Establishment of a Panel by the European Communities (12 November 1998) WT/DS114/5 177
Cannington v Nuttall (1871) LR 5 HL 205 (HL) 47, 59
Canterbury Agricultural College for LP's Application [1958] RPC 85 (SE) 81
Cappelini & Bloomberg's Applications [2007] EWHC 476 318
Carpenter v Smith (1841) Web Pat Cas 530 (NP); (1842) Web Pat Cas 540 (Ex) 36, 46
Cartwright v Eamer (1800) G 112 (LC) 27, 41
CFPH's Application [2005] EWHC 1623 .... 6, 8, 110, 251, 308, 309, 310, 312, 315, 316, 318, 321, 322
Chiron Corporation v Murex Diagnostics Ltd; Chiron Corporation v Organon Teknika Ltd [1996] RPC 535 (CA) 291, 294, 295, 297, 298, 300
Christian Franceries/Traffic Regulation T0016/83 [1988] EPOR 65 .... 196, 198, 200, 201, 205, 206, 208, 209, 210, 214, 218, 219, 220, 254, 264, 275, 276, 277, 288, 289
Ciba-Geigy AG (Durr's) Applications [1977] RPC 83 (PAT, CA) 104

Ciba-Geigy/Propagating Material [1979–85] EPOR C758 221
Citibank/Integrated Account T0368/05 [2007] EPOR 54 240
Clark v Adie (1873) LR 10 Ch 667 (CA); (1877) LR 2 App C 315 (HL) 61, 67
Clothworkers of Ipswich Case (1615) 11 Co Rep 53a, 77 ER 1218 Godb 252, 78 ER 147, 1 Rolle 4, 81 ER 285, Show K B 266, 89 ER 988 (KB) 20, 30
Cochrane v Smethurst (1816) G 124, 2 Coop G 57, 47 ER 1048 (LC); (1816) 1 Stark 205, 171 ER 448 (NP) 41, 43, 45
Commercial Solvents Corp v Synthetic Products Company Ltd (1926) 43 RPC 185 (HC) 80
Compagnies Reunies des Glaces et Verres Speciaux Du Nord De La France's Application (1933) 48 RPC 185 (SG) 75
Commissioner of Patents v Microcell Ltd (1959) 102 CLR 232 (HCA) 75
Comvik/Two Identities T0641/00 [2004] EPOR 10 .... 240, 333
Conor Medsystems Inc v Angiotech Pharmaceuticals Inc [2008] UKHL 49 .... 320, 321
Cooper's Application (1901) 19 RPC 53 (AG) 73, 75, 76, 77, 104, 316, 329
Cornish v Keene (1835) Web Pat Cas 497 (NP); (1837) Web Pat Cas 513, 3 Bing NC 570, 132 ER 530 (CP) 34
Crane v Price (1842) Web Pat Cas 393 (CP) 36, 46, 52, 58, 59, 61, 62, 66, 92, 327
Crawford's Application [2005] EWHC 2417 314
Cropper v Smith (1884) 1 RPC 81 (CA) 73
Crossley v Potter (1853) Macr 239 (NP) 40, 48, 328
Curtis v Platt (1866) 35 LJ Ch 852 (HL) 47, 55, 66

D A and K's Application (1925) 43 RPC 154 (SG) 74, 76, 102
D Company's Application (1921) 38 RPC 897 (SG) 76
Dangerfield v Jones (1866) 13 LT 142 (VC) 55
Darcy v Allin (Case of Monopolies) (1599) Noy 173, 74 ER 1131 (KB); (1601) 11 Co Rep 84(b), Web Pat Cas 1, 1 Carp Pat Cas 26; 77 ER 1260 (KB) 19, 30, 325
Data Access v Powerflex Services (1999) 202 CLR 1 (HCA) 332
Designers Guild Ltd v Russell Williams (Textiles) Ltd [2000] UKHL 58 114
Dixon's Application [1978] RPC 687 (PAT) 104
Dolland v Champneys (1766) G 175, Web Pat Cas 42, 1 Carp Pat Cas 28, 2 H Bl 469, 126 ER 651 (NP) 27
Duns Licensing Associates/Estimating sales activity T0154/04 [2007] EPOR 38 .... 1, 5, 111, 122, 236, 237, 238, 240, 247, 317, 318, 319, 324, 332, 334
Dyson Appliances Ltd v Hoover Ltd [2001] EWCA Civ 1440 251

E M's Application (1924) 41 RPC 590 (SG) 85
E S P's Application (1944) 62 RPC 87 (SE) 74, 75
Eades v Starbuck Wagon Co (1881) WN 160 (Ch) 63
Eastman Kodak Company's Application [1970] FSR 393 (PAT) 97, 296
Edgar Rhodes' Application [1973] RPC 243 (PAT) 75, 103, 104
Eisai/Second Medical Indication G0005/83 [1979-85] EPOR B241 110
Electric Telegraph Company v Brett (1851) 10 CB 838, 138 ER 331 (CP) 39, 45
Eli Lilly & Company's Application [1975] RPC 438 (PAT) 336
Elias v Grovesend Tinplate Company (1890) 7 RPC 455 (QB, CA) 72, 73
Elton and Leda Chemical Ld's Application for a Patent [1957] RPC 267 (PAT) 79
Esau's Application (1932) 44 RPC 85 (SG) 76
Esswein/Automatic Programmer T0579/88 [1991] EPOR 120 210
Ex parte Allen 2 USPQ2d (BNA) 1425, 1427 (BPAI 1987); 846 F2d 77 (Fed Cir 1988) 221
Ex parte Fox (1812) 1 Ves & Bea 67, 35 ER 26 (Lincoln's Inn Hall) 32

Ex parte Hibberd 227 USPQ (BNA) 443 (BPAI 1985) ... 221
Exxon/Fuel Oils T0409/91[1994] EPOR 149 ... 292, 297, 300, 301, 304, 305, 306, 308
Exxon v Exxon Insurance Corp [1981] 3 WLR 541 (CA) ... 332
F M M's Application (1940) 58 RPC 115 (PAT) ... 75
F's Application (1920) 37 RPC 112 (SG) ... 74
F's Application (1954) 72 RPC 127 (SE) ... 75, 79
Fishburn's Application (1938) 57 RPC 245 (PAT) ... 73, 74
Forsyth v Manton (1846) 3 Hay Pat Cas 705 (LC) ... 43
Fox v Dellestable (1866) G 198 (VC) ... 56
Foxwell v Bostock (1864) 46 ER 934 (LC) ... 46, 58
Fujitsu Limited's Application [1996] RPC 511 (PC); [1997] RPC 608 (CA) ... 201, 265, 275, 277, 280, 281, 282, 283, 284, 285, 287, 290, 301, 312, 317
Fujitsu/Coloured Disk Jacket T0119/88 [1990] EPOR 615 ... 212, 214

G E C's Application (1942) 60 RPC 1 (PAT) ... 76, 77, 92, 93, 329
Gadd v The Mayor, &c., of Manchester (1892) 9 RPC 516 (CA) ... 72, 73, 76
Gale's Application [1991] RPC 305 (CA) ... 121, 251, 265, 266, 267, 268, 269, 270, 271, 272, 273, 274, 275, 276, 277, 278, 282, 285, 287, 289, 296, 297, 317
Genentech Inc's Patent [1987] RPC 553 (Pat); [1989] RPC 147 (CA) ... 4, 253, 258, 259, 260, 261, 262, 264, 265, 266, 267, 268, 271, 272, 282, 288, 294, 297, 299, 301
General Electric Co Ltd's Application (Patent) [1961] RPC 21 (SE, PAT) ... 95, 184
Generics (UK) Ltd v Lundbeck A/S [2007] EWHC 1040 (Pat); [2008] EWCA Civ 311; [2009] UKHL 12 ... 84, 303, 305, 306, 307, 308, 311, 332, 333, 334
Gevers' Application [1970] RPC 91 (PAT) ... 101
Gibson v Brand (1841) Web Pat Cas 627 (NP); (1842) Web Pat Cas 631 (CP) ... 36
Griffin v Isaacs (1938) 12 ALJ 169 (HCA) ... 75, 102, 104
Gum v Stevens (1923) 33 CLR 267 (HCA) ... 75

Hall v Boot (1822) G 217, 1 Carp Pat Cas 423, Web Pat Cas 97 (NP) ... 34, 43, 46
Hallen Co v Brabantia (UK) Ltd [1991] RPC 195 (CA) ... 306
Halliburton Energy Services Inc v Smith International Inc [2005] EWHC 1623 ... 1, 304, 305, 306, 312, 314, 315, 321, 322
Hare v Harford (1803) 1 Carp Pat Cas 180 (NP) ... 44
Hartley's Patent (1777) Web Pat Cas 54 ... 30
Harvard/Onco-Mouse Application [1991] EPOR 525 (Examining Division) ... 221, 222
Harwood v Great Northern Railway Co (1860) 121 ER 1044 (QB); (1862) 121 ER 1056 (ExCh); 9 ER 1488 (HL) ... 47, 51, 66, 67, 327
Hasting's Patent (c1578) Noy 184, 74 ER 1139, Web Pat Cas 5, 6 (ExCh) ... 23
Hayward v Hamilton (1879–81) 1 Gr 115 (CA) ... 47, 64, 66
Heath v Unwin (1852) 2 Web Pat Cas 236, 12 CB 522, 138 ER 1010 (ExCh); (1855) 5 HL Cas 505, 10 ER 997 (HL) ... 38
Hickton's Patent Syndicate v Patents and Machine Improvements Company Ltd (1909) 26 RPC 71 (HC); [1909] RPC 339 (CA) ... 72, 259
Higgs v Goodwin (1858) 120 ER 607 (QB) ... 50
Hill v Thomas (1907) 24 RPC 415 (CA) ... 75
Hill v Thompson (1817) G 241, 3 Mer 622, 36 ER 239, Web Pat Cas 225, 1 Carp Pat Cas 377 (LC); (1817) Holt NP 636, 171 ER 367 (NP); (1817) 3 Mer 626, 36 ER 239, Web Pat Cas 235 (LC); (1818) 8 Taunt 375, 129 ER 427, Web Pat Cas 239 (CP) ... 27, 33, 41, 42, 43, 45, 61, 66, 326

Hill v Tombs (1881) JPM 82 (MR) 65, 67
Hiller's Application [1969] RPC 267 (PAT) 100
Hinks v Safety Lighting Company (1876) LR 4 ChD 608 (MR) 56, 66
Hitachi Ltd's Application [1991] RPC 415 (PO) 201, 274, 275, 280, 281, 289, 299
Hitachi/Auction Method T0258/03 [2004] OJ EPO 12/575 .... 316, 321
Hoermann's Application [1996] RPC 341 (PO) 236, 237
Hornblower v Boulton (1799) G 263, 8 TR 95, 101 ER 1285, Dav Pat Cas 221, 1 Carp Pat Cas 156 (KB) 27, 28, 32
Horton v Mabon (1862) 142 ER 1213 (CP) 53, 66, 67
Househill Iron Co v Neilson (1842) Web Pat Cas 676 (NP); (1843) Web Pat Cas 673 (HL) 37, 44, 104, 327
Howard Florey/Relaxin, Application No. 83 307 553.4 [1995] EPOR 541 (Opposition Division) 221, 222, 224, 225, 227, 248, 288
Huddart v Grimshaw (1803) G 266, 1 Carp Pat Cas 200, Web Pat Cas 85, Dav Pat Cas 265 (NP) 32, 43, 44

IBM/Card Reader T0854/90 [1994] EPOR 89 215
IBM/Computer Programs T0935/97 [1999] EPOR 301 .... 201, 228, 233, 234, 236
IBM/Computer-Related Invention T0115/85 [1990] EPOR 107 .... 197, 205, 206, 207, 208, 209, 210, 269, 273
IBM/Database Back-Solving T-1449/06 [2010] EPOR 3 238
IBM/Data Processor Network T0006/83 [1990] EPOR 91 .... 212, 269, 273
IBM/Document Abstracting and Retrieving T0022/85 [1990] EPOR 98 .... 201, 210, 211, 212, 215, 216, 269, 270, 273
IBM/Document Composition T0167/92 [2003] EPOR 20 .... 212, 217
IBM/Editable Document Form T0110/90 [1995] EPOR 185 .... 201, 212
IBM/Editing Business Charts T0790/92 (Unreported Technical Board of Appeal, 29 October 1993) 201, 212, 214, 217
IBM/External Interface Simulation T0833/91 [1998] EPOR 431 .... 211, 212, 217
IBM/Homophone Checker T0065/86 [1990] EPOR 181 .... 201, 211, 212, 216, 280, 282
IBM/Method for physical VLSI-chip design T453/91Unreported, 31 May 1994 313
IBM/Rotating Displayed Objects T0059/93 (Unreported Technical Board of Appeal, 20 April 1994) 212, 214, 282
IBM/Semantically-Related Expressions T0052/85 [1989] EPOR 454 .... 201, 211, 212, 216
IBM/Spatially Related Data T0186/86 (Unreported Technical Board of Appeal, 5 December 1989) 212, 216
IBM/Text Clarity Processing T0038/86 [1990] EPOR 606 .... 201, 211, 212, 216
IBM's Application [1980] FSR 564 .... 102, 104, 331
IGT/Acres Gaming Inc's Application [2008] EWHC 568 316
IGT/Computer Implemented Game Process T1023/06 [2007] EPOR 36 240
ITS Rubber Ltd's Application [1979] RPC 318 (Pat) 103
In re Bilski 545 F 3d 943 (Fed Cir 2008) 9, 228, 247, 249
In re Martin and Hyam's Patent (1855) 3 WR 433 (LC) 48
In re May & Baker Ltd (1948) 45 RPC 255 (Ch) 86

Jacques Kessler's Application [1973] RPC 413 (PAT) 75, 104
Johnson v Rylands (1873) 1 Gr 138 (ExCh) 61
Johnson's Application (1901) 19 RPC 56 (SG) 74, 75
Johnson's Application (1930) 47 RPC 178 (SG) 74, 75, 76
Johnson's Application (1930) 47 RPC 361 (SG) 75, 76
Jones v Pearce (1832) Web Pat Cas 123 (NP) 34, 44, 45
Joos v Commissioner of Patents [1973] RPC 59 (HCA) 62 .... 94, 99, 100, 101, 336

Jordan v Moore (1866) LR 1 CP 624 (CP) 47, 56, 66, 67
Jupe v Pratt (1837) Web Pat Cas 145 (Ex) 35, 44, 327

Kay v Marshall (1839) 2 Web Pat Cas 71, 5 Bing NC 492, 132 ER 1189 (CP); (1841) 2 Web Pat Cas 79, 8 Cl&F 245, 8 ER 96 (HL) 35, 44
Kearney/Computer-Related Invention T0042/87 [1997] EPOR 236 .... 205, 212
Kelvin & Hughes Ld's Application (1953) 71 RPC 103 (PAT) 74, 75
Kirin-Amgen Inc v Hoechst Marion Roussel Ltd [2002] EWCA 1096; [2004] UKHL 46 .... 293, 294, 298, 301, 302, 303, 306, 331, 332, 333, 337, 338
Klaber's Patent (1906) 23 RPC 461 (HL) 75
Koch & Sterzel/X-Ray Apparatus T0026/86 [1988] EPOR 72 .... 197, 202, 203, 204, 205, 206, 207, 208, 209, 216, 228, 230, 238
Kopp v Rosenwald Brothers (1900) 19 RPC 205 (HC) 83, 105, 165, 329

L & G's Application (1940) 58 RPC 21 (PAT) 75, 76
L'Oreal's Application [1970] RPC 565 (PAT) 75, 104
Lane-Fox v The Kensington and Knightsbridge Electric Lighting Company (Limited) (1892) 9 RPC 413 (CA) 72, 73
Lee Pharmaceuticals' Applications [1978] RPC 51 (SE) 99
Leland Stanford/Modified Animal Application 2002 [EPOR] 2 (Opposition Division) 221, 222, 223
Lenard's Application (1954) 71 RPC 190 (PAT) 79, 80
Lewis v Davis (1829) Web Pat Cas 488 (NP) 34, 41, 43
Lewis v Marling (1829) Web Pat Cas 490 (NP); Web Pat Cas 493 (KB) 27, 41, 44
Liardet v Johnson (1778) 1 Carp Pat Cas 35, Web Pat Cas 52, Bull NP 78 (NP); (1780) 1 Y & C 527n, 62 ER 1000 (Ch) 27, 41, 67
Lister v Leather (1855–1858) G 283, 120 ER 373 (QB); G 283, 120 ER 384 (ExCh) 50, 61
Littlewoods Mail Order Stores Ld's Application (1954) 71 RPC 185 (SE) 74
London Borough of Lewisham v Malcolm [2008] UKHL 43 .... 177, 178
London Rubber Industries Limited's Patent [1968] RPC 31 (PAT) 98, 99, 336
Losh v Hague (1838) Web Pat Cas 202 (NP) 35, 45, 46, 47, 66, 74, 327, 329
Lubrizol/Hybrid Plants T0320/87 [1990] EPOR 173 221
Lucent Technologies/Structured Voicemail Messages T0858/02 [2006] EPOR 2 234
Lux Traffic Controls Limited v Pike Signals Limited [1993] RPC 107 (Pat) 265, 275, 276, 277, 279, 281, 285, 287, 289

M's Application (1922) 39 RPC 261 (SG) 85
M's Application (1924) 41 RPC 159 (SG) 74, 75
Macfarlane v Price (1816) G 294, 1 Stark 199, 171 ER 446, 1 Carp Pat Cas 309 (NP) 41, 45
Mackelean v Rennie (1862) 143 ER 21 (CP) 53
Maeder v Busch (1938) 59 CLR 684 (HCA) 78, 81, 92, 93
Manton v Parker (1814) G 297, 1 Carp Pat Cas 274, Dav Pat Cas 327 (NP) 27, 41, 45
Markem Corporation v Zipher Ltd [2005] EWCA Civ 267 312
Matthey's Patent (c1580) Noy 184, 74 ER 1139, 1140, Web Pat Cas 5, 6-7 (PC) 23, 325
Merrell Dow Pharmaceuticals Inc v H N Norton & Co Ltd [1996] RPC 76 (HL) 8, 251, 303, 306, 319
Merrill Lynch Inc's Application [1988] RPC 1 (Pat); [1989] RPC 561 (CA) 201, 210, 253, 255, 256, 257, 258, 259, 260, 261, 262, 263, 264, 265, 266, 267, 268, 270, 271, 273, 275, 276, 277, 278, 279, 281, 282, 283, 285, 287, 288, 302, 315, 317, 320

Microsoft/Data transfer T424/03 (Unreported, Technical Board of Appeal, 23 February 2006) 316, 317, 320
Minter v Wells (1834) 1 CM&R 505, 149 ER 1180 (Exch) 34
Mitchell v Reynolds (1711) 1 P Wms 181 24 ER 347 (Ch) 30
Mölnlycke AB v Procter & Gamble Ltd [1992] FSR 549 (Pat) 306
Monier Industries Ltd v Hume Pipe Co (Australia) Ltd (1943) 13 AOJP 575 (HCA) 75
Monsanto Co v Merck & Co Inc [2000] RPC 709 (PAT) 300, 301, 307
Morgan v Seaward (1836) Web Pat Cas 170 (NP); (1837) Web Pat Cas 187, 2 M&W 544, 150 ER 874 (Ex) 27, 29, 35, 44, 67, 100
Morgan v Windover (1888) 5 RPC 295 (CA); (1890) 7 RPC 131 (HL) 72, 73, 76
Morris v Bramsom (1776) G 311, 1 Carp Pat Cas 30, Web Pat Cas 50, Bull NP 76 (NP) 28, 30, 326
Morton v Middleton (1863) 1 Macph 718 (CS) 54
Mullard Radio Valve Co Ltd v Philco Radio & Television Corp [1936] 2 All ER 920 (HL) 86, 88, 97, 98, 173, 299, 300, 301, 303, 329, 333
Muntz's Patent (1846) 3 Hay Pat Cas 705 (PC) 44, 45
Murray v Clayton (1869) LR 7 Ch 570 (CA) 46, 47, 58, 63

N V Philips' Gloeilampenfabrieken's Application (1954) 71 RPC 192 (PAT) 80, 81, 93
Naamloose Vennootschap de Bataafsche Petroleum Maatschappij's Application (1939) 57 RPC 341 (PAT) 76
NAT/Bagging Plant T0636/88 [1993] EPOR 517 214
National Research Development Co. v Commissioner of Patents (1959) 102 CLR 252 (HCA) 3, 90, 92, 93, 94, 95, 96, 97, 98, 99, 100, 101, 102, 104, 105, 183, 184, 191, 246, 252, 281, 330
Natural Colour Kinematograph Co Ld v Bioscemes Ld (1915) 32 RPC 256 (HL) 83
Neilson v Harford (1841) Web Pat Cas 295 (NP); (1841) Web Pat Cas 331 (Ex) 35, 44
Nelson's Application [1980] RPC 173 (HC) 104
Neva Corp's Application [1968] FSR 537 (SE) 97, 98
Newall v Elliot (1864) 10 Jur 954 (Ex) 47, 55
Newton v Vaucher (1851) 6 Ex 859, 155 ER 794 (Ex) 39, 44
Nova Productions Limited v Mazooma Games [2007] EWCA Civ 219 114
Novartis/Transgenic Plant G0001/98 [2000] EPOR 303 221

Olin Mathieson Chemical Corp v Biorex Laboratories Ltd [1970] RPC 157 (Ch) 97, 105, 178, 300, 301, 331
Oneida Indian Nation's Application [2007] EWHC 954 (Pat) 319
Organon Laboratories Limited's Application [1970] RPC 574 (PAT) 104
Ormsom v Clarke (1862) 143 ER 134 (CP); (1863) 143 ER 531 (ExCh) 54, 66, 67
Otto v Linford (1882) 46 LT 35 (CA) 65, 67

P's Application (1924) 41 RPC 201 (SG) 74, 75
Palmer's Application [1970] RPC 597 (PAT) 99
Parkes v Stevens (1869) LR 8 Eq 359 (VC); (1869) LR 5 Ch 36 (LC) 47, 57, 66, 67
Partington v The Hartlepools Pulp and Paper Company Ld (1895) 12 RPC 295 (HC) 72
Patterson v The Gas Light and Coke Company (1875) G 362, LR 2 ChD 812 (CA); (1876) G 362, 3 App Cas 239 (HL) 62, 67, 328
PBS Partnership/Controlling pension benefits system T0931/95 [2002] EPOR 52 .... 201, 235, 236, 237, 238, 241, 247, 248, 316
Penn v Bibby (1866) LR 2 Ch 127 (VC) 47, 56, 66

Pettersson/Queuing System T1002/92 [1996] EPOR 1 .... 218, 219, 220, 233, 236, 286
Philips Electrical Industries Ld's Application [1959] RPC 341 (HC) 77, 79
Philips/Record Carrier T1194/97 [2001] EPOR 25 .... 234, 235, 236
Pitman's Application [1969] RPC 646 (PAT) 104
Pitney Bowes/A system for processing mail T0767/99 (Unreported Technical Board of
Appeal, 13 March 2002) 236
Plant Genetic Systems/Glutamine Synthetase Inhibitors T0356/93 [1995]
EPOR 357 .... 221, 222, 223
Pottier's Application [1967] RPC 170 (PAT) 97, 299
Pow v Taunton (1845) G 391 (QB) 37
Pozzoli SPA v BDMO SA [2007] FSR 37 177
President's Reference/Computer Program Exclusion G0003/08 [2009] EPOR 9 6
Puharich and Lawrence's Application [1965] RPC 395 (PAT) 98

R H F's Application (1944) 61 RPC 49 (PAT) 76
R L Huber's Application for a Patent [1956] RPC 50 (SE) 79
R v Arkwright (1785) G 15, Web Pat Cas 64, 1 Carp Pat Cas 53,
Dav Pat Cas 61 (NP); (1785) Web Pat Cas 73, 1 Carp Pat Cas 101,
Dav Pat Cas 141 (KB) 27, 28, 30, 43, 44, 326
R v Cutler (1816) G 152, 1 Stark 354, 171 ER 495, 1 Carp Pat Cas 351 (NP) 41, 44, 45, 47
R v Cutler (1847) Macr 124, 3 Car & Kir 215, 175 ER 526 (NP); (1849) Macr 137,
3 Car & Kir 232, 175 ER 534 (QB); (1849) Macr 138, 3 Car & Kir 233,
175 ER 534 (NP) 38
R v Else (1785) G 190, Web Pat Cas 75, 1 Carp Pat Cas 103 (KB) 28
R v Fussell (1826) 1 Carp Pat Cas 449; Godson 274 (KB) 41
R v Wheeler (1819) G 497, 2 B & Ald 345, 106 ER 392 (KB) 33, 41, 43, 66, 99
Ralston v Smith (1860) 142 ER 45 (CP); (1861) 142 ER 880 (ExCh); (1865)
11 ER 1318 (HL) 47, 50, 67, 72
Rantzen's Application (1946) 64 RPC 63 (PAT) 77, 79
Rau GmbH's Applications (1935) 52 RPC 362 (PAT) 80
Raytheon Co's Application [1993] RPC 427 (PC) 265, 275, 277, 278, 279,
280, 281, 284, 287
Research in Motion UK Ltd v Inpro Licensing SARL [2006] EWHC 70 .... 6, 308
Rogers v The Commissioner of Patents (1910) 10 CLR 701 (HCA) 76
Rolls-Royce Ltd's Application [1963] RPC 231 (PAT) 100, 101, 102
Rudolfe Battig's Application (1929) 49 RPC 415 (AC) 75
Rushton v Crawley (1870) LR 10 Eq 522 (VC) 46, 59, 67
Russell v Cowley (1834) Web Pat Cas 459 (NP); (1834) Web Pat Cas 463,
1 Cr M & R 864, 149 ER 1331 (Ex) 34, 44

Saunders v Aston (1832) 1 Carp Pat Cas 510, 3 B & Ald 881, 110 ER 324 (KB) 34, 45
Schering AG's Application [1971] RPC 337 (PAT) 99, 331
Sharp & Dohme Inc v Boots Pure Drug Company Ltd (1928) 45 RPC 153 (CA) 85
Sharp v Brauer (1886) 3 RPC 193 (VC) 76
Shell Development Co's Application (1947) 44 RPC 151 (PAT) 87
Shopalotto.com Ltd's Application [2005] EWHC 2416 .... 6, 308, 309, 314, 318, 319, 321
Siemens/Character Form T0158/88 [1992] EPOR 69 .... 211, 212, 217
Sigma/Classifying Areas T0162/82 [1987] OJ EPO 533 122
Simon-Carves Limited's Application (1928) 45 RPC 407 (SG) 76
Slee & Harris's Applications [1966] RPC 194 (SE) 101
Sohei/General-Purpose Management System T0769/92 [1996] EPOR 253 .... 212, 285

Spencer v Jack (1863) 11 LT 242 (CA) 47, 54, 67
Stahl and Larsson's Application [1965] RPC 596 (PAT) 101
Standard Oil Development Company's Application (1951) 68 RPC 114 (PAT) 80
State Street Bank & Trust Co v Signature Financial Group 149 F 3d 1368 (Fed Cir 1998) 228
Stead v Williams (1848) 5 CB 528, 136 ER 985 (CP) 37
Steiner v Heald (1851) 6 Exch 697, 155 ER 686 (Ex) 39
Sternheimer/Harmonic Vibrations T0366/87 [1989] EPOR 131 .... 197, 206, 207, 208, 209, 220
Stevens v Keating (1847) 2 Web Pat Cas 176 (VC); (1847) 2 Web Pat Cas 181 (NP); (1848) 2 Web Pat Cas 192 (Ex) 38, 44
Stockburger/Coded Distinctive Mark T0051/84 [1986] EPOR 229 .... 197, 198, 200, 202, 208, 209, 210, 214, 219, 220, 254
Sturtz v De la Rue (1828) G 448 5 Russ 322, 38 ER 1048, 1 Carp Pat Cas 463 (LC) 41, 45
Swift & Co's Application [1961] RPC 129 (PAT); [1962] RPC 37 (Divisional Court) 98
Swift v Commissioner of Patents [1960] NZLR 775, (NZSC) 779 .... 95, 98, 184
Sykes v Howarth (1878) LR 12 ChD 826 (Ch) 63
Symbian v Comptroller General of Patents [2008] EWCA Civ 1066 .... 5, 9, 114, 176, 306, 310, 318, 319, 320, 321, 322, 334
Synthon BV v SmithKline Beecham plc [2005] UKHL 59 .... 4, 304
Szuecs's Application [1956] RPC 25 (SE) 81, 93

T S's Application (1924) 41 RPC 530 (SG) 89, 100
T's Application (1919) 37 RPC 109 (SG) 73, 76, 103
Tatham v Dania (1869) 1 Gr 213 (CP) 47, 57, 67
Teckelpenny v Army and Navy Co-operative Society (Limited) (1888) 5 RPC 405 (CA) 76
Tetley v Easton (1852) Macr 33 (NP); (1853) Macr 78 (NP); (1853) Macr 82, 2 E&B 956, 118 ER 1024 (QB) 40, 48
Texas Instruments/Language Understanding System T0236/91 [2000] EPOR 156 212
The Badische Anilin und Soda Fabrik v Levinstein (1885) 2 RPC 73 (CA) 83
The Cementation Company Ltd's Applications (1945) 62 RPC 151 (PAT) 77, 81, 329
The Dow Chemical Company's Application for a Patent [1956] RPC 247 (SE) 77, 80
The Patent Bottle Envelope Company v Seymour (1858) 141 ER 65 (CP) 47, 50, 67
The Upjohn Company (Robert's) Application [1977] RPC 94 (CA) 336
Thermos Ltd v Isola Ld [1910] RPC 388 (HC) 75
Thierry v Riekmann (1895) 12 RPC 412 (CA) 73
Thompson v James (1863) 55 ER 224 (MR) 47, 54, 67
Thomson v American Braided Wire Co (1889) 6 RPC 518 (HL) 82, 83
Turner v Winter (1787) G 470, 1 TR 602, 99 ER 1274, Web Pat Cas 77 (NP) 27, 28

United States Rubber Company's Application [1964] RPC 104 (SE) 98

Vicom/Computer-Related Invention T0208/84 [1987] EPOR 74 .... 197, 199, 201, 202, 203, 204, 205, 206, 207, 208, 209, 216, 220, 231, 236, 254, 256, 257, 258, 262, 263, 264, 269, 270, 271, 274, 275, 277, 278, 279, 280, 281, 282, 283, 284, 286, 287, 289, 290, 320
Virginia-Carolina Chemical Corporation's Application [1958] RPC 38 (PAT) 76, 80
Vittorio Cobianchi's Application (1953) 70 RPC 199 (PAT) 74
Vorwerk v Evans (1890) 7 RPC 265 (CA) 82

W R's Application (1924) 41 RPC 216 (SG) 75
W's Application (1914) 31 RPC 141 (SG) 74, 75
W's Application (1922) 39 RPC 263 (SG) 85
Walton v Bateman (1841) Web Pat Cas 613 (NP) 36, 44, 45, 67
Walton v Potter (1841) Web Pat Cas 585 (NP); (1842) Web Pat Cas 597 (CP) 35, 44, 45
Wang Laboratories Inc's Application [1991] RPC 463 (PC) 265, 272, 273, 274, 276, 278, 279, 280, 284, 285, 286, 287, 289
Ward's Application (1911) 29 RPC 79 (SG) 74, 75
Waterhouses's Patent (1906) 23 RPC 470 (CA) 73
Webster v Uther (1824) Godson 232, Web Pat Cas 128n(b) (NP) 43, 45
Welcome Real-Time SA v Catuity Inc [2001] FCA 445 228
West v James Keeves & Sons Ld (1911) 28 RPC 474 (HC) 76
White v Toms (1867) 37 LJ Ch 204 (VC) 57, 66, 67
Williams v Nye (1890) 7 RPC 62 (CA) 76
Willis v Davison (1863) 1 NR 234 (QB) 54, 67
Wilson v Wilson Bros Bobbin Company Ld [1911] RPC 612 (HC) 76
Wood v Raphael (1896) 13 RPC 730 (HC) 72, 73
Wood v Zimmer (1815) G 502, Holt NP 58, 171 ER 161, 1 Carp Pat Cas 290 (NP) 41
Wright v Hitchcock (1870) LR 5 Ex 37 (Ex) 59

Young v Fernie (1864) 66 ER 836 (LC) 55, 67
Young v Rosenthal (1884) 1 RPC 29 HC (QB) 72

Table of Statutes and Official Papers

STATUTES

An Act to Amend the Law Touching Letters Patent for Inventions 1835 (5 & 6 Wm IV, c 83)
s II 2
Disability Discrimination Act 1995
s 24 177
European Communities Act 1972
s 1(2) 177
Foreign Clothworkers Act 1337 (11 Edw III, c 5) 14
Patent Law Amendment Act 1852
(15 & 16 Vict, c 83) 328
s LV 2, 25, 47, 71, 327
Patents Act 1949 .... 4, 96, 180, 183, 300
s 4(4) 88, 97, 329
s 7 71
s 10(1)(c) 182
s 14 71
s 32(1) 97, 98
(i) 300
s 37 88
s 101 .... 2, 71, 323
Patents Act 1977 .... 2, 91, 110, 177, 190, 251, 253, 254, 301, 307, 308
s 1 .... 114, 189, 251, 252, 338
(1)-(4) 334
(1) 1, 8, 123, 176, 253, 256, 257, 270, 296, 298, 299, 306, 307
(2) 4, 5, 179, 190, 253, 254, 255, 256, 257, 258, 259, 260, 261, 262, 263, 264, 265, 266, 268, 269, 270, 271, 272, 273, 274, 275, 276, 278, 279, 281, 282, 283, 285, 287, 290, 292, 295, 296, 302, 309, 312, 314, 332
(a) 258, 259, 261, 262, 266, 269, 270, 274, 275, 283, 294, 297, 301
(c) 254, 255, 256, 257, 262, 263, 266, 267, 269, 270, 271, 272, 274, 277, 278, 279, 280, 281, 283, 285, 271, 308
(4) 123
(5) 298
ss 2-6 .... 123, 334
s 2 1
s 3 1
s 4 1
s 14(2)(b) 295
(3) 123, 295, 334
(5) 123, 334
(a) 1
(b) 1
(c) 1, 291, 295, 296, 304
(6) 334
s 37(5) 334
s 54 .... 123, 334
s 60 .... 1, 123, 176, 334
(1) 305
s 69 .... 123, 334
s 72 .... 123, 303
(1) 123, 304, 334
(c) 291, 295, 300, 304, 305
(2) 123, 334
s 74(4) 123, 334
s 82 .... 123, 334
s 83 .... 123, 334
s 100 .... 123, 334
s 125 .... 1, 123, 295, 334, 338
(1) 192, 332
s 130(7) 114, 123, 189, 193 251, 268, 298, 312, 334
Patents and Designs Act 1907 75
s 25 71
s 26 71
s 32(1)(i) 88
s 38A(1) 85, 87, 88, 105, 176, 329, 330
s 75 .... 89, 336
s 93 .... 2, 71, 323
Patents and Designs Act 1919
s 11(1) 85, 329
Patents and Designs Act 1932
s 3(d)-(g) 323
(d) 71
s 8 85
s 11 .... 89, 336

Patents, Designs & Trade Marks Act 1883 (46 & 47 Vict, c 57) 25, 69, 71, 82, 83, 328, 329
s 5(2) 42
s 11 71
s 22 69
s 46 .... 2, 71, 316, 323, 328
s 86 .... 89, 336
Statute of Monopolies 1623 (21 Jac I, c 3) 2, 4, 13, 19, 20, 26, 100, 325, 326
s 6 .... 2, 3, 21, 22, 24, 25, 43, 46, 47, 72, 73, 88, 90, 91, 92, 100, 105, 106, 252, 323, 330, 336

DIRECTIVES AND REGULATIONS

Council Regulation (EC) 2100/94 of 27 July 1994 on Community plant variety rights [1994] OJ L227/1 113
Directive 98/44 of the European Parliament and of the Council of 6 July 1998 on the legal protection of biotechnological inventions (Biotech Directive) [1998] OJ L213/13 .... 113, 126, 128

INTERNATIONAL AGREEMENTS

Act Revising the Convention on the Grant of European Patents (European Patent Convention) of 5 October 1973, last revised on 17 December 1991, Special Edition No 4 OJ EPO (2001) 232
Agreement on Trade-Related Aspects of Intellectual Property Rights (TRIPS Agreement) Marrakesh Agreement Establishing the World Trade Organization, Annex 1C, Legal Instruments—Results of the Uruguay Round, vol 31 (15 April 1994, 33 ILM 81) 112, 114, 124, 177, 227, 228, 229,231, 249, 335
Art 7 178
Art 8.1 178
Art 27.1 .... 9, 123, 124, 176, 177, 178, 196, 226, 232, 233, 240, 248, 249, 335
Art 65 124
Art 70 124
Convention Establishing the World Intellectual Property Organization (WIPO) (Stockholm, 14 July 1967, 6 ILM 782) 112, 113, 114
Convention for the European Patent for the Common Market (Community Patent Convention) (CPC) (Luxembourg, 15 December 1975, 15 ILM 5) 113, 115, 124, 186, 334
Convention on the Grant of European Patents (European Patent Convention) (Munich, 5 October 1973, 13 ILM 268) 1, 6, 9, 10, 106, 110, 111, 121, 123, 126, 127, 159, 160, 162, 163, 181, 186, 189, 190, 191, 195, 203, 226, 251, 253, 324, 325, 331, 334, 337
Pt V 123
Art 1 122
Art 9 .... 147, 148, 149
(2) 145, 147, 148, 149, 151, 152, 154, 156, 159, 160, 161, 163, 164
(a)-(e) 147
(a) 147, 148, 149, 150, 151, 152, 160
(b) 150
(c) 156
(d) 149, 154, 160, 161
(e) 147, 148, 161
(f) 148, 149
(g) 148, 149
Art 10 .... 147, 148, 149, 155, 161
(b) 148, 161
Art 18 239
Art 23(3) 122
Art 46 1
Arts 52-57 .... 110, 122
Art 52 .... 176, 190, 195, 196, 206, 223, 229, 231, 251, 298, 312, 313, 314, 317, 320, 332
(1) 1, 8, 9, 106, 114, 123, 176, 177, 195, 199, 202, 206, 207, 209, 210, 218, 220, 232, 233, 235, 236, 237, 238, 239, 240, 244, 247, 249, 250, 255, 323, 324, 332, 334
(2) 4, 5, 6, 7, 106, 110, 111, 114, 118, 119, 143, 146, 147, 159, 178, 179, 190, 191, 192, 195, 196, 197, 198, 199, 202, 203, 204, 205, 206, 207, 208, 209, 210, 211, 212, 213, 217, 219, 220, 223, 224, 225, 227, 228, 231, 232, 233, 234, 235,

236, 237, 238, 239, 240, 241, 242, 244, 245, 246, 247, 248, 249, 252, 254, 255, 257, 258, 309, 310, 312, 313, 314, 315, 316, 317, 318, 319, 320, 321, 332, 333, 334
(a) ... 199, 202, 206, 221, 222, 223, 257, 271
(b) ... 206, 212
(c) ... 154, 195, 197, 198, 199, 200, 201, 202, 205, 206, 208, 209, 211, 215, 216, 218, 219, 228, 229, 231, 232, 233, 234, 235, 236, 241, 247, 313, 319, 320, 321
(d) ... 204, 212, 213, 214, 217, 229, 234, 235, 313, 337
(3) ... 4, 5, 6, 106, 110, 111, 114, 119, 143, 146, 147, 152, 159, 179, 190, 191, 192, 195, 196, 199, 206, 209, 210, 211, 220, 227, 230, 231, 232, 233, 234, 236, 237, 238, 239, 240, 241, 242, 244, 245, 246, 248, 249, 252, 257, 309, 312, 314, 315, 316, 317, 318, 319, 320, 321, 333, 334
(4) ... 119, 195, 196
Art 53(a) ... 221, 222, 223, 337
(b) ... 119, 167, 169, 170, 196, 221, 226, 249
(c) ... 119, 337
Art 54 ... 1, 317
Art 56 ... 239, 317
Art 57 ... 1
Art 60 ... 332
Art 65 ... 113, 126
Art 69 ... 123, 295, 332
Art 71 ... 173
Art 84 ... 1
Art 112 ... 317
(1)(a) ... 317
(b) ... 318
Art 138 ... 123
Art 167(1)(a) ... 171
Art 172 ... 146
Convention on the Unification of Certain Points of Substantive Law on Patents for Inventions (Strasbourg Convention) (Strasbourg, 27 November 1963, ETS 47) ... 5, 10, 94, 114, 115, 126, 149, 155, 159, 160, 162, 167, 169, 180, 181, 182, 183, 184, 190, 195, 196
Art 1 ... 115, 141, 143, 162, 195
Art 2 ... 168
Art 2(a) ... 180
Art 2(b) ... 119, 195
Art 3 ... 162
Art 9(2)(b) ... 147
European Convention on the International Classification of Patents for Inventions (Paris, 19 December 1954, CETS No 017)
European Convention relating to the formalities required for Patent Applications (Paris, 11 December 1953, CETS No 016)
Implementing Regulations to the Convention on the Grant of European Patents (5 October 1973, as amended)
rr 27-29 ... 210
r 27 ... 225, 226
(1) ... 239
r 29 ... 225, 226
(1) ... 214, 239
International Convention for the Protection of New Varieties of Plants (UPOV Convention) (Paris, 2 December 1961, 33 UST 2703), opened for signature 2 December 1961, 33 UST 2703 ... 113
Paris Convention for the Protection of Industrial Property (Paris Convention, as amended) (Paris, 20 March 1883, 13 UST 1) ... 112, 113, 115, 124
Art 1 ... 9, 115, 132, 133, 162
(2) ... 115
(3) ... 115
Art 6 ... 133
Art 19 ... 115
Patent Cooperation Treaty (Washington, 19 June 1970, 9 ILM 978) ... 112, 115, 117, 118, 120, 123, 149, 150, 153, 155, 159, 160, 161, 179, 190, 195, 196, 234, 334
r 39 ... 148
r 39.1 ... 5, 118, 119, 120, 121, 159, 160, 195
r 67 ... 148
r 67.1 ... 5, 118, 119, 120, 121
Patent Cooperation Treaty International Search and Preliminary Examination Guidelines, PCT/GL/ISPE/1 ... 118, 121
Patent Law Treaty (Cm 6779, 2000) ... 112

Resolution on the Adjustment of National Patent Law (annexed to the Community Patent Convention) 123
Treaty Establishing the European Community [2002] OJ C325
Art 251 241
Treaty Establishing the European Economic Community (EEC Treaty) (Rome, 25 March 1957, 37 ILM 56) 116
Art 249 114
Treaty of Lisbon Amending the Treaty on European Union and the Treaty Establishing the European Community (Lisbon Treaty) (Lisbon, 13 December 2007) 113
United Nations (UN) General Assembly Resolution 217 A (III), Universal Declaration of Human Rights (Paris, 10 December 1948) 113
Vienna Convention on the Law of Treaties (Vienna Convention) (Vienna, 23 May 1969, UNTS 1155, 321) 229, 230
Arts 31-33 315
Art 31 .... 111, 230
(4) 231
Art 32 .... 110, 111

OFFICIAL PAPERS OF THE UNITED KINGDOM

United Kingdom, Explanatory and Financial Memorandum to the Patents HL Bill 1976–1977 .... 180, 189, 192
United Kingdom, Hansard HC (series 5) vol 118 85
United Kingdom, Hansard HC (series 5) vol 910 244
United Kingdom, Hansard HL (series 5) vol 37 85
United Kingdom, Hansard HL (series 5) vol 379 .... 4, 89, 154, 179, 180, 183, 192, 252, 256, 322, 329, 337
United Kingdom, Hansard HL (series 5) vol 381 .... 154, 192, 251
United Kingdom, Hansard HL (series 5) vol 383 .... 154, 155
United Kingdom, Parl Debs (series 4) vol 63 84
United Kingdom, Memorandum (Patents and Designs Bill) (Cmnd 4067, 1931–1932) 85
United Kingdom Memorandum on German Use of our Patent Law (1916) 84
United Kingdom, Patent Law Reform (Cmnd 6000, 1975) 180
United Kingdom, Patents and Designs Acts; Final Report of the Departmental Committee (Cmnd 7206, 1947) 87, 94, 180, 191
United Kingdom, Promoting Innovation through patents—The follow-up to the Green Paper on the Community patent and the patent system in Europe, Com (1999) 42 final (3 February 1999) 241
United Kingdom, Report from the Select Committee on Letters Patent (C 193, 1872) 68
United Kingdom, Report of the Commissioners appointed to Inquire into the Working of the Law Relating to Letters Patent for Inventions ([3419], 1864) 68, 191, 328
United Kingdom, Report of the Department Committee on the Patents and Designs Acts and on the practice of the Patent Office (Patent Office Practice) (Cmnd 3829, 1930–1931) 84
United Kingdom, The British Patent System: Report of the Committee to Examine the Patent System and Patent Law (Cmnd 4407, 1970) 84, 119, 180
United Kingdom, United Kingdom Patent Law: The Effects of the Strasbourg Convention of 1963 (Cmnd 2835, 1965) 87, 94, 95, 112, 330

OFFICIAL PAPERS OF EUROPEAN INSTITUTIONS AND ORGANIZATIONS

European Patent Organisation, Basic Proposal for the Revision of the European Patent Convention, MR/2/00 (2000) 232, 241

European Patent Organisation
Agreement on the Application of Article 65 of the Convention on the Grant of European Patents (London Agreement) (London, 17 October 2000, [2001] OJ EPO 550) 113, 126
European Patent Organisation Basic Proposal for the Revision of the European Patent Convention, CA/100/00 e 37 176
Bureau of the Committee of Experts, Report of the meeting held in Paris on 16th and 17th March 1961, EXP/Brev B (61) 3 (24 March 1961) 140
Committee of Experts, Article published in Europe's Marché commune (No 560 of 24 November 1959): The Six EEC Countries Plan to Establish a Uniform European Law on Industrial Property, EXP/Brev (59) 1 (3 December 1959) 116
Committee of Experts, Comments by the Dutch experts on questions raised in the Resolution adopted by the Committee of Experts on 7th October 1955, EXP/Brev (57) 1 (14 June 1957) 136
Committee of Experts, Comments by the Luxembourg expert on questions dealt with in the Resolution adopted by the Committee of Experts, dated 7th October, 1955 EXP/Brev (56) 9 (26 October 1956) 135, 136
Committee of Experts, Comparative Study of Substantive Law in Force in the Countries Represented on the Committee of Experts on Patents presented by the Secretariat-General, EXP/Brev (53) 18 (7 November 1953) 86, 132, 138, 167, 172
Committee of Experts, Criteria of Novelty and Patentability, CM/WP IV (51) 9 (undated) 86, 131, 164
Committee of Experts, Draft by delegations of Germany and United Kingdom, EXP/Brev (52) 3 (16 January 1952) 167
Committee of Experts, Draft of a European Convention relating to Patents of Invention, EXP/Brev (53) 19 (10 November 1953) 133, 167, 173
Committee of Experts, Draft Official Report of the meeting of 2nd July 1951, CM/WP IV (51) PV4 (4 July 1951) 131, 140, 172, 175
Committee of Experts, Draft Official Report of the meeting of 5th July 1951, CM/WP IV (51) PV7 (6 July 1951) 165
Committee of Experts, Experience of the effects of UK legislation relating to licensing of patents, with particular reference to patents for chemical products, food, medicine and surgical or curative devices; Report presented by the UK Experts, EXP/Brev (52) 10 (6 March 1952) 88
Committee of Experts, Memorandum concerning the establishment of a European Patents Office presented to the Committee of Experts on Patents by the Experts of the Federal Republic of Germany, CM/WP IV (51) 25 (June 1951) 130
Committee of Experts, Memorandum on the unification of legislation (Item 4 of the Agenda for the meeting of 28th November 1960) by Mr Finniss, Rapporteur-General, EXP/Brev (60) 7 (28 November 1960) 121
Committee of Experts, Note by the Secretariat General on the Functions and Powers of a European Patents Office, CM/WP IV (51) 8 (18 June 1951) 130
Committee of Experts, Note on the Standardisation of Laws, included in the Agenda for the Fourth Meeting of the Committee of Experts on Patents, EXP/Brev (52) 23 (26 June 1952) 167, 172
Committee of Experts, Observations and proposals of the German experts on the unification of general conditions of patentability, EXP/Brev (56) 8 (26 October 1956) 137, 144
Committee of Experts, Observations of the United Kingdom Delegation, EXP/Brev (56) 2 (11 June 1956) 88
Committee of Experts, Observations of the Scandinavian Experts: Note on the question of the role and scope of descriptions and claims by the Scandinavian Experts, EXP/Brev (56) 1 (14 April 1956) 134

Committee of Experts, Observations of the United Kingdom Delegation; Industrial Character, EXP/Brev (56) 3 (3 May 1956)....................................86, 87, 135, 329
Committee of Experts, Observations of the United Kingdom Delegation; Technical Progress and Creative Effort, EXP/Brev (56) 4 (11 June 1956).................................. 140
Committee of Experts, Observations on the proposed European Convention on Patents proposed during the meeting of the Bureau of the Committee of Experts held at Strasbourg from the 2nd to 6th February, 1954. Presented by M Reimer, EXP/Brev B (54) 2 (10 March 1954).................................. 133
Committee of Experts, Preliminary draft Convention on the unification of certain points of substantive law on patents for inventions, CM (61) 97 (12 June 1961) Appendix V 168, 173
Committee of Experts, Proposal for the institution of a European patent with the possible creation of a European Court of Justice. Made by M Reimer, EXP/Brev B (55) 1 (14 February 1955)........................133, 134
Committee of Experts, Questionnaire drawn up by the Bureau of the Committee of Experts of the Council of Europe, EXP/Brev (53) 3 (15 January 1953)...................... 132
Committee of Experts, Reply of the Italian Delegation to the Questionnaire Adopted by the Committee of Experts, EXP/Brev (56) 7 (10 August 1956)................................ 135
Committee of Experts, Reply to the Questionnaire drawn up by the Bureau of the Committee of Experts of the Council of Europe from the point of view of legislation in the United Kingdom, EXP/Brev (53) 7 (17 March 1953)........................87, 88, 329
Committee of Experts, Reply to the Questionnaire drawn up by the Bureau of the Committee of Experts of the Council of Europe from the Point of view of the German legislation, EXP/Brev (53) 1 (12 January 1953)......................136, 167
Committee of Experts, Report by Mr Herbert Kuhnemann, German Expert, on the unification of European patent law in respect of the requirements of technical progress and creative effort as conditions of patentability, EXP/Brev (60) 5 (15 November 1960)..................138, 175
Committee of Experts, Report by Mr Nicolas Vouropoulos, Greek Expert, on the questions relating to the unification of European patent law selected by the Committee of Experts in October 1955, EXP/Brev (60) 3 (16 August 1960) 134
Committee of Experts, Report of the Committee of Experts on Patents to the Committee of Ministers, EXP/ Brev (52) 8 rev (22 January 1952)... 167
Committee of Experts, Report of the Committee of Experts to the Committee of Ministers on the meeting held Strasbourg from 10th to 13th July 1962, CM (62) 160 (8 August 1962)............................168, 173
Committee of Experts, Report of the Committee of Experts to the Committee of Ministers on the meeting held Strasbourg from 14th to 17th May 1963: Annex II (Declaration by the Austrian delegation) CM (63) 101 (27 May 1963) 168
Committee of Experts, Report of the French Delegation on the Unification of European Laws in respect of Patents, EXP/Brev (57) 3 (13 May 1957)............136, 164, 167
Committee of Experts, Report on the Creation of a European Patents Office presented by MJC de Haan, Vice-Chairman of the Committee of Experts, CM/WP IV (51) 26 (30 November 1951) 131, 133
Committee of Experts, Report on the form and content of patent specifications; presented by the Experts of the UK of Great Britain and Northern Ireland, EXP/Brev (52) 11 (18 March 1952) 88
Committee of Experts, Report on the questions inscribed on the Agenda for the 2nd session of the Committee of Experts presented by Mr Finniss, Rapporteur-General, CM/WP IV (51) 11 (25 June 1951)................................ 130

Committee of Experts, Report on the unification of certain points of patent law in the countries represented by the Committee of Experts by Mr Finniss, Rapporteur-General, EXP/Brev B (61) 1 (1 March 1961) 140
Committee of Experts, Resolution, CM/WP IV (51) 4 (16 January 1951) 130
Committee of Experts, Resolutions adopted by the session of the Committee of Experts on Patents held at The Hague on 2nd to 9th July 1951, CM/WP IV (51) 17 final (7 July 1951) 165
Committee of Experts, Second Preliminary Draft of a Convention Establishing a European System for the Grant of Patents, 1971 169
Committee of Experts, The Patentability of Products, CM/WP IV (51) 27 (30 November 1951) 165
Committee of Experts, The Preliminary Report by the Joint Scandinavian Patent Committees: Draft Patent Law and Extracts from Explanatory Notes, EXP/Brev (62) 2 (2 May 1962) 141
Committee of Experts, The Programme for the Committee of Experts on Patents, statement presented by the Scandinavian Experts, Stockholm, September 28th, 1955, EXP/Brev (55) 3 (3 October 1955) 133
Committee of Experts, Unification of Laws Convention; Amendment suggested by the Austrian Delegation, EXP/Brev (63) 7 (10 May 1963) 168
Committee of Ministers, 12th Session, Report to the Committee of Ministers presented by the Committee of Experts on Patents, CM/12 (52) 89 (12 July 1952) 131
Committee of Ministers, Eighth Meeting of the Ministers' Deputies (Paris, 18th December, 1952): Report to the Committee of Ministers submitted by the Bureau of the Committee of Experts on Patents, CM/12 (52) 149 (17 November 1952) 132
Committee of Ministers, Report by the Committee of Experts to the Committee of Ministers on the meeting held at Strasbourg from 2nd to 5th May 1961 CM (61) 97 (12 June 1961) 141, 168, 173
Committee of Ministers, Report to the Committee of Experts on Patents to the Committee of Ministers (on the meeting at Strasbourg from the 3rd to the 7th October 1955) CM (55) 137 (7 October 1953) 134
Common Position (EC) 20/2005 adopted by the Council on 7th March 2005 with a view to adopting Directive (EC) 2005/... of the European Parliament and of the Council on the patentability of computer-implemented inventions [2005] OJ C 144 E/9 243
Common Position (EC) No 19/98 adopted by the Council of 26 February 1998 with a view to adopting Directive 981/../EC of the European Parliament and of the Council on the legal protection of biotechnological inventions [1998] OJ C110/17 224
Consultative Assembly of the Council of Europe, Doc 75 Appendix I (Introductory Note to the Study of the Problem of the Creation of a European Patents Office) 129
Consultative Assembly of the Council of Europe, First Ordinary Session, Sixteenth sitting, 6th September 1949, Doc 75 (Creation of a European Patents Office: Report proposed by M Longchambon on behalf of the Committee on Economic Questions) 109
Consultative Assembly of the Council of Europe, First Ordinary Session, Sixteenth sitting, 6th September 1949, Doc 110 (Creation of a European Patents Office: Recommendation to the Committee of Ministers adopted 8th September 1949 on the conclusion of the Debates) Art 1 129
Consultative Assembly of the Council of Europe, Memorandum from the Swedish Government on the Draft Proposal for a Convention on the Creation of a European Patent Office, SG/D (50) 3 (undated) 129
Consultative Assembly of the Council of Europe, Observations of Governments or of Experts on the Draft Proposal for a Convention on the Creation of a European Patent Office proposed by the Consultative Assembly of the Council of Europe, SG/D (50) 3 (20 July 1950) 129

Consultative Assembly of the Council of Europe, Proposal for Establishment of a Western European Patent Office, SG/D (50) 33 (23 May 1950) 129
Consultative Assembly of the Council of Europe, Remarks by the Luxembourg Official Expert on the draft on the possible creation of a European Patents Office, SG/D (50) 3 (undated) 129
Consultative Assembly of the Council of Europe, Report on a European Patent Law (Mr A Borels) (15 January 1964) II (Explanatory Memorandum) 130
Consultative Assembly of the Council of Europe, Report of the Netherlands Experts, SG/D (50) 3 (4 May 1950) 129
Council Decision 94/800/EC (of 22 December 1994) concerning the conclusion on behalf of the European Community, as regards matters within its competence, of the agreements reached in the Uruguay Round multilateral negotiations (1986–1994) [1994] OJ L336/1 .... 9, 124
Council of the European Union, Working Party on Intellectual Property (Patents), Working Document 5072/09 (8 January 2009) (Draft Agreement on the European and Community Patents Court and Draft Statute) 113, 125
Directive 98/44/EC of the European Parliament and of the Council of 6 July 1998 on the legal protection of biotechnological inventions [1998] OJ L213/13 [13–21] 113, 224, 225, 226, 227, 231, 241, 335
Art 1 226
Art 2 226
Art 3 226
Art 4 226
Art 5 226
Art 6 226
Art 7 226
Art 10 226
Art 11 226
Art 12 226
Art 14 226
Art 15 .... 226, 335
Art 16 .... 226, 335
Art 19 .... 226, 335
Art 22 226
Recital 12 227
Recital 14 227
Recital 16 227
Recital 19 227
European Convention on the International Classification of Patents for Inventions (Paris, 19 December 1954, CETS 17) 126
European Convention relating to the Formalities required for Patent Applications (Paris, 11 December 1953, CETS 16) 126
European Parliament legislative resolution on the Council common position with a view to the adoption of a directive of the European Parliament and of the Council on the patentability of computer-implemented inventions (11979/1/2004—C6-0058/2005—2002/0047 (COD)) [2006] OJ C157/265 243
European Patent Organisation Decision of the Administrative Council of 16 June 1999 Amending the Implementing Regulations to the European Patent Convention [1999] OJ EPO 437 225
European Patent Organisation, Diplomatic Conference for the Revision of the EPC, <http://www.epo.org/patents/law/legal-texts/epc2000/diplomatic-conference-nov2000.html> 123, 239
European Patent Organisation, Notice from the President of the EPO, 24 March 2000, (2000) 4 OJ EPO 173 231
German Association of Comparative Law Mannheim Conference on Fundamental Questions of Patent Protection for Chemical Inventions (1973) 3 IIC 357 .... 174, 175, 176
Group of Advisers on the Ethical Implications of Biotechnology to the European Commission, Opinion No 8 (Ethical Aspects of Patenting Inventions Involving Elements of Human Origin) 25 September 1996

(<http://ec.europa.eu/european_group_ethics/docs/opinion8_en.pdf>)................................226, 227, 335
Letter from the Norwegian Ministry of Foreign Affairs, SG/D (50) 3 (23 May 1950)...................................... 129
Luxembourg Conference, Draft Convention Establishing a European System for the Grant of Patents (Stage reached on 20th May 1972) BR/199/72 (25 May 1972) 146, 154
Luxembourg Conference, First Preliminary Draft of a Convention Establishing a European System for the Grant of Patents, BR/70/70 (1970) 169
Luxembourg Conference, Minutes of the meeting of Working Party I (Luxembourg 8–11 July 1969) BR/7/69 (29 July 1969) 116
Luxembourg Conference, Minutes of the meeting of Working Party I, Luxembourg, 7th–11th September 1970, BR/49/70 (26 October 1970).................169, 170, 173
Luxembourg Conference, Minutes of the 2nd Meeting held at Luxembourg on 13th to 16th January 1970, BR/26/70 (30 January 1970)................................ 153
Luxembourg Conference, Minutes of the 3rd meeting of the Co-ordinating Committee (Luxembourg, 23rd, 24th and 27th June 1972) BR/218/72 (26 September 1972) 151, 169, 170
Luxembourg Conference, Minutes of the 4th Meeting of the Inter-Governmental Conference for the Setting up of a European System for the Grant of Patents (Luxembourg, 20th to 28th April 1971) BR/125/71 (7 July 1971)..................148, 149, 169, 170
Luxembourg Conference, Minutes of the 5th meeting of the Inter-Governmental Conference for the setting up of a European System for the Grant of Patents (Luxembourg Conference, 26th January to 1st February 1972) BR/169/72 (1972).........................119, 126
Luxembourg Conference, Minutes of the 5th Meeting of the Inter-Governmental Conference for the Setting up of a European System for the Grant of Patents Parts 1 and 3 (Luxembourg, 24th–25th January and 2nd–4th February 1972) BR/168/72 (15 March 1972).................................. 147
Luxembourg Conference, Minutes of the 5th Plenary Meeting (Part II: Hearing of the non-governmental international organizations on the Second Preliminary Draft of a Convention establishing a European System for the Grant of Patents Minutes) (Luxembourg 26th January–1st February 1972) BR/169/72 (15 March 1972)................................. 147, 148, 149, 150, 153, 170
Luxembourg Conference, Minutes of the 6th Meeting of the Inter-Governmental Conference for the Setting up of a European System for the Grant of Patents (Luxembourg, 19th to 30th June 1972) BR/219/72 (26 September 1972) 151, 156, 170
Luxembourg Conference, Minutes of the 7th Meeting of Working Party I held at Luxembourg from 26th–29th January 1971, BR/94/71 (6 April 1971)...............................148, 154
Luxembourg Conference, Minutes of the 9th Meeting of Working Party I, Inter-Governmental Conference for the Setting up of a European System for the Grant of Patents (Luxembourg 12th–22nd October 1971) BR/135/71 (17 November 1971) 119, 147, 148, 149, 150, 153, 154, 155, 156
Luxembourg Conference, Minutes of the meeting of Working Party I (Luxembourg 8th–11th July 1969) BR/7/69 (29 July 1969) 116
Luxembourg Conference, Preliminary Draft Convention for a European System for the Grant of Patents, Arts 1 to 41 prepared by Working Party I (8th to 11th July 1969) BR/6/69 (25 July 1969) 145, 146
Luxembourg Conference, Report on the 11th meeting of Working Party I held in Luxembourg from 28th February to 3rd March 1972, BR/177/72 (13 April 1972)................... 148, 149, 150, 151, 154, 155, 156, 164, 234

Luxembourg Conference, Working Document for the 11th meeting of Working Party I, to be held from 28th February to 3rd March 1972, BR/GT I/145/72 (14 February 1972) 148

Luxembourg Inter-Governmental Conference for the Setting up of a European System for the Grant of Patents (1969 to 1972) 252

Memorandum of the Council of the European Communities, R/360/69 (28 February 1969) 117

Memorandum on the unification of legislation (Item 4 of the Agenda for the meeting of 28th November 1960) by Mr Finniss, Rapporteur-General, EXP/Brev (60) 7 (undated) 140

Munich Conference, Comments by CNIPA Committee of National Institutes of Patent Agents, M/20 (2 April 1973) 146, 147, 156

Munich Conference, Comments by CPCCI Standing Conference of the Chambers of Commerce and Industry of the European Economic Community, M/18 (2 April 1973) 147

Munich Conference, Comments by the Government of the Federal Republic of Germany, M/11 (29 March 1973) 151, 152, 154, 161

Munich Conference, Comments by the United Kingdom Government, M/10 (29 March 1973) 156

Munich Conference, Conference Document, M/66/I (11 September 1973) (proposal by the International Association for the Protection Industrial Property (IAPIP)) 152, 170

Munich Conference, Minutes of the Munich Diplomatic Conference for the Setting up of a European System for the Grant of Patents (Munich 10 September–5 October 1973), M/PR/1 (1973) 147, 151, 152, 155, 156, 170

Munich Conference, Minutes of the Conference, Munich 10 September to 5 October 1973: Minutes of the Proceedings of the Main Committee I, M/PR/II 147

Opinion of the Economic and Social Committee on the Proposal for a Directive of the European Parliament and of the Council on the patentability of computer-implemented inventions [2003] OJ C61/154 .... 6, 242

Patents Working Party, Amendments to the Preliminary Draft Convention relating to a European Patent Law (Arts 1 to 185) 2335/IV/65 (22 January 1965) 120, 145

Patents Working Party Comments on the First Preliminary Draft Convention relating to a European patent law of 14 March 1961 (Art 11 to 29) IV/2071/61–E (undated) 129, 144

Patents Working Party First Preliminary Draft Convention relating to a European Patent Law, Arts 11 to 29, IV/2071/61, Section 13 (14 March 1961) 129, 144

Patents Working Party, Joint proposal by the German and Netherlands delegations on Art 9 regarding exceptions to the concept of invention, 9663/IV/64 120

Patents Working Party, Proceedings of the 12th meeting of the Patents Working Party held at Brussels from 26th February–6th March 1964, 2632/IV/64–E–Final (15 April 1964) 144

Patents Working Party, Proceedings of the 14th meeting of the Patents Working Party held at Munich from 1st to 12th June 1964, 6498/IV/64–E–Final (1 August 1964) 145

Patents Working Party, Proceedings of the 15th meeting of the Patents Working Party held at Brussels from 19th to 26th October 1964, 11821/IV/64–E–Final (7 December 1964) 120, 145, 152

Proposal for a Council Decision establishing the Community Patent Court and concerning appeals before the Court of First Instance—Opinion of the Court of Justice of the European

Communities (8 November 2004) Document No 14349/4 125
Proposal (EC) 88/496 on the legal protection of biotechnological inventions for a Council Directive [1989] OJ C10/3 224
Proposal (EC) 2000/412 for a Council Regulation on the Community Patent [2000] OJ C337 E/278 125
Proposal (EC) 2002/92 for a Directive of the European Parliament and of the Council on the patentability of computer-implemented inventions [2002] OJ C151/129 ....6, 241
Proposed Software Directive 2002 241
Plenary Debate on the Commission Proposal for a Directive of the European Parliament and of the Council on the patentability of computer-implemented inventions, 23 September 2003 243
Position of the European Parliament adopted at first reading on 24 September 2003 with a view to adopting Directive (EC) 2003/...EC of the European Parliament and of the Council on the patentability of computer-implemented inventions [2004] OJ C77/E/230 242
Proposal for a Council Decision conferring jurisdiction on the Court of Justice in disputes relating to a Community patent, COM [2003] 827 final 125
Proposal for a Council Decision establishing the Community Patent Court and concerning appeals before the Court of First Instance, COM [2003] 828 final 125
Proposal for a Directive of the European Parliament and of the Council on the patentability of computer-implemented inventions-Political agreement on the Council's common position, Council Doc No 97/13/04, 2002/0047 (COD) (24 May 2004) 243
Proposal (EC) 2000/412 for a Council Regulation on the Community patent [2000] OJ C337 E/278 125
Proposal (EC) 2002/92 for a Directive of the European Parliament and of the Council on the patentability of computer implemented inventions [2002] OJ C151/129 ....6, 241
Report from the Commission to the European Parliament and the Council—Development and implications of patent law in the field of biotechnology and genetic engineering, COM (2002) 545 final (7 October 2002) 225, 227
Report from the Select Committee on European Communities, The Community Patent and the Patent System in Europe, HL (1998) 121
Results of the Competitiveness Council of Ministers, Brussels, 11th March 2004 Internal Market, Enterprise and Consumer Protection Issues, MEMO/04/58 (12 March 2004) 125
Rules of Procedure of the Boards of Appeal [2007] OJ EPO 536
Art 20 122
(1) 122
(2) 123
Scandinavian Conference, 13th, on the Protection of Industrial Property 150

OTHER OFFICIAL PAPERS

Australian Government's Advisory Committee on Intellectual Property, Review of Patentable Subject Matter (<http://www.acip.gov.au/reviews.html#%23subject>) 228
Patent Act of 1790 (USA) 15
United States Senate Subcommittee on Patents, Trademarks, and Copyrights, '"To Promote the Progress of ... Useful Arts" in an Age of Exploding Technology', Report of the President's Commission on the Patent System (17 November 1966) 90th Congress 1st Session SENATE, Doc No 5 119
Washington Diplomatic Conference on the Patent Cooperation Treaty (Washington 25th May–19th June 1970) Post-Conference Documents, History of the Patent Cooperation Treaty, PCT/PCD/1 (16 October 1970) 117

Washington Diplomatic Conference on the Patent Cooperation Treaty (Washington 25 May–19 June 1970), Summary Minutes of the 16th Meeting of Main Committee I (3 June 1970) 118

Washington Diplomatic Conference on the Patent Cooperation Treaty (Washington 25th May–19th June 1970) Verbatim and Summary Minutes, Verbatim Minutes of the First Plenary Meeting (25 May 1970).................................118, 120

List of Abbreviations

Courts and Jurisdictions

AC	Acting Comptroller-General of Patents (Patent Office)
AG	Attorney-General
BPAI	Board of Patent Appeals and Interferences
CA	Court of Appeal
Ch	Chancery
CP	Common Pleas
EBA	Enlarged Board of Appeal (European Patent Office)
ECJ	European Court of Justice
EWCA Civ	England and Wales Court of Appeal, Civil Division
EWHC	High Court of England and Wales
Ex	Exchequer
Ex Ch	Exchequer Chamber
FCA	Federal Court of Australia
Fed Cir	Federal Circuit
HC	High Court of Justice
HCA	High Court of Australia
HL	House of Lords
KB	King's Bench
LC	Lord Chancellor (Chancery)
MR	Master of the Rolls (Chancery)
NP	Nisi Prius
NZSC	New Zealand Supreme Court
PAT	Patents Appeal Tribunal
Pat	Patents Court (High Court of England and Wales)
PC	Privy Council
PE	Principal Examiner (Patent Office)
PO	Patent Office
QB	Queen's Bench
SE	Superintending Examiner (Patent Office)
SG	Solicitor General
TBA	Technical Board of Appeal (European Patent Office)
VC	Vice Chancellor (Chancery)

Notes on Citation

In footnotes to the text, cases are given one citation only. For UK cases decided before 11 January 2001, this is generally to the Reports of Patent Cases (RPC),

English Reports (ER), Webster's Patent Cases (Web Pat Cas), or another standard nominate report. UK cases decided after 11 January 2001 are cited in neutral form consistent with the Practice Direction on the Form of Judgments, Paragraph Marking and Neutral Citation (2001) 1 WLR 194, extended in effect to the High Court by Practice Direction (Neutral Citations) (2002) 1 WLR 346. European Patent Office citations that begin with the letter 'T' are to decisions of the Technical Board of Appeal, while European Patent Office citations that begin with the letter 'G' are to decisions or other documents of the Enlarged Board of Appeal. A more complete list of citations is included in the Table of Cases, prepared with assistance from the English Reports, the All England Reports Reprint, and Hayward's Patent Cases (Hay Pat Cas).

With respect to the European patent archival material, CM documents are of the Council of Europe's Committee of Ministers, EXP/Brev and CM/WP documents are of the Council of Europe's Committee of Experts on Patents, BR documents are of the Inter-Governmental Conference for the Setting Up of a European System for the Grant of Patents (Luxembourg Conference), and M documents are of the Munich Diplomatic Conference for the Setting up of a European System for the Grant of Patents (Munich Conference). In the footnotes to the text, archival material is cited in full for its first use, and in brief for all subsequent uses.

Key Sources, Institutions, and Agreements

ALJ	Australian Law Journal
All ER	All England Law Reports
AOJP	Australian Official Journal of Patents, Trade Marks and Designs
App Cas	Law Reports, Appeal Cases
Atk	Atkyns' Chancery Reports
B & Ald	Barnewall & Alderson's King's Bench Reports
Banks Report	United Kingdom, *The British Patent System: Report of the Committee to Examine the Patent System and Patent Law* (Cmnd 4407, 1970)
BingNC	Bingham's New Cases, English Common Pleas
Biotech Directive	Directive 98/44 of the European Parliament and of the Council of 6 July 1998 on the legal protection of biotechnological inventions (Biotech Directive) [1998] OJ L213/13
BIRPI	Bureaux Internationaux Réunis pour la Protection de la Propriété Intellectuelle
BNA	Bureau of National Affairs
Brod & Bing	Broderip & Bingham's Common Pleas Reports
Bull NP	Buller's Law of Nisi Prius
CB	Common Bench Reports

C&K	Carrington & Kirwan's Nisi Prius Reports
Carp Pat Cas	Carpmael's Patent Cases
Ch	Law Reports, Chancery
Ch App	Law Reports, Chancery Appeals
Ch D	Law Reports, Chancery Division
Cl&F	Clark & Finnelly's House of Lords Cases
CLR	Commonwealth Law Reports
CM	Council of Ministers (Council of Europe)
CM&R	Crompton, Meeson & Roscoe's Exchequer Reports
CofE	Council of Europe
Committee of Experts	Committee of Experts on Patents of the Council of Europe (1950 to 1969)
Coop	Cooper's Chancery Reports Tempore Eldon
Co Rep	Coke's King's Bench Reports
CPC	Convention for the European Patent for the Common Market (Community Patent Convention) (Luxembourg, 15 December 1975, 15 ILM 5)
D & E	Durnford & East's King's Bench Reports
Dav Pat Cas	Davies' Patent Cases
Dick	Dicken's Chancery Reports
E&B	Ellis & Blackburn's Queen's Bench Reports
EC	European Community
EEC	European Economic Community
EIPR	European Intellectual Property Review
EPC	Convention on the Grant of European Patents (European Patent Convention, as amended) (Munich, 5 October 1973, 13 ILM 268)
EPO	European Patent Office
EPOR	European Patent Office Reports
Eq Ca Abr	Equity Cases Abridged
ER	English Reports
ETS	European Treaty Series
EU	European Union
FSR	Fleet Street Reports
G	Gale's Exchequer Reports
GATT	The General Agreement on Tariffs and Trade
Godb	Godbolt's King's Bench Reports
Grif Pat Cas	Griffin's Patent Cases

Hay Pat Cas	Hayward's Patent Cases
H Bl	Henry Blackstone's Common Plea Reports
Holt KB	Holt's King's Bench Reports
Holt NP	Holt's Nisi Prius Reports
IIC	International Review of Intellectual Property and Competition Law
ILM	International Legal Materials
JPM	J Johnson & J Johnson, *The patentee's manual; a treatise on the law and practice of patents for inventions, with an Appendix of statutes, rules, and foreign and colonial patent laws, International convention and protocol* (6th edn, New York: Longmans, Green and Co, 1890)
J Pat & Trademark Off Soc'y	Journal of the Patent & Trademark Office Society
Jur	Jurist Reports
Keb	Keble's King's Bench Reports
LJ Ch	Law Journal Reports, Chancery New Series
LJ QB	Law Journal Reports, Queen's Bench New Series
LQR	Law Quarterly Review
LR Ch D	Law Reports, Chancery Division
LR CP	Law Reports, Common Pleas
LR Eq	Law Reports, Equity Cases
LT	Law Times Reports
Luxembourg Conference	Inter-Governmental Conference for the Setting Up of a European System for the Grant of Patents (1969 to 1972)
M&W	Meeson & Welsby's Exchequer Reports
Macph	Macpherson's Session Cases
Macr	Macrory's Patent Cases
Man & G	Manning & Granger's Common Pleas Reports
Mer	Merivale's Chancery Reports
MLR	Modern Law Review
Mod	Modern Reports
Moore	Moore's King's Bench Cases
Munich Conference	Munich Diplomatic Conference for the Setting up of a European System for the Grant of Patents (1973)
Newton's Lond J	W Newton, *The London Journal of Arts and Sciences; and Repertory of Patent Inventions* (London: Sherwood, Gilbert, and Piper, Paternoster Row, 1835)

Noy	Noy's Kings Bench Reports
NR	New Reports
NZLR	New Zealand Law Reports
OJ EPO	Official Journal of the European Patent Office
P Wms	Peere-Williams' Chancery & King's Bench Cases
PA	Patents Act 1977, as amended
PA 1977	Patents Act 1977, as enacted
Paris Convention	Paris Convention for the Protection of Industrial Property (Paris Convention, as amended) (Paris, 20 March 1883, 13 UST 1)
Parl Rep	Series 1 to 4 of the Offcial Report (Hansard) of the Parliamentary Debates
PAT	Patent Appeals Tribunal
Patents Working Party	Patents Working Party of the European Economic Community (1961 to 1965)
PCT	Patent Cooperation Treaty (Washington, 19 June 1970, 9 ILM 978)
PLG	Patents Liaison Group
PLG Report	*United Kingdom Patent Law: The Effects of the Strasbourg Convention of 1963* (Cmnd 2835, 1965)
Rolle	Rolle's King's Bench Reports
RPC	Reports of Patent Cases
Russ	Russell's Chancery Reports tempore Eldon
Salk	Salkeld's King's Bench Reports
Show K B	Shower's King's Bench Reports
SPC/Strasbourg Convention	Convention on the Unification of Certain Points of Substantive Law on Patents for Inventions (Strasbourg Convention) (Strasbourg, 27 November 1963, ETS 47)
Stark	Starkie's Nisi Prius Reports
TRIPS Agreement	Agreement on Trade-Related Aspects of Intellectual Property Rights (TRIPS Agreement) Marrakesh Agreement Establishing the World Trade Organization, Annex IC, Legal Instruments—Results of the Uruguay Round, vol 31 (15 April 1994, 33 ILM 81)
UKIPO	United Kingdom Intellectual Property Office
UNTS	United Nations Treaty Series
USPQ	United States Patents Quarterly
UST	United States Treaties and Other International Agreements

Ves & B	Vesey & Beames' Chancery Reports
Ves Jr	Vesey Junior's Chancery Reports
Ves Sup	Hovenden's Supplement to Vesey Junior's Chancery Reports
Vienna Convention	Vienna Convention on the Law of Treaties (Vienna Convention) (Vienna, 23 May 1969, UNTS 1155, 321)
Web Pat Cas	Webster's Patent Cases
White Paper	United Kingdom, *Patent Law Reform* (Cmnd 6000, 1975)
WIPO	World Intellectual Property Organization
WLR	Weekly Law Reports
WN	Weekly Notes of Cases
WR	Weekly Reporter
WTO	World Trade Organization
Y & C	Younge and Collyer's Chancery Cases

What is an 'invention' . . . is a topic bedevilled by verbal formulae – and by the sweeping of problems under the carpet. So, before I go any further I want to bring some of those problems out into the light of day. But first: does it really matter? Is it merely a sterile argument about the meaning of words? To which I answer that whoever controls the meaning of 'invention' controls what can be patented and hence an important aspect of industrial policy.

Mr Prescott QC[1]

1 *CFPH's Application* [2005] EWHC 1623, Pat, [9]–[10].

Introduction

A. The Requirement for an Invention and the Concept of Inherent Patentability

Under the European Patent Convention[1] (EPC), a patent may only be granted for an 'invention'.[2] This immediately raises the question, '*what is an invention?*' The answer, I suggest, is something the constitutive properties of which make it suitable for a patent; hence my use of the expression 'inherent patentability'[3] to denote the legal requirement for an invention.

Establishing the existence of an invention does not guarantee an entitlement to a patent. Rather, such an entitlement requires that the invention be 'patentable' in the sense of involving an 'inventive step', and being 'new' and 'susceptible of industrial application'.[4] An applicant must also satisfy certain other conditions, including that the subject matter for which a patent is sought is defined in claims that are 'clear and concise and . . . supported by the description' in the specification.[5]

It follows that the existence of an invention by itself is a necessary but not a sufficient condition for entitlement to a patent. It is nevertheless an important condition;[6] the threshold requirement for monopoly protection against which the 'relative'[7] criteria of novelty, inventive step, and industrial applicability are assessed,[8] and the resulting patent property defined.[9] This is critical, for it underlines the two aspects of inherent patentability properly construed, namely, the determination of the subject matter to which the patent system is capable of applying, and the restriction of protection to such subject matter when specifically conceived *qua* invention.

[1] Convention on the Grant of European Patents (European Patent Convention, as amended) (Munich, 5 October 1973, 13 ILM 268).

[2] EPC, Art 52(1). The corresponding UK provision is PA, s 1(1).

[3] For an instance of judicial use of this term, see *Halliburton Energy Services Inc v Smith International Inc* [2005] EWHC 1623, Pat per Pumfrey J.

[4] See EPC, Art 52(1). See also PA, s 1(1).

[5] See EPC, Art 84. See also PA, ss 14(5)(b), 14(5)(c).

[6] cf *Biogen Inc v Medeva plc* [1997] RPC 1, HL, 41 per Lord Hoffmann, discussed below at pages 6, 296–299.

[7] T_154/04 (*DUNS LICENSING ASSOCIATES/Estimating Sales Activity*) [2007] EPOR 38, [10].

[8] See EPC, Arts 54, 46, 57; also PA, ss 2, 3, 4.

[9] See EPC, Arts 69, 84; also PA, ss 60, 14(5)(a), 125.

B. The Term 'Invention'

A natural place to begin thinking about inherent patentability is with the term 'invention' by which it is denoted. That term has a long association with patent law, reflected in its use in the earliest Patent Rolls and literature of England.[10] While not included in the first English patent legislation, the Statute of Monopolies 1623,[11] it did appear in all subsequent legislation prior to 1977.[12] According to one commentator, this long association reflects the early common language meaning of the verb 'to invent', which was sufficiently wide that use of the word 'invention' to denote patentable subject matter 'was of great utility, for it shortened expression and imported no theory'.[13] The early meaning of the verb 'to invent' was 'to come upon, to find, to reach', and only derivatively 'to think out, to *excogitare*';[14] a meaning that ignored modern distinctions between original conception, discovery, and the acquisition of knowledge by other means.[15] In so doing, it also explains the use of the term 'invention' in early English patent custom to mean first introduction including importation.[16]

Consistent with this, on the eve of the introduction of the Patents Act 1977 (PA) it was widely accepted that conceptions of the invention depended not on the meaning of the term 'invention', but rather on the meaning of section 6 of the Statute of Monopolies. This was consistent with the reference to that section in all statutory definitions of the 'invention' from 1852.[17] Further, those definitions evolved little beyond the original definition, contained in section LV of the Patent Law Amendment Act 1852, as follows:

The Expression 'Invention' shall mean any Manner of new Manufacture the Subject of Letters Patent and Grant of Privilege within the Meaning of the Act of the Twenty-first Year of the Reign of King *James* the First, Chapter Three [the Statute of Monopolies].

[10] See WH Price, *The English Patents of Monopoly* (Cambridge: Harvard University Press, 1913) 8. Compare the situation regarding the term 'technology', which had no currency in UK patent law prior to the EPC, and limited non-legal currency before the 1829 publication of the first edition of J Bigelow, *Elements of Technology* (2nd edn, Boston: Hilliard, Gray, Little and Wilkins, 1831). On the latter, see HA Meier, 'Technology and Democracy, 1800–1860' (1957) 43 Mississippi Valley Historical Rev 618.

[11] 21 Jac I c 3.

[12] See An Act to Amend the Law Touching Letters Patent for Inventions 1835 (5 & 6 Wm IV c 83), s II; Patent Law Amendment Act 1852 (15 & 16 Vict c 83) s LV; Patents, Designs, and Trade Marks Act 1883 (46 & 47 Vict c 57) s 46; Patents and Designs Act 1907, s 93; Patents Act 1949, s 101. Section 6 of the Statute of Monopolies did include the terms 'inventor' and 'inventors'.

[13] A Daniell, 'Inventions and Invention' (1899) 11 Juridical Rev 151, 151.

[14] ibid; see also the Oxford English Dictionary definition of 'invention'.

[15] EW Hulme, 'The History of the Patent System under the Prerogative and at Common Law. A Sequel' (1900) LQR 44, 52.

[16] Price, *The English Patents of Monopoly*, 8.

[17] See Patent Law Amendment Act 1852, s LV; Patents, Designs, and Trade Marks Act 1883, s 46; Patents and Designs Act 1907, s 93; Patents Act 1949, s 101.

In interpreting such definitions, the UK courts tended often to focus on the language of section 6 of the Statute of Monopolies, and the requirement for a 'manner of new manufacture'.[18] Implicit in that phrase was a concern with advancing the manufacturing arts by rewarding ingenious methods for producing mechanical and chemical artefacts.[19] That concern was tested in the nineteenth century by the grant of patents for uses of such artefacts, resulting in the recognition of a threshold exclusion for manifestly non-inventive categories of subject matter. It was further tested in the twentieth century by the grant of patents for a range of non-mechanical and non-chemical subject matter, resulting in a conception of the invention as a human action of mechanical or chemical utility directed to advancing the manufacturing arts, and requiring to that end the production of a physical (and vendible) product. Finally, the concern was tested in the nineteenth and twentieth centuries by the grant of patents for products as such, resulting in expressions of judicial doubt regarding the inherent patentability of those products, independent of the methods by which they were made.

Two notable pre-modern exceptions to the focus of the courts on the statutory language were Eyre and Rooke CJJ in *Boulton v Bull*,[20] for whom the meaning of the requirement for an invention derived from the history and 'spirit' of the legislation.[21] Still, and despite the terms of the statutory definition above, that did not become the accepted view in modern UK law until the 1960s, following the High Court of Australia's decision in *National Research Development Corp v Commissioner of Patents*.[22] In that case, the Court held the Statute of Monopolies-based definition of 'invention' to depend on neither the meaning of the term 'invention', nor the existence of a 'manner of new manufacture', but rather the principles developed by the courts 'for the application of s. 6 of the Statute of Monopolies'.[23] The effect, on its face, was two-fold. First, it suggested a philosophical and policy-based conception of the invention, focused on what had previously been regarded as suitable for a patent.[24] And second, it confirmed the importance of judicial law-making in the area, notwithstanding the system's longstanding

[18] This is in the cases in which they construed the Statute, rather than merely assuming the issue and focusing (for example) on the sufficiency of the specification.

[19] For a discussion of conceptions of *techne* as 'purposive action' (a means to an end) and 'artefacts' (nuts and bolts) and the tension between them, see R Laudan, 'Natural Alliance or Forced Marriage? Changing Relations Between the Histories of Science and Technology' (1995) 36 Technology and Culture S17–S28.

[20] (1795) 126 ER 651, CP.

[21] ibid, per Rooke J, 666 per Eyre CJ.

[22] (1959) 102 CLR 252, HCA.

[23] ibid, 269.

[24] These conceptions are informed by the definitions of technology formulated by Misa in TJ Misa, 'Theories of Technological Change: Parameters and Purposes' (1992) 17 Science, Technology, & Human Values 3, 4–7.

statutory basis. That importance was recently noted by Lord Walker in *Synthon BV v SmithKline Beecham plc*, as follows:[25]

The law of patents is wholly statutory, and has a surprisingly long history. It has been wholly statutory since the Statute of Monopolies 1623, an important landmark in constitutional history because of its effect in curbing the royal prerogative. It is a field of law in which statutory change (although important in its cumulative effect, especially in reflecting international treaty obligations) has been a process, not of revolution, but of slow evolution. Bankruptcy law and, until recently, the law of rating are comparable in this respect. In these fields the courts have shown an inclination to enrich the bare simplicity of the statutory text with their own explanatory commentary. But from time to time this process is interrupted by periods of reappraisal. . . . This sort of judicial exposition is more than what the old cases called glossing the statute. In the interpretation and application of patent statutes judge-made doctrine has over the years done much to clarify the abstract generalities of the statutes and to secure uniformity in their application.

With the introduction of the current Patents Act, the 1949 version of the 1852 definition above has been repealed, and the definition of the 'invention' contained in Article 52(2) and (3) of the EPC transposed in its place.[26] The result is a substantially different provision. For a start, the definition does not refer to any prior legislation, suggesting that history is no longer its interpretive base. And second, it is not expressed in positive or purposive terms, but rather as a formal, albeit inclusive list of subject matter not capable of constituting an 'invention'.[27] That list is contained in section 1(2) of the PA, which reads as follows:

It is hereby declared that the following (among other things) are not inventions for the purposes of this Act, that is to say, anything which consists of –
(a) a discovery, scientific theory or mathematical method;
(b) a literary, dramatic, musical or artistic work or any other aesthetic creation whatsoever;
(c) a scheme, rule or method for performing a mental act, playing a game or doing business, or a program for a computer;
(d) the presentation of information;
but the foregoing provision shall prevent anything from being treated as an invention for the purposes of this Act only to the extent that a patent or application for a patent relates to that thing as such.

[25] [2005] UKHL 59, [56]–[57]. cf *Genentech Inc's Patent* [1989] RPC 147, CA, 258–259 per Mustill LJ.

[26] On the purpose of PA, s 1(2) in transposing EPC, Art 52(2), (3), see Hansard HL (series 5) vol 379, cols 1402–1411 (15 February 1977).

[27] On the tension between the definition's inclusiveness and expression as a formal list or 'code'. see ibid, col 1407 per Lord Lloyd.

C. The Origins of the Contemporary (EPC) Definition of the 'Invention'

The difference between section 1(2) of the PA and pre-1977 definitions of the 'invention' raises the question as to the origins and reasons for Article 52(2) and (3) of the EPC: where did they come from and why were they adopted? Their origins lie in a comparative study of national European laws that was undertaken in 1953 by the Secretariat-General of the Council of Europe's Committee of Experts on Patents, in the course of considering a 1949 proposal for the creation of a European patent office. In 1956, certain headings from that study were proposed by Germany as the basis for a possible definition of the 'invention' for inclusion in (what became) the Strasbourg Convention (SPC).[28] That proposal was rejected. In 1964, however, it was the basis for a further proposal by Germany and the Netherlands to a Working Party of the European Economic Community that was formed to create a Common Market patent. While that Party was subsequently disbanded, the proposal itself survived to influence the drafting of Rules 67.1 and 39.1 of the Patent Cooperation Treaty[29] (PCT), which Rules subsequently became the basis for the drafting of Article 52(2) and (3) of the EPC themselves.

Less complex is the reason for the Articles' adoption. The primary reason was to facilitate harmonization, including by ensuring the literal identity of the Articles with the text of Rules 39.1 and 67.1 of the PCT, notwithstanding the different purpose of those provisions.

D. The Uncertainty and Controversy Surrounding the Requirement for an Invention

Since the introduction of section 1(2) of the PA, the law of inherent patentability has been mired in legal and political controversy. Legally, the UK courts and EPO Boards of Appeal have each been 'rather deprecatory about the approach of the other', as the Court of Appeal in *Symbian v Comptroller-General*[30] expressed it.[31] Indeed, in *Aerotel Ltd v Telco Holdings Ltd*[32] the Court of Appeal described the approaches of the Boards as 'mutually contradictory',[33] and one of their approaches as 'not intellectually honest'.[34] Similarly in *DUNS LICENSING ASSOCIATES/Estimating*

[28] Convention on the Unification of Certain Points of Substantive Law on Patents for Inventions (Strasbourg Convention) (Strasbourg, 27 November 1963, ETS 47).

[29] Patent Cooperation Treaty (Washington, 19 June 1970, 9 ILM 978).

[30] [2008] EWCA Civ 1066.

[31] See ibid, [10].

[32] [2006] EWCA Civ 1371.

[33] ibid, [25].

[34] ibid, [27].

Sales Activity,[35] the EPO Technical Board described the UK approach as *mala fides*: not 'consistent with a good faith interpretation of the EPC'.[36] Following a request by the *Aerotel* Court,[37] the matter has now been referred by the EPO President to the Enlarged Board of Appeal for clarification.[38] At the time of writing that decision is yet to be published.

The controversy over the EPC requirement for an invention has not been confined to the courts. Indeed, even the European Community institutions have conflicted over the proper approach to Article 52(2) and (3), amid their concern to regulate the law of patentability, and rectify their exclusion from the EPC system.[39]

Underlying the legal uncertainty which such conflicts have generated remains a further issue and object of disagreement: the importance of the requirement for an invention at all. According to Lord Hoffmann in *Biogen Inc v Medeva plc*,[40] for example, the requirement will invariably be an 'academic' one, which is why (he suggested) the EPC drafters failed to reach agreement on the meaning of 'invention'.[41] By contrast, Lord Mustill believed that it might sometimes require 'close conceptual analysis', particularly in the context of new technologies, 'far distant from the mechanical and chemical inventions to which so much of traditional patent law relates.'[42] As a member of the Patents Court, Pumfrey LJ took a similar view,[43] as has Mr Prescott QC:[44]

> What is an 'invention' . . . is a topic bedevilled by verbal formulae – and by the sweeping of problems under the carpet. So, before I go any further I want to bring some of those problems out into the light of day. But first: does it really matter? Is it merely a sterile argument about the meaning of words? To which I answer that whoever controls the meaning of 'invention' controls what can be patented and hence an important aspect of industrial policy. There can be but one justification for having a patent system, and that is that it is good for the people of the country. If the patenting of certain things does more

[35] [2007] EPOR 38.

[36] ibid, [41].

[37] [2006] EWCA Civ 1371, [25].

[38] G_3/08 (*PRESIDENT'S REFERENCE/Computer Program Exclusion*) [2009] EPOR 9. This case will be the first of an Enlarged Board on the meaning of Art 52(2) and (3).

[39] See, eg *Opinion of the Economic and Social Committee on the Proposal for a Directive of the European Parliament and of the Council on the patentability of computer-implemented inventions* COM(2002) 92 final—2002/0047(COD) (19 September 2002) (criticizing the EPO's interpretation of EPC, Art 52(2) and (3) and *Proposal for a Directive of the European Parliament and of the Council on the patentability of computer-implemented inventions* COM(2002) 92 final (20 February 2002), both discussed in J Pila, 'Dispute Over the Meaning of "Invention" in Article 52(2) EPC: The Patentability of Computer-Implemented Inventions in Europe' (2005) 36 IIC 173–191.

[40] [1997] RPC 1, HL.

[41] ibid, 41.

[42] ibid, 31–32.

[43] *Shopalotto.com Ltd's Application* [2005] EWHC 2416, Pat, [6]–[7]; see also *Research in Motion UK Ltd v Inpro Licensing SARL* [2006] EWHC 70, Pat, [185] per Pumfrey J.

[44] *CFPH's Application* [2005] EWHC 1623, [9]–[10] per Mr Prescott QC.

harm than good, it matters. Patents that are wrongly granted can be very expensive to challenge and may deter small and medium enterprises.

E. The Need for a Conception and Ontology of the Invention . . .

As the last two sentences of Mr Prescott QC's statement above suggest, beyond the uncertainty regarding the importance of the requirement for invention lies one matter on which contemporary courts seem to agree: that the key to understanding that requirement does not depend on any exhaustive, positive definition of the 'invention', but rather on a list of things that are not 'inventions'. And yet, despite a large number of cases and extensive judicial discussion of the list that exists in Article 52(2) of the EPC, the law is still in a 'depressingly' uncertain state.[45] Further, in focusing on what the requirement for an invention excludes from the system, the courts have all but ignored what I suggest to be the second aspect of that requirement, namely, the restriction of patentability to subject matter *qua* invention. Consequently, they have all but ignored the fundamental question, '*what does it mean for a subject matter to be patentable* qua *invention?*' It is also in respect of this particular question that 'close conceptual analysis' is required. Hence my suggestion, that the resolution of the uncertainty that exists in this area depends on two things. The first is successfully articulating a positive conception of the invention. And the second is determining an ontology of the invention, and what it means to conceive a subject matter *qua* invention.

The need for a positive conception of the invention is underlined by the failure, since 1977, of negative definitions to offer a method by which to resolve the morass of hard cases that lie on the periphery of the current patent system.[46] While the existence of such cases is not of itself a cause for alarm, the absence of any coherent underlying theory of inherent patentability means that they assume a significance beyond that of cases in the usual penumbra of uncertainty common to most substantive legal concepts.[47] In particular, they illustrate the difficulty not only in identifying what makes a subject matter (for example) technical, but also in stating with confidence that inventions *are* technical, and *why* the properties constitutive of a technical process or artefact establish a subject matter's inherent suitability for a patent. In other words, the uncertainty relates to both the normative basis of the requirement for inherent patentability and the nature of the requirement itself. In this way, the hard cases demonstrate that the uncertainty lies at the heart of the law, and not merely at its boundaries.

The same applies with respect to the conception of subject matter *qua* invention, the importance of which has only recently been alluded to by the courts,

[45] See *AT&T Knowledge Ventures LP* [2009] EWHC 343, Pat, [15] per Lewison J.

[46] See HLA Hart, *The Rule of Law* (Oxford: OUP Clarendon, 1961) 14–15.

[47] ibid, 12.

along with its essentially philosophical nature. In *CFPH's Application*, for example, Mr Prescott QC emphasized the need for a subject matter to be novel and non-obvious '*under the description* "invention"'.[48] Similarly in *Merrell Dow Pharmaceuticals Inc v H N Norton & Co Ltd*,[49] Lord Hoffmann acknowledged that legal definitions of novelty depend on the law's answer to the question, '*what does it mean to know something, so that it can be part of the state of the art?*', and that the question is an epistemological one.[50] Consistent with this, my suggestion is that it is only by understanding what it means for something to exist *qua* invention, and ensuring that legal conceptions of individual subject matter reflect that understanding, that protection can be properly restricted to '*patentable inventions*', and the requirement for inherent patentability fulfilled.[51]

F. . . . Formulated in Light of the Purpose of the System, the Meaning of Europeanization, and Insights from the History and Philosophy of Science and Technology

The basis of the existing legal uncertainty regarding the requirement for an invention is philosophical and policy-based, and pertains in part at least to the role of the patent system, and the implications of that role for the requirement itself. Historically, the function of the patent system was to advance the industrial arts by granting exceptional and discretionary Crown monopolies for the first introduction of trades and devices capable of supporting a new industry. Central was the notion of industrial progress, or rather, social progress through the advancement of the industrial arts. Given this, it is not surprising that the century in which the Statute of Monopolies was enacted was also the century of Francis Bacon, René Descartes, and Isaac Newton. This is because it was their connection of applied science with utilitarian ends, as well as their ideas regarding physical laws and the need for an *arts mechanic* supported by a history of individual trades,[52] that were the basis for Western ideas of social progress, along with the Western project of industrial technology itself.[53] That connection is important, among other things,

[48] [2005] EWHC 1589, Pat, [93], emphasis in original.

[49] [1996] RPC 76, HL.

[50] ibid, 88, emphasis added.

[51] The requirement for a 'patentable invention' is contained in EPC, Art 52(1) and PA, s1(1); see also PA, ss 21(1), 72(1)(a), 74A(1)(b).

[52] cf, eg E Zilsel, 'The Sociological Roots of Science' (1942) 47 Am J of Sociology 544–562 (locating the origins of technology in the separation of the liberal and mechanical arts in around 1550); L Edelstein, 'Review: Recent Trends in the Interpretation of Ancient Science' (1952) 13 J History of Ideas 573–604 (discussing technological applications of ancient science).

[53] See generally J Bury, *The Idea of Progress: An Inquiry into its Growth and Origin* (New York: Dover Publications, 1932), especially ch 2; WE Houghton, 'The History of Trades: Its Relation to Seventeenth-Century Thought as Seen in Bacon, Petty, Evelyn, and Boyle' (1941) 2 J History of

for implying a non-legal conception of the invention as including a progressive and sociological component.

The reference here to technology is apposite, for in today's jurisprudence it is this concept which plays the central role in delimiting the threshold limits of the patent system, displacing, one might say, the role previously played by the concept of industry.[54] In particular, it pervades the decisions of the European Patent Office (EPO), and since the 2000 revision of the EPC, has been expressly reflected in the text of the EPC itself.[55] As revised, Article 52(1) of the EPC requires that patents be available for inventions 'in all fields of technology', consistent with Article 27.1 of the TRIPS Agreement.[56] In combination with the TRIPS requirement that patents 'be available and patent rights enjoyable without discrimination as to . . . the field of technology', and its definition of the aim of the patent system as being to 'encourage and reward technological innovation', this provision of Article 27.1 puts technology at the heart of the international patent system. Since the revision of Article 52(1) of the EPC above, and the EC's adoption of the TRIPS Agreement in 1994,[57] the concept of 'technology' is also at the heart of the European system.

However, and as the UK courts have all but recognized, the focus on 'technology' as informing and delimiting the requirement for an invention begs more questions than it has so far answered.[58] One reason is alluded to in Lord Walker's description of patent law above, as a history of 'judge-made doctrine . . . clarify[ing] the abstract generalities of the statutes . . . to secure uniformity in their application'. That reason is the normative vacuum in which patent doctrine has tended to be made. Indeed, and with some important exceptions, the contemporary law of inherent patentability has lacked principled theoretical analysis, either of the purpose of the

Ideas 33–60; KH Ochs, 'The Royal Society of London's History of Trades Programme: An Early Episode in Applied Science' (1985) 39 Notes and Records of the Royal Society of London 129–158.

[54] See, eg Paris Convention for the Protection of Industrial Property (Paris Convention, as amended) (Paris, 20 March 1883, 13 UST 1) Art 1 (defining patents as 'industrial property'), discussed at page 115; chs 1–3.

[55] Though note that r 43(1) of the EPC Implementing Regulations (requiring that the claims define the matter for which protection is sought in terms of the technical features of the invention) predates (as r 29) the 2000 EPC Revision, and that the UK patent system was described on the eve of the PA as 'an instrument for making known new technology' (see Hansard HL (series 5) vol 379, cols 234–235 (24 January 1977) per Lord-Elwyn Jones (LC)).

[56] Agreement on Trade-Related Aspects of Intellectual Property Rights (TRIPS Agreement), Marrakesh Agreement Establishing the World Trade Organization, Annex 1C, Legal Instruments—Results of the Uruguay Round, vol 31 (15 April 1994, 33 ILM 81).

[57] See Council Decision (of 22 December 1994) concerning the conclusion on behalf of the European Community, as regards matters within its competence, of the agreements reached in the Uruguay Round multilateral negotiations (1986–1994) [1994] OJ L336/1.

[58] See, eg *Symbian v Comptroller-General* [2008] EWCA Civ 1066, [13], [21]; *AT&T Knowledge Ventures LP* [2009] EWHC 343, [15] per Lewison J. See also *Re Bilski* 545 F 3d 943 (Fed Cir, 2008) (rejecting a 'technological arts' test of inherent patentability on the grounds, among others, of the ambiguity surrounding 'technology' and related concepts), in relation to which a decision of the US Supreme Court is currently pending.

patent system and the role of inherent patentability in furthering that purpose,[59] the implications for that purpose of harmonization,[60] or the substantial non-legal learning on the history and philosophy of science and technology. Instead, it has reflected the same pragmatic, legalistic, and casuistic preoccupations that characterized aspects of the development of UK law in this area prior to 1977, and that have been described as having characterized the development of other regimes at common law as well, including other intangible property regimes.[61]

The purpose of this book is to consider these issues through a study of the UK requirement for an invention in its historical and international statutory context. My interest in doing so is with what might loosely be described as the law's intellectual history: the way in which legal ideas are constructed. In the context of inherent patentability, legal ideas have been constructed primarily by the courts, the Boards of the EPO, and the drafters of the Strasbourg and European Patent Conventions. It is therefore primarily with these sources that my study is concerned. While I do not suggest that complete understanding can come from this method, and my study does not purport to be exhaustive, I do believe that reading the judgments of the courts and the records of the discussions and decisions of the other makers of our laws is essential to attaining that understanding. In particular, I believe that it is only by close analysis of the documentary record of the decisions and discussions in which the ideas of the law are expressed that we can begin to understand the intellectual history of our system—the way in which its central ideas have been created—as an essential part of contemplating its future.

[59] See PA David, 'Reflections on the Patent System and IPR Protection in the Past, Present and Future' in *Interviews for the Future* (Munich: European Patent Office, 2006) 115–140 (criticizing the patent system for having acquired a life of its own, independent of its original purpose).

[60] See S Weatherill, 'Why Object to the Harmonization of Private Law by the EC?' (2004) 12 Eur Rev Private L 633–660 (regarding the importance of considering the meaning of Europeanization in different legal contexts).

[61] See, eg J Getzler, *A History of Water Rights at Common Law* (Oxford: OUP, 2005) 7.

PART I

THE REQUIREMENT FOR AN INVENTION IN HISTORICAL PERSPECTIVE

1

The Origins of Patents for Inventions, and Early Conceptions of the Invention: Pre-1623

A. Introduction

The origins of UK patents for inventions lie in the patent custom[1] of fourteenth century England, and its evolution over three centuries into a body of discrete legal principles that was given legislative expression in section 6 of the Statute of Monopolies 1623.[2] The historical significance of that provision is twofold. First, it established the exception to the general prohibition against monopoly grants on which the UK patent system was based until 1977. And second, it provided the express basis for all statutory definitions of the 'invention' from 1852 to 1977. At the time of its enactment, however, the aim of the Statute of Monopolies was

[1] 'Custom' is used here in accordance with its accepted legal definition of '[a] rule of conduct, obligatory on those within its scope, established by long usage' (R Bird, *Osborn's Concise Law Dictionary* (7th edn, London: Sweet & Maxwell, 1983)). Its use reflects the general agreement among patent historians that it is inappropriate to speak of an English patent system or law before the seventeenth or eighteenth century, and is consistent with the reported recognition by the High Courts of England in the early seventeenth century that the Crown's power to issue 'monopoly patents to inventors' derived from 'prescription' or 'ancient custom'. See FD Prager, 'Historic Background and Foundation of American Patent Law' (1961) 5 Am J Leg Hist 309, 309–314; W Holdsworth, *A History of English Law* (3rd edn, London: Sweet & Maxwell, 1945) vol iv, 343 (describing the early English patent grants as justified by custom); DS Davies, 'Further Light on the Case of Monopolies' (1932) 48 LQR 394, 396 (dating the commencement of the English patent system as a system to the second year of Elizabeth I's reign); C MacLeod, *Inventing the Industrial Revolution: The English Patent System, 1600–1800* (Cambridge: Cambridge University Press, 1988) 1 (describing the 'patent system' prior to 1800 as a 'misnomer'); TM Meshbesher, 'The Role of History in Comparative Patent Law' (1996) 78 J Pat & Trademark Off Soc'y 594, 602 (describing the evolution of patent law in England and other European countries from a set of customs to a settled body of principles or rules of law, and defining 'custom' in that context to mean 'a system, generally in the hands of the sovereign, in which rewards were granted to inventors or importers of inventions in the form of temporary and exclusive rights to exploit the subject matter of the grant').

[2] 21 Jac I c 3. For a list of important recent studies of this evolution, see EC Walterscheid, 'The Early Evolution of the United States Patent Law: Antecedents (Pt 2)' (1994) 76 J Pat & Trademark Off Soc'y 849, 849 n 2. This list might reasonably be extended to include several of the further works cited above and below, including the first two parts of Walterscheid's own five-part study of the early evolution of US Patent Law, published in (1994) 76 J Pat & Trademark Off Soc'y 697–715, 849–880. On the earlier historical roots of patents, see E Kaufer, *The Economics of the Patent System* (Chur: Harwood Academic Press, 1989) 1–3; M Frumkin, 'The Origin of Patents' (1945) 27 J Pat & Trademark Off Soc'y 143, 143–144; RA Klitzke, 'Historical Background of the English Patent Law' (1959) 41 J Pat & Trademark Off Soc'y 615, 617–618.

not to give rise to a new body of law, but rather to codify existing principles of English law. It follows that any history of legal conceptions of the invention in the UK must begin with a history of those principles, and the interpretive context of the Statute in which they were expressed.

B. The Royal Custom of Encouraging the Establishment of New Industries

The practice of granting patents in England originated with the feudal custom of conferring privileges on the early merchant and craft guilds in order to reserve to them exclusive control over the manufacture of goods and the provision of services.[3] The guilds were essentially group monopolies controlled by the Crown,[4] and dominated all aspects of domestic English trade throughout the twelfth and thirteenth centuries.[5] By the fourteenth century, however, England's economy was trailing behind that of Europe, and its industry had outgrown the jurisdiction and capacity of the guilds.[6] There was therefore an increasing recognition of the need for some form of special stimulus to encourage the creation of new local industries and the greater exploitation of local resources, whether through innovation or the importation of successful industries from abroad.[7] Initially, importation was preferred and in 1326 Edward II formulated a policy of encouraging skilled foreign workers to settle and introduce their trades in England.[8] Under that policy, open 'letters of protection' were offered promising exemption from the strict guild rules and other forms of royal patronage in exchange for an undertaking to work a trade within the realm and apprentice locals in its working.[9] The first grant under this policy[10] was reportedly made in 1331 to a wool weaver, John Kempe of Flanders.[11]

[3] See generally HG Fox, *Monopolies and Patents: A Study of the History and Future of the Patents Monopoly* (Toronto: University of Toronto Press, 1947) 35–38.

[4] FL Vaughan, *The United States Patent System: Legal and Economic Conflicts in American Patent History* (3rd edn, Norman: University of Oklahoma Press, 1977) 13.

[5] See Walterscheid, 'The Early Evolution of the United States Patent Law: Antecedents (Pt 2)', 851–852.

[6] WH Price, *The English Patents of Monopoly*, (Cambridge: Harvard University Press, 1913) 16.

[7] EC Walterscheid, 'The Early Evolution of the United States Patent Law: Antecedents (Pt 1)' (1994) 76 J Pat & Trademark Off Soc'y 697, 706.

[8] ibid. cf MacLeod, Inventing the Industrial Revolution, 10; PE Schaafsma, 'An Economic Overview of Patents' (1997) 79 J Pat & Trademark Off Soc'y 241, 245, both referring to the policy as having been introduced by Edward III.

[9] ibid.

[10] Compare the first grant still traceable on the Patent Rolls, which was to John of Utynam in 1449, who undertook to make coloured glass required for the windows of Eton and other college chapels within the realm, and to 'instruct divers lieges of the King' in glass making and other arts theretofore unknown in England, in exchange for the exclusive right to practice those arts for 20 years. For a discussion, see P Meinhardt, *Inventions, Patents and Trade Marks* (London: Gower Press, 1971) 21; Klitzke, 'Historical Background of the English Patent Law', 627.

[11] Vaughan, *The United States Patent System: Legal and Economic Conflicts in American Patent History*, 13. For the text of the grant, see EW Hulme, 'The History of the Patent System under the Prerogative and at Common Law' (1896) 12 LQR 141, 142. Foreign Clothworkers Act 1337 (11 Edw III c 5)

The patronage offered in early letters of protection did not generally include monopoly grants.[12] The making of such grants to encourage the importation of new industries did, however, become usual at around the same time in the city-states of Italy.[13] In 1421, for example, the Italian engineer and architect Filippo Brunelleschi was given the exclusive right for three years to manufacture and use a device of his own conception for transporting heavy loads on water.[14] Fifty-three years later, the first general patent statute was enacted in Venice, promising anyone who built a new and useful device the exclusive right to manufacture it for ten years.[15]

With the accession of the Tudor dynasty, the mediaeval practice of granting letters of protection in England had given way to a system of secret negotiations conducted by the Crown to attract skilled foreign artisans into its service.[16] During the sixteenth century, however, knowledge of the Italian practice spread throughout England and the continent.[17] The earliest evidence of this for England is a letter dated 1537 from a Venetian silk maker, Antonio Guidotti, to Thomas Cromwell, informing him that Guidotti had persuaded some Italian silk weavers to go to England and practice their craft, and asking that he intercede with the King to secure Guidotti a privilege for fifteen or twenty years to prevent others from doing the same.[18] This letter has been seen as evidence of the influence of the Italian patent practices on the English Crown during the sixteenth century,[19] coming as it did fifteen years before Edward VI granted a twenty-year monopoly to Henry Smyth in consideration for undertaking to introduce into England

confirmed the grant by providing that all cloth workers of other countries would be given special franchises and privileges if they settled in England and practiced and taught their trades (PJ Frederico, 'Operation of the Patent Act of 1790' (1936) 18 J Pat & Trademark Off Soc'y 237, 293).

[12] Prager, 'Historic Background and Foundation of American Patent Law', 309.

[13] Schaafsma, 'An Economic Overview of Patents', 243. See also G Mandich, 'Venetian Origins of Inventors' Rights' (1960) 42 J Pat & Trademark Off Soc'y 378 (translated by FD Prager) 379; M Silberstein, 'The Patents of Marini, 1443 to 1457' (1955) 37 J Pat & Trademark Off Soc'y 674–676.

[14] The purpose of the patent was to facilitate Brunelleschi's construction of the cupola of the cathedral of Florence, and the terms of the grant included that the work of anyone imitating his invention would be burned. For a discussion, see AA Gomme, *Patents of Invention, Origin and Growth of the Patent System in Britain* (London: Longmans Green and Co, 1946) 6.

[15] See Walterscheid, 'The Early Evolution of the United States Patent Law: Antecedents (Pt 1)', 708–709 (reproducing the Statute in full). On the origins of the Statute, see Mandich, 'Venetian Origins of Inventors' Rights'. On its impact, see Kaufer, *The Economics of the Patent System*, 6.

[16] Hulme, 'The History of the Patent System under the Prerogative and at Common Law', 144 (describing the negotiations as a 'perversion of the mediaeval policy of the encouragement of industry').

[17] Prager has argued that the widespread reception of identical patent forms in Germany, France, the Netherlands, and England in the period immediately after 1550 indicates that a supra national driving force emanating from the idea of intellectual property was also at work (see FD Prager, 'The Early Growth and Influence of Intellectual Property' (1952) 34 J Pat & Trademark Off Soc'y 106, 108).

[18] Klitzke, 'Historical Background of the English Patent Law', 629.

[19] See, eg Gomme, *Patents of Invention, Origin and Growth of the Patent System in Britain*, 8–9; cf Davies, 'Further Light on the Case of Monopolies', 397.

foreign workmen 'mete and experte' in the making of Normandy glass.[20] The grant to Henry Smyth was the first of its kind in over a century,[21] and marks the time when the English custom of granting patents for inventions began to regularize.[22] That custom received a firmer basis in 1561, when Elizabeth I, three years after acceding to the throne, pursued an industrial patent policy in the hope of making England self-sufficient.[23]

C. The Elizabethan Policy of Granting Monopolies to Encourage 'Others to Study and Seke for the Knowledge of . . . Good Engines and Devyses' and Other 'Things Useful for the Public'

As with the 1552 grant to Henry Smyth, the Elizabethan policy of 1561 has been traced to a petition by an Italian manufacturer. Writing in 1559, Jacobus Acontius sought protection against the unauthorized use of furnaces and 'wheel machines' he had invented,[24] supporting his request with the following plea:[25]

> Nothing is more honest than that those who by searching have found out things useful to the public should have some fruits of their rights and labors, as meanwhile they abandon all other modes of gain, are at much expense in experiments, and often sustain much loss, as has happened to me. I have discovered most useful things, new kinds of wheel machines, and of furnaces for dyers and brewers, which when known, will be used without my consent, except there be a penalty, and I, poor with expenses and labor, shall have no returns. Therefore I beg a prohibition against using any wheel machines, either for grinding or bruising, or any furnaces like mine without my consent.

The idea that a person who introduced 'a new kind of [useful] machine' should be rewarded by a monopoly in respect of its use seems to have influenced the Queen and informed her subsequent patent policy. Pursuant to that policy, patents were granted to reward the investment of those who introduced new things of use to the public, and thereby encourage others to do the same. This can be seen from the text of an early Elizabethan patent for the importation of a dredging machine, dated 26 May 1562, which was conferred in the hope that it would 'give courage to others to study and seke for the knowledge of like good engines and devyses.'[26]

[20] Fox, *Monopolies and Patents: A Study of the History and Future of the Patents Monopoly*, 60–61.

[21] The last patent to have been granted was that to John of Utynam 103 years earlier; see n 10.

[22] Klitzke, 'Historical Background of the English Patent Law', 630.

[23] ibid, 632–633.

[24] Frederico, 'Operation of the Patent Act of 1790', 296–297.

[25] Fox, *Monopolies and Patents: A Study of the History and Future of the Patents Monopoly*, 27. A patent was granted to Acontius in 1565; see Patent No XVI (dated 7 September 1565) listed in Hulme, 'The History of the Patent System under the Prerogative and at Common Law', 148.

[26] Hulme, 'The History of the Patent System under the Prerogative and at Common Law', 146.

As with the mediaeval letters of protection, the objective of the Elizabethan policy was industrial growth: the stimulation of English industry through the increase in native industries and the greater exploitation of native resources.[27] Also in line with the mediaeval practice was the initial focus of the Crown in pursuing this objective, which was on the importation of continental manufactures as a way of expanding local industry without disrupting existing employment or practice. This concern accounts for three related aspects of the Elizabethan custom. First, patents were granted in exchange for the introduction within the realm of a manufacture capable of supporting a new industry, with 'introduction' in this sense requiring that the manufacture be worked and locals apprenticed in its working.[28] Second, the manufacture had to be new, in the sense of not having been worked within the realm in recent memory.[29] And third, patents were not granted for any manufacture the establishment of which would 'inconvenience' the public, including by impairing local employment.[30] The last of these is of particular interest, as it suggests an early recognition that the industrial interests furthered by the policy might be overridden by competing public concerns, and perhaps even a recognition of social benefit as a constitutive element of the early invention.[31]

The procedures for obtaining patents under the old English patent custom were elaborate and involved several layers of checks and balances to ensure that no document acquired the Great Seal without the Crown's knowledge and sanction.[32] Broadly speaking they involved the submission of a petition and supporting declaration to the Crown through one of its Law Officers, the examination of those documents by the Officer for compliance with certain formalities, the preparation

[27] On the proportion of native and foreign inventions during this period, see EW Hulme, 'The History of the Patent System under the Prerogative and at Common Law: A Sequel' (1900) 16 LQR 44, 52; C MacLeod, *Inventing the Industrial Revolution* 13.

[28] MacLeod, *Inventing the Industrial Revolution*, 12.

[29] See Hulme, 'The History of the Patent System under the Prerogative and at Common Law: A Sequel', 56 (quoting from the Patent Rolls of 1562).

[30] MacLeod, *Inventing the Industrial Revolution*, 13; Hulme, 'The History of the Patent System under the Prerogative and at Common Law', 152 (citing the Elizabethan concern with protecting existing industry as a reason for its unwillingness to grant patents for 'first-finders' or 'inventors'). See also Walterscheid, 'The Early Evolution of the United States Patent Law: Antecedents (Pt 2)', 859; E Coke, *Institutes on the Laws of England* (London, 1628) vol 3, 181 ('a mans trade is accounted his life, because it maintaineth his life; and therefore the monopolist that taketh away a mans trade, taketh away his life, and therefore is so much the more odious'), quoted in A Mossoff, 'Rethinking the Development of Patents: An Intellectual History, 1550–1800' (2001) 52 Hastings LJ 1255, 1263. Clauses of revocation for 'inconveniency' appear in patent grants from 1575; see DS Davies, 'The Early History of the Patent Specification' (1934) 50 LQR 86, 102.

[31] While in practice 'inconveniency' clauses were invoked for the central purpose of protecting local workers, in theory the clauses provided the Crown and Privy Council with an unfettered discretion in respect of patentability. See Davies, 'The Early History of the Patent Specification' 102–104. The fact of their limited application is thus best viewed as reflecting the social preoccupations of Elizabethan England rather than any restrictive interpretation of the 'inconveniency' clauses themselves. See MacLeod, *Inventing the Industrial Revolution*, 13.

[32] N Davenport, *The United Kingdom Patent System: A Brief History with Bibliography* (Portsmouth: Mason, 1979) 15.

by the Officer of a report for the Crown recommending that a patent be granted or denied, and the rejection of the petition (without reasons or avenues for appeal) or, if successful, the preparation of several Crown Bills and eventually of letters patent communicating conferral of the grant. The formalities for which petitions were examined included the provision of undertakings with respect to the trade or device's novelty and public utility, and the petitioner's intention to introduce it within the realm.[33] While patents were generally granted on the strength of those undertakings alone,[34] and without inquiry into the nature of the trade or device, there is some evidence of the Crown taking practical steps to ascertain a petitioner's ingenuousness before approving a grant.[35] In addition, from 1575 patents regularly contained a clause invalidating the grant if the invention was not put into practice, and in 1639 a proclamation provided for the invalidation of any patent the subject matter of which was not introduced within three years of its conferral.[36]

D. The Abuse of the Elizabethan Policy and Devolution of Patent Jurisdiction to the Common Law Courts

In the first two years of the Elizabethan policy, patents were granted for new manufactures only and for a limited term of ten years.[37] Soon, however, a depleted Crown exchequer and a desire to patronize favoured courtiers led to a series of renewed or extended-term patents for existing trades and known devices,[38] including some patents for the manufacture of basic commodities such as starch, salt, and vinegar, for which there were already well-established local industries.[39]

33 EW Hulme, 'On the Consideration of the Patent Grant, Past and Present' (1897) 13 LQR 313, 315.

34 Klitzke, 'Historical Background of the English Patent Law', 639; also Davies, 'The Early History of the Patent Specification', 100–106; Davenport, *The United Kingdom Patent System*, 46. Hulme describes the early patent grants as including a clause stating that the patent should be favourably construed at law 'notwithstanding the not full and certain describing of the nature and quality of the said invention, or of the materials thereunto conducing and belonging' (see Hulme, 'On the Consideration of the Patent Grant, Past and Present', 313).

35 Davies, 'The Early History of the Patent Specification', 106–107.

36 Davenport, *The United Kingdom Patent System*, 46.

37 See Patents No I, II, and III (dated 3 January 1561, 8 August 1561, and 26 May 1562 respectively), reproduced in Hulme, 'The History of the Patent System under the Prerogative and at Common Law', 145–146. Patents No III and IV (dated 31 December 1562) were granted for 20 years each. See Hulme, 146.

38 See Price, *The English Patents of Monopoly*, 8–9 (noting that by the second decade of the Elizabethan patent policy, grants for 20, 21, and 30 years had become more usual than the initial 10-year grants).

39 See Patents No XXXVI, XXXXVII, and XL for the manufacture of salt (dated 26 June 1582, 22 September 1582, and 1 September 1585 respectively), XLIII for the manufacture of starch (dated 15 April 1588), and LI for the manufacture of vinegar (dated 23 March 1594), reproduced in Hulme, 'The History of the Patent System under the Prerogative and at Common Law: A Sequel', 47, 49, 50.

The grant of these 'odious monopolies' was devastating for the public,[40] and provoked intense political opposition that led, in 1601, to a debate between the Crown and the House of Commons over the scope of the royal prerogative.[41] In the course of that debate, the Attorney-General Sir Francis Bacon provided a list of monopolies which the Crown maintained were within its power to grant. Included on his list were monopolies of limited duration granted 'in regard of [the] Pains, Travel and Charge' incurred by 'any man [who] out of his own Wit, Industry, or Endeavour, find out anything Beneficial for the CommonWealth, or bring any New Invention, which every subject of this Realm may use'.[42] The debate was eventually resolved and threatened legislative action forestalled by a royal commitment to submit the worst monopolies to the Exchequer, and all monopolies to 'Tryal... in the *Common Pleas*'.[43] Thus began the devolution of the Crown courts' exclusive jurisdiction in respect of patents for inventions to the common law courts.

The first occasion for the exercise of that jurisdiction came the following year in the *Case of Monopolies*.[44] At issue in that case was the validity of a patent for the importation, manufacture, and sale of playing cards.[45] According to the report—which was likely drafted by the rumoured author of the Statute of Monopolies itself, the then Attorney-General, Sir Edward Coke—the Court invalidated the patent as a monopoly grant, explaining its decision as follows:[46]

> [T]here are three inseparable incidents to every monopoly against the commonwealth[:] [t]hat the price of the same commodity will be raised, for he who has the sole selling of any commodity, may and will make the price as he pleases[;] that after the monopoly granted, the commodity is not so good and merchantable as it was before: for the patentee having the sole trade, regards only his private benefit, and not the common wealth[; that] it tends to the impoverishment of divers artificers and others, who before, by the labour of their

[40] See generally Holdsworth, *A History of English Law*, 347. As memorably put by Vaughan, the odious monopolies meant that '[f]or a small pittance most of the life necessities were farmed out as private monopolies'. Vaughan, *The United States Patent System: Legal and Economic Conflicts in American Patent History*, 14. For a classification of the Elizabethan monopolies, see Davies, 'Further Light on the Case of Monopolies', 397–398, quoting from E Lipson, *The Economic History of England* (London: A & C Black Ltd, 1929) vol III, 352–356.

[41] The focus of the debate was the proposed Act for the Explanation of the Common Law in Certain Cases of Letters Patent, which sought to clarify the common law limits of the Crown's power to grant monopolies (see 'Proceedings in the Commons, 1601: November 16th – 20th', *Historical Collections: An exact Account of the Proceedings of the Four last Parliaments of Q. Elizabeth* (1680) pp 216–236; available at <http://www.british-history.ac.uk>).

[42] ibid.

[43] ibid.

[44] *Darcy v Allin* (*Case of Monopolies*) (1602) 77 ER 1260, KB. For a discussion, see Davies, 'Further Light on the Case of Monopolies'; JI Corre, 'The Argument, Decision, and Reports of Darcy v Allen' (1996) 45 Emory LJ 1261–1327.

[45] See Patent No LIV (dated 11 August 1598), reproduced in Hulme, 'The History of the Patent System under the Prerogative and at Common Law: A Sequel', 51.

[46] 77 ER 1260, 1263 (footnotes omitted).

hands in their art or trade, had maintained themselves and their families, who now will of necessity be constrained to live in idleness and beggary.

Aside from this passage, the *Case of Monopolies* is significant for the report it contains of the argument of Mr Fuller, the defendant's counsel, in favour of allowing patents for inventions:[47]

Now therefore I will shew you how the Judges have theretofore allowed of monopoly patents, which is, that where any man by his own charge and industry, or by his own wit or invention doth bring any new trade into the realm, or any engine tending to the furtherance of a trade that never was used before: and that for the good of the realm: that in such cases the King may grant to him a monopoly patent for some reasonable time, until the subjects may learn the same, in consideration of the good that he doth bring by his invention to the commonwealth: otherwise not.

This passage, which drew on the language used by Bacon in the House of Commons, appears again in the report of the *Clothworkers of Ipswich Case* of 1615.[48] The effect of its repetition was to support both the presumptive illegality of monopoly patents, and the recognition of an exception 'for the good of the realm' in respect of certain new trades and devices.

The *Case of Monopolies* was decided in 1602, the year before Elizabeth I's death and the accession of James I to the throne. Contrary to an early undertaking,[49] James I soon revived the practice of granting monopolies in respect of established trades, sparking new petitions of grievance. Those petitions resulted, in 1610, in the publication of the King's *Book of Bounty*,[50] in which all patents of monopoly were declared to be against the law and the King's royal pleasure, with a single exception for monopolies in respect of 'projects of new invention, so they be not contrary to the Law, nor mischievous to the State, by raising prices of commodities at home, or hurt of trade, or otherwise inconvenient'.[51] Despite this declaration, the Crown abuses of the patent custom continued, and in 1620 Parliament was forced to act by undertaking an extensive review of public grievances in respect of monopolies.[52] That review resulted in the enactment of the Statute of Monopolies,

[47] 74 ER 1131, 1139.

[48] (1615) 77 ER 1218, KB.

[49] See his proclamation of 7 May 1603, reproduced in Price, *The English Patents of Monopoly*, 163.

[50] King James I, *A Declaration of His Majesties Royall pleasure, in what sort He thinketh fit to enlarge, or reserve Himselfe in matter of Bountie* (London: Robert Barker, 1610), republished as an appendix in W Gordon, *Monopolies by Patents* (London: Stevens and Sons, 1897), and available at <http://www.archive.org/details/declarationofhis00greauoft>.

[51] ibid, 21.

[52] Walterscheid, 'The Early Evolution of the United States Patent Law: Antecedents (Pt 2)', 872–873.

by which all monopoly grants were declared illegal, with a limited exception in section 6 as follows:

Provided also, and be it declared and enacted, that any declaration, before-mentioned, shall not extend to any letters patent and grants of privilege for the term of fourteen years, or under, hereafter to be made, of the sole working or making of any manner of new manufactures within this realm, to the true and first inventor and inventors of such manufactures, which others at the time of making such letters patent and grants shall not use, so as also they be not contrary to the law, nor mischievous to the state, by raising prices of commodities at home, or hurt of trade, or generally inconvenient. The said fourteen years to be accounted from the date of the first letters patent of grants of such privilege hereafter to be made, but that the same shall be of such force as they should be, if this act had never been made, and of none other.

E. Section 6 of the Statute of Monopolies: The Common Law Invention in its Original Form

According to section 6 of the Statute of Monopolies, patents could only be granted by the Crown for 'the sole working or making of any manner of new manufactures within this realm, to the true and first inventor and inventors of such manufactures . . . so as also they be not contrary to the law, nor mischievous to the state . . . or generally inconvenient'. Writing in the 1840s, Hindmarch read the phrase 'manner of new manufacture' as including both the art of making something by workmanship and the thing resulting from the practice of that art; an interpretation he attributed to common language meaning.[53] The question remained as to the nature of inventions; a question which Hindmarch answered as follows:[54]

The patent . . . purports to grant . . . a privilege in the *invention*, and the question is, what is meant by the word invention, which according to the patent the patentee is to have the sole use of?

It is important to distinguish between the use of an invention which a patentee has thus granted to him, and the articles or results which may be produced by means of the invention. . . . If any person, except the patentee, make articles according to the patentee's invention, he commits an infringement of the patent, and yet the privilege granted by the patent did not vest any right in the patentee to make the particular articles which have been made in violation of the patent, for the materials of which they are composed were never his property, and therefore he could not exercise any control over them. But the person who thus makes such articles must necessarily use the art invented by the patentee, and so infringes the patent.

[53] WM Hindmarch, *A Treatise on the Law Relating to Patent Privileges for the Sole Use of Inventions; and the Practice of Obtaining Letters Patents for Inventions: With an Appendix of Statutes, Rules, Forms, &c. &c.* (London: Stevens and Norton, 1846) 78.

[54] ibid, 56–57.

The only thing therefore which can be the subject of the privilege granted by a patent, is the ART by means of which the articles are made or manufactured . . . The word invention then in a patent grant means the art which the patentee alleges that he has invented, and therefore that art is the subject of the privilege which is granted by a patent.

The conception of the invention in this passage as an art of manufacture, and not a product resulting from the practice of such an art, nor the practice of any other industrial art, is consistent with the language of section 6 and the majority of sixteenth and seventeenth century patents. On the other hand, it is more restrictive than the conception suggested by the *Case of Monopolies*, and does not explain the few sixteenth century grants for such non-manufacturing (and even non-industrial) methods as teaching shorthand,[55] drainage,[56] mining,[57] and conserving fuel.[58]

Ultimately, and as Hindmarch himself emphasized, the best insight into the meaning of section 6 at the time of its enactment is to be found in Coke's contemporaneous legislative commentary, and the jurisprudence which section 6 purportedly codified. Written in 1628, the former described the validity of a patent as depending on the following specific properties:[59]

First, it must be for twenty-one years or under. Secondly, it must be granted to the first and true inventor. Thirdly, it must be of such manufactures, which any other at the making of such letters patent did not use. . . . Fourthly, the privilege must not be contrary to law. . . . Fifthly, nor mischievous to the state, by raising the prices of commodities at home. In every such new manufacture that deserves a privilege, there must be *urgens necessitas et evidens utilitas*. Sixthly, nor to the hurt of trade. . . . Seventhly, nor generally inconvenient.

With respect to the nature of 'new manufactures', Coke stated as follows:[60]

[A new manufacture] must be substantially and essentially newly invented; but if the substance was *in esse* before, and a new addition thereunto, though that addition made the former more profitable, yet it is not a new manufacture in law. And so it was resolved in the Exchequer Chamber (E T 15, E 4), in Bircot's Case, for a privilege concerning the preparing and melting &c of lead ore; for there it was said, that that was to put but a new button to an old coat, and it is much easier to add than to invent. And there it was also resolved, that if the new manufacture be substantially invented according to law, yet no old manufacture in use before can be prohibited.

[55] See, eg Patent No XLIV (dated 26 July 1588) 'to teach, print, and publish works in shorthand', reproduced in Hulme, 'The History of the Patent System under the Prerogative and at Common Law: A Sequel', 49.

[56] See, eg Patent No XXVIII (dated 8 July 1573) to 'put into practice an instrument for land drainage, and for the stopping of breaches in dams', reproduced in ibid, 46.

[57] See, eg Patents No IX (dated 10 October 1564) 'for mining in eight English Counties' and XXI (dated 20 April 1569) 'to use the arte of myninge', reproduced in Hulme, 'The History of the Patent System under the Prerogative and at Common Law', 147, 149.

[58] See, eg Patent No XVIII (dated 26 August 1567) 'to cut iron, save fuel and extract oil', reproduced in ibid, 149.

[59] Coke, *Institutes on the Laws of England*, vol 3, 181 (reproduced in Web Pat Cas 31, n (p)).

[60] ibid.

The implication of this passage, that a new improvement to an existing device would lack the invention required to justify a patent, even if commercially significant, suggests an early requirement for a minimum degree of ingenuity as inherent in the concept of a new manufacture.[61] Aside from *Bircot's Case*, support for such a requirement can be found in two decisions from the sixteenth century. The first concerned a patent for knives made 'with bone hafts and plates of lattin' that differed only 'lightly' from knives previously made in England,[62] which was reportedly invalidated on the ground that 'a light difference or invention should be no cause to restrain [previous manufacturers], whereupon [the patentee] could never have benefit of this patent, although he laboured very greatly therein.'[63] And the second concerned a patent for a particular type of cloth, which was similarly invalidated on the ground that the cloth was 'very like to' cloth previously manufactured.[64] Both cases support the principle suggested by Coke, that a subject matter of light or insubstantial invention would not justify the restraint on 'old manufacture' that a monopoly for it would entail. In so doing they underline the exceptional nature of early patents as consideration for the establishment of new trades and industries, and suggest an alternative or additional basis for the early principle of law that a mere improvement to an existing trade or industry[65] was not something for which a patent might validly be granted. That basis, supported by Hulme, was the requirement for utility, and the view that any subject matter 'prejudicially affecting an existing industry' would inconvenience rather than benefit the realm. That the social impact of a subject matter might be relevant to its status as an invention implies a conception of the latter as having an ethical component, consistent with the historical (seventeenth century) connection between the ideas of the practical arts and of social progress.[66]

[61] See FD Prager, 'Standards of Patentable Invention from 1474 to 1952' (1952–53) U Chicago L Rev 69, 71; cf A Daniell, 'Inventions and Invention' (1899) 11 Juridical Rev 151, 160; Hulme, 'The History of the Patent System under the Prerogative and at Common Law: A Sequel', 56.

[62] *Matthey's Patent*, (*c*1580) Web Pat Cas 5, 6–7, PC; see also Hulme, 'The History of the Patent System under the Prerogative and at Common Law: A Sequel', 45; TM Stevens, *Lewis Edmunds, The Law and Practice of Letters Patent for Inventions* (2nd edn, London: Stevens and Sons, 1897) 6, 885. The patent was granted in 1571, and *Hayward's Patent Cases* gives 1580 as the year of the decision; cf Mossoff, 'Rethinking the Development of Patents: An Intellectual History, 1550–1800', n 30.

[63] 74 ER 1139, 1140.

[64] *Hasting's Patent* (*c*1578) Web Pat Cas 5, 6, Ex Ch; see also Hulme, 'The History of the Patent System under the Prerogative and at Common Law', 149; Stevens, *Lewis Edmunds, The Law and Practice of Letters Patent for Inventions*, 883. The patent was granted in 1569, and *Hayward's Patent Cases* gives 1578 as the year of the decision.

[65] See Hulme, 'The History of the Patent System under the Prerogative and at Common Law: A Sequel', 152.

[66] On this connection, see page 8.

F. Conclusion

The origins of the UK patent system are closely related to one of the most dramatic periods in English constitutional history. The Crown practice of rewarding favours with monopolies in respect of established trades and manufactures helped precipitate the crisis that led to the devolution of judicial authority from the Crown in the early seventeenth century, and to the enactment of legislation prohibiting monopoly grants. As Hulme first argued, however, such rewards were themselves a perversion of the original social policy informing the fourteenth century grants of letters of protection. That policy was to encourage the establishment of new industries by supporting the introduction of new trades and devices. It was based on the early recognition of the social value of practical knowledge, and the consequential desirability of a system of granting limited exclusive rights for the exploitation of such knowledge as a means of advancing the industrial arts. Thus understood, section 6 of the Statute of Monopolies did not introduce a new industrial policy previously unknown to English law, but rather constituted a renewal and reaffirmation of an earlier policy that had theretofore been repeatedly neglected.

There remains the nature of the invention itself. According to the earliest common law statement, a patentable invention was 'any new trade . . . or any engine tending to the furtherance of a trade that never was used before: and that for the good of the realm'.[67] Thus conceived, and notwithstanding the more restrictive language of section 6 of the Statute of Monopolies, an invention in 1623 is best understood to have been an ingenious method of working pre-existing materials to produce a useful result in the industrial arts. Put differently, it was an industrial art, with 'art' in this context requiring an ingenious purposive human action on the physical world.

[67] *Case of Monopolies* (1602) 74 ER 1131, 1139.

2

The Requirement for an Invention in the Pre-Modern Era of Mechanical and Chemical Manufacture: 1623 to 1882

A. Introduction

The history of the law of inherent patentability in the United Kingdom prior to the introduction of the first modern patent legislation—the Patents, Designs, and Trade Marks Act 1883[1]—can be divided into four main periods defined with reference to significant legislative and judicial developments. They are the periods from: (1) the enactment of section 6 of the Statute of Monopolies 1623[2] to 1794; (2) the first detailed consideration by an English court of the nature of an invention within the meaning of the Statute of Monopolies in 1795[3] to 1829; (3) 1830 to 1851, representing the start of a new era of patent law decision-making marked by declining judicial hostility to patents; and (4) the introduction of the first express definition of the 'invention' into English statutory law in 1852[4] to 1882. The purpose of the current chapter is to trace the evolution of legal conceptions of the invention during these periods, which represent collectively the era of early mechanical and chemical manufacture in which the basic principles of patent validity were formulated. What emerges is a view of the invention from the introduction of the specification in the eighteenth century that it was any method of sufficient mechanical or chemical ingenuity directed to advancing the industrial arts. The two central issues of uncertainty within this broad conception were: (a) whether the action needed to result in a physical (and vendible) product, such that the art advanced would be a manufacturing art, and (b) whether a patent could protect such a product independent of the method by which it was made.

1 46 & 47 Vict c 57.
2 21 Jac I c 3.
3 See *Boulton v Bull* (1795) 126 ER 651, CP.
4 See Patent Law Amendment Act 1852 (15 & 16 Vict c 83), s LV, reproduced at page 2.

B. 1623 to 1794: The Post-Specification Invention as an Idea or Discovery

Enactment of the Statute of Monopolies led the vibrant if controversial patent custom of Elizabethan and Jacobean England into a one hundred-year abeyance under the Stuarts who, still wary of the damage caused by the odious monopolies, saw the custom neither as a source of revenue nor as an instrument of industrial policy.[5] This lack of Crown interest in patents, combined with the complexity and expense of the procedures for obtaining them,[6] accounts for the relatively few patents that were granted during the seventeenth century.[7] In addition, there was virtually no development in the common law principles of patentability until the second half of the eighteenth century.[8] One reason for this was that jurisdiction in respect of patent matters was exercised almost exclusively by the conciliar courts until 1752, when it was formally divested from (and by) the Privy Council.[9]

Soon after receiving jurisdiction in respect of patent matters, the common law courts effected a change that altered fundamentally the principles of patent law, including conceptions of the invention.[10] That change was the recognition of a requirement that patentees provide a written specification describing their invention as a condition of their grant. This requirement had been introduced by the Law Officers in the early eighteenth century,[11] but was only given a common

[5] C MacLeod, *Inventing the Industrial Revolution: the English Patent System, 1660–1800* (Cambridge: Cambridge University Press, 1988) 34.

[6] N Davenport, *The United Kingdom Patent System: A Brief History with Bibliography* (Portsmouth: Mason 1979) 17–18.

[7] See EC Walterscheid, 'The Early Evolution of the United States Patent Law: Antecedents (Pt 4)' (1996) 78 J Pat & Trademark Off Soc'y 77, 80.

[8] Before 1800 there were very few cases and those that existed were rarely reported and generally not discussed in the law books. See HI Dutton, *The Patent System and Inventive Activity During the Industrial Revolution, 1750–1852* (Dover: Manchester University Press, 1984) 70–71 (noting that in the second half of the eighteenth century no more than 22 cases came before the superior courts of London).

[9] See EC Walterscheid, 'The Early Evolution of the United States Patent Law: Antecedents (Pt 3)' (1993) 77 J Pat & Trademark Off Soc'y 771, 774 (suggesting that jurisdiction over patent matters had remained with the conciliar courts partly out of a genuine uncertainty as to whether the common law courts had jurisdiction and partly out of respect for the Crown).

[10] cf JN Adams and G Averley, 'The Patent Specification: The Role of Liardet v Johnson' (1986) 7 J Leg History 156–177.

[11] See generally DS Davies, 'The Early History of the Patent Specification' (1934) 50 LQR 86–109, 260–274. The requirement for a written specification arose from the Law Officers' practice of inserting a clause into the patents they granted requiring the filing of a detailed description of the patentee's invention in the High Court of Chancery within one (and later six) months to secure the grant's validity. The earliest surviving specification was filed by John Nasmith for a patent granted in 1711 in respect of a method of fermenting wash from sugar and molasses.

law footing in *Liardet v Johnson*,[12] where Mansfield LJ directed the jury as follows:[13]

The third point is whether the specification is such as instructs others to make it. For the condition of giving encouragement is this: that you must specify upon record your invention in such a way as shall teach an artist, when your term is out, to make it – and to make it as well as you by your directions; for then at the end of the term, the public have the benefit of it. The inventor has the benefit during the term, and the public have the benefit after.

The emergence of the specification as a condition of the grant had two further consequences of jurisprudential importance. The first was a change in the nature of the consideration for which patents were granted, from the introduction of an invention within the realm to the disclosure of an invention to the realm.[14] And the second was a change in the conception of patents, from rewards for the establishment of a new industry, to contracts between an inventor and the Crown recorded in the body of the specification.[15] These changes impacted on the principles of patentability by precipitating an amendment of the pre-1623 tests of novelty and utility. The test of novelty went from being whether the invention had been worked within the realm in recent memory to whether it had been previously disclosed to the realm,[16] and the test of utility went from being whether the invention was capable of successful introduction within the realm to whether it was workable in the manner and to achieve the results described in the specification.[17] They also impacted on the nature of inventions, which went from being a trade or device capable of supporting new industry to an idea or discovery with respect to the same. This shift in the conception of the invention was consistent with a shift in the type of subject matter for which patents were sought after 1623, from the entire trades and devices of the pre-Statute of Monopolies era to improvements to existing trades and devices. Aside from being more abstract in nature, such

[12] (1778) 1 Carp Pat Cas 35, NP; see also *The Morning Post* (London, 23 February 1778) 2, reproduced in EW Hulme, 'On the History of the Patent Law in the Seventeenth and Eighteenth Centuries' (1902) 18 LQR 280, 285; Adams and Averley, 'The Patent Specification', 162–167.

[13] Hulme, ibid.

[14] Adams and Averley, 'The Patent Specification', 156. On the test of adequate disclosure, see Hulme, 'On the History of the Patent Law in the Seventeenth and Eighteenth Centuries', 285; *Dolland v Champneys* (1766) Web Pat Cas 42, NP.

[15] See, eg *Ex p O'Reilly* (1790) 34 ER 686, LC, 686. On the specification as contract, see *Arkwright v Nightingale* (1785) 1 Carp Pat Cas 38, CP, 49–50; *Hornblower v Boulton* (1799) 101 ER 1285, KB, 1286 per Ashurst J, 1288–1289 per Grose J; *Cartwright v Eamer* (1800) G 112, LC, 112.

[16] See, eg *Dolland v Champneys* (1766) Web Pat Cas 42, NP, 43.

[17] See, eg *Turner v Winter* (1787) 99 ER 1274, NP, 1276. On the role of utility, see also *R v Arkwright* (1785) Web Pat Cas 64, NP, 66; *Morgan v Seaward* (1837) Web Pat Cas 187, Ex, 196–197; TM Stevens, *Lewis Edmunds, The Law and Practice of Letters Patent for Inventions* (2nd edn, London: Stevens and Sons, 1897) 100; *Liardet v Johnson* (1778) 1 Carp Pat Cas 35, NP, in Hulme, 'On the History of the Patent Law in the Seventeenth and Eighteenth Centuries', 286–287; *Manton v Parker* (1814) G 297, NP; *Hill v Thompson* (1817) 36 ER 239, LC; *Lewis v Marling* (1829) Web Pat Cas 490, NP, verdict aff'd (1829) Web Pat Cas 493, KB.

improvements were increasingly complex and difficult to differentiate,[18] making the task of determining their entitlement to a patent difficult for the Officers charged with their assessment and with advising the Crown with respect to their merit.[19] Hence the value of the specification as a record of subject matter claimed for protection against which future disputes could be determined by the courts.[20]

From 1778, scrutiny of the specification dominated the courts' approach to determining patent validity,[21] applying to the detriment of many of the patents challenged before 1794,[22] when the courts were still hostile to patents because of their links with the odious monopolies.[23] That approach involved asking whether the subject matter described in the specification was both useful and novel; the former in the sense of being capable of reproduction from the description, and the latter in the sense of extending beyond what was previously available.[24] Further, while the nature of the invention was not discussed in detail by the courts until 1795, some consideration of it was inevitable because of the change in patenting practices described above. As a result of that change, patents for improvements were being granted notwithstanding their invalidity on the principles of pre-1623 cases.

Those principles were successfully challenged for the first time in 1776,[25] resulting in the recognition of the ability of an addition to support a patent on the pragmatic ground that to conclude otherwise 'would go to repeal almost every patent that ever was granted'.[26] Nine years later that decision was affirmed in *R v Arkwright*,[27] where the patentability of additions to a known machine was supported on condition that they be 'material and new', and reflect 'an improvement

[18] Hulme, 'On the Consideration of the Patent Grant, Past and Present' (1897) 13 LQR 313, 317; N Davenport, *The United Kingdom Patent System: A Brief History with Bibliography*, 29. See also Adams and Averley, 'The Patent Specification', 158.

[19] MacLeod, *Inventing the Industrial Revolution*, 51–53. cf Hulme, ibid; Davies, 'The Early History of the Patent Specification', 93.

[20] MacLeod, *Inventing the Industrial Revolution*, 51.

[21] *Turner v Winter* (1787) 99 ER 1274, 1277 per Buller J; *Boulton v Bull* (1795) 126 ER 651, 665 per Eyre CJ.

[22] See, eg *R v Else* (1785) G 190, KB; *Bramah v Hardcastle* (1789) 1 Carp Pat Cas 168, NP.

[23] See, eg *Blanchard v Hill* (1742) 26 ER 692, LC, 693; *Hornblower v Boulton* (1799) 101 ER 1285, KB, 1287–1288 per Kenyon LC.

[24] *Turner v Winter* (1787) 99 ER 1274, 1277.

[25] *Morris v Bramsom* (1776) G 311, NP (set of needles to be applied to a stocking frame for making eyelet-holes).

[26] ibid. See also *Boulton v Bull* (1795) 126 ER 651, 667 per Eyre CJ (describing the argument that methods of production were inherently unpatentable as contradicted by the evidence of the patents granted since 1623, 'three-fourths' of which were likely to have been for methods of operating and manufacture 'producing no new substances and employing no new machinery').

[27] (1785) Web Pat Cas 64, NP.

of the trade'.[28] With respect to the requirement for materiality, the Court directed the jury as follows:[29]

[T]he question is, whether it is material or useful? The witnesses upon the part of the prosecution say it is of no use at all. In the first place, they had that before which answered the same purpose, though not made exactly in the same form . . . So if it was new it is of no use, but they say it is not new, for though it was not precisely the same shape, in substance it was the same thing.

By implication from this passage, for an addition to merit a patent it needed to support a new result, and to be new in substance and not merely in form.[30] Unclear, however, was the scope of the resulting patent, and whether the invention it protected was the machine or merely the addition.[31]

C. 1795 to 1829: The Invention in and after *Boulton v Bull*

1. The Invention as a Manufacture

The most detailed consideration of the post-specification invention before 1883 took place in *Boulton v Bull*.[32] That case involved a challenge to a patent for a new method of using an old steam engine in a more beneficial way, described in the specification as consisting of certain principles applied in a particular mode to the purposes of the invention.[33] The question before the Court was whether the alleged invention so described was a method or principle and, if it was, whether this rendered the patent invalid for lack of patentable subject matter. The judges differed greatly in approach and conception, dividing 2:2 on the issue of the patent's validity.

An initial point of disagreement concerned the method of interpretation to be applied in construing the Statute. Three of the Judges, Eyre CJ, Buller J, and Heath J, regarded inherent patentability as depending on whether the subject matter could be categorized as a 'manufacture' within the meaning of section 6. The fourth Judge, Rooke J, relied on the spirit rather than the meaning of the legislation. In his judgment, the spirit of section 6 supported the patent, there being a new and useful improvement in fire engines sufficiently described

[28] ibid, 71.

[29] ibid, 73.

[30] ibid.

[31] ibid, 71–72. cf *Boulton v Bull* (1795) 126 ER 651, 664 per Buller J. See also *Morgan v Seaward* (1837) Web Pat Cas 187, Ex, 193–194.

[32] (1795) 126 ER 651, CP.

[33] The 'principles' of the invention concerned the nature of steam, the 'particular mode' involved keeping the steam vessel in the engine as hot as the steam during the time the engine was at work, and the 'purposes' were to lessen the engine's consumption of fuel (see ibid). For a documentary history of James Watt and his patents, including that litigated in *Boulton*, see E Robinson and AE Musson, *James Watt and the Steam Revolution* (London: Adams and Dart, 1969).

Diagram A Some cases of note regarding patentable subject matter, 1600 to 1851

Case	Judicial understandings of key concepts	Examples of subject matter regarded as appropriate/inappropriate for a patent
Darcy v Allin (Case of Monopolies) (1602) 11 Co Rep 84, 77 ER 1260, Noy 173, 74 ER 1131, Moore 671, 72 ER 830, Web Pat Cas 1, 1 Carp Pat Cas 26 (KB)	A patent may be granted for any new trade, or any new engine tending to the furtherance of a trade, brought into the realm by a man's own charge and industry, or wit and invention.	
Clothworkers of Ipswich Case (1615) 11 Co R 53a, 77 ER 1218, Godb 252, 78 ER 147, 1 Rolle 4, 81 ER 285, Shower 266, 89 ER 988 (KB)	A patent may be granted for a new invention, and a new trade, brought within the kingdom in peril of a man's life and consumption of his estate or stock; or a new discovery of anything.	
Mitchell v Reynolds (1711) 1 P Wms 181, 24 ER 347 (Ch)	A patent may be granted for a trade newly invented.	
Morris v Bramsom (1776) G 311, 1 Carp Pat Cas 30, Web Pat Cas 50, Bull NP 76 (NP)		An addition to an old machine is patentable.
Hartley's Patent (1777) Web Pat Cas 54		A method of disposing plates of iron to produce a useful effect is patentable.
R v Arkwright (1785) G 15, Web Pat Cas 64, 1 CPC 53, Dav Pat Cas 61 (NP); (1785) Web Pat Cas 73, 1 Carp Pat Cas 101, Dav Pat Cas 141 (KB)		Any material new part of an old machine that is an improvement of the trade will support a patent. Whether the patent protects the addition only or the whole machine is a separate question.

Boulton v Bull (1795) G 70, 2 H Bl 463, 126 ER 651, Dav Pat Cas 162, 1 Carp Pat Cas 117 (CP); (1796) 3 Ves 140, 30 ER 937, 1 Ves Sup 345, 34 ER 820, 1 Carp Pat Cas 155 (LC)	Heath J—Manufactures are reducible to two classes: vendible machinery, and vendible substances formed by chemical and other processes. The subject of the patent is in each case the machinery, and the substance produced. Eyre CJ—Manufactures are things made (including new compositions of things, and all mechanical inventions whether made to produce old or new effects), and practices of making (including all new artificial manners of operating with the hand or instruments in common use, and new processes in any art producing effects useful to the public). The subject of the patent granted for the second type of manufacture is (where the effect produced is a new substance or composition of things) the substance or composition produced or (where the effect is not a new substance or composition of things) either the mechanism (if a new mechanism is used) or the process (if it is a new method of operating) with or without the old mechanism, by which the effect is produced. All—Manufactures include mechanical and chemical discoveries, including new compounds of known and used articles. Rooke J—The term 'principle' is equivocal; it may denote either the radical elementary truths of a science, or those consequential axioms that are founded on radical truths but used as fundamental truths by those who do not find it expedient to have recourse to first principles. Heath J—A patent must be for a vendible matter, and not for a principle. The application of a principle to produce an old machine is not a proper subject of a patent otherwise all subsequent applications of that principle would be an infringement.	An addition to an old machine is patentable, but the patent must be for the addition only, and not for the old machine.

(Cont'd)

Diagram A Some cases of note regarding patentable subject matter, 1600 to 1851 (*Cont'd*)

Case	Judicial understandings of key concepts	Examples of subject matter regarded as appropriate/inappropriate for a patent
	Buller J—A principle is the first ground and rule for arts and sciences, or in other words the elements and rudiments of them. A patent must be for some new production from those elements, and not for the elements themselves.	
Hornblower v Boulton (1799) G 263, 8 D & E 95, 101 ER 1285, Dav Pat Cas 221, 1 Carp Pat Cas 156 (KB)	Kenyon CJ—A manufacture is something made by the hands of man. Grose J—A patent may be for the manufacture of a thing, where the mode of making the manufacture is described. The patent is then for the manufacture of the thing so made, and not merely for the principle upon which it is made. Lawrence J—'Engine' is synonymous with 'method', and both can be the subject of a patent. 'Method' is placing several things and performing several operations in the most convenient order; but may also signify a contrivance or device, as may an engine. So 'Principle' may mean a mere elementary truth, but it may also mean constituent parts.	An addition is patentable.
Huddart v Grimshaw (1803) G 266, 1 Carp Pat Cas 200, Web Pat Cas 85, Dav Pat Cas 265 (NP)		A new combination of old materials so as to produce a new effect is patentable. A method of obtaining an old result that is not identical to any existing method is patentable.
Ex parte Fox (1812) 1 Ves & B 67, 35 ER 26 (Lincoln's Inn Hall)		Improvements to an engine for which a patent has been granted are patentable, even where those improvements could not be used without the original invention (but the substratum of another patent cannot be used until the patent has expired).

Hill v Thompson (1817) G 241, 3 Mer 622, 36 ER 239, Web Pat Cas 225, 1 Carp Pat Cas 377 (LC); (1817) Holt NP 636, 171 ER 367 (NP); (1817) 3 Mer 626, 36 ER 239, Web Pat Cas 235 (LC); (1818) 8 Taunt 375, 129 ER 427, Web Pat Cas 239 (CP)		A new combination of materials previously in use for the same purpose is patentable, as is a new method of applying such materials.
R v Wheeler (1819) G 497, 2 B & Ald 345, 106 ER 392 (KB)	A 'manufacture' is a thing made that is useful and vendible for its own sake; or an engine or instrument (or some part of an engine or instrument) to be employed in making a previously known article or in some other useful purpose; or possibly a new process (ie method) to be carried on by known implements or elements, acting upon a known substance, but producing it in a cheaper or more expeditious manner, or of a better and more useful kind. It must be something of a corporeal and substantial nature, and something that can be made by man from the matters subject to his art and skill, or at the least some new mode of employing practically his art and skill. It does not include a merely philosophical or abstract principle.	
Brunton v Hawkes (1820) G 97, 1 Carp Pat Cas 405 (NP); (1821) 4 B & Ald 541, 106 ER 1034, 1 Carp Pat Cas 410 (KB)		A machine comprised of a new combination of old parts that produces a new result is patentable. The union of two old pieces in a new mode for an old purpose is patentable.

(*Cont'd*)

Diagram A Some cases of note regarding patentable subject matter, 1600 to 1851 (*Cont'd*)

Case	Judicial understandings of key concepts	Examples of subject matter regarded as appropriate/inappropriate for a patent
Hall v Boot (1822) G 217, 1 Carp Pat Cas 423, Web Pat Cas 97 (NP)		A method of causing lace to pass through a flame of gas to singe off the projecting fibres is patentable, notwithstanding that lace has previously been passed through other flames for the same purpose.
Lewis v Davis (1829) Web Pat Cas 488 (NP)		A machine constructed to shear from list to list by means of rotary cutters is patentable, notwithstanding that shearing from list to list, and shearing from end to end by means of rotary cutters, was previously known and in use.
Jones v Pearce (1832) Web Pat Cas 123 (NP)		A method of making wheels on the principle of suspension is patentable, and not a mere principle.
Saunders v Aston (1832) 1 Carp Pat Cas 510, 3 B & Ald 881, 110 ER 324 (KB)		Merely putting together an old button and flexible shaft does not create an invention as can support a patent.
Russell v Cowley (1834) Web Pat Cas 459 (NP); (1834) Web Pat Cas 463, 1 CM&R 864, 149 ER 1331 (Ex)		A process of manufacturing iron tubes to produce longer, more uniform and cheaper tubes is patentable.
Minter v Wells (1834) 1 CM&R 505, 149 ER 1180 (Exch)		A claim that is not for a principle, but for the application of a principle so as to produce a specified effect in a specified manner, is valid.
Cornish v Keene (1835) Web Pat Cas 497 (NP); (1837) Web Pat Cas 513, 3 BingNC 570, 132 ER 530 (CP)	'New manufacture' has a very wide and extended meaning—you may call it almost an invention. A 'manufacture' is a vendible article produced by the art and hand of man.	

Losh v Hague (1838) Web Pat Cas 202 (NP)		The application of a well-known thing to an operation analogous to that for which it has already been used is not a proper subject matter for a patent.
Kay v Marshall (1839) 2 Web Pat Cas 71, 5 BingNC 492, 132 ER 1189 (CP); (1841) 2 Web Pat Cas 79, 8 Cl&F 245, 8 ER 96 (HL)		A new means of using an old machine is patentable. The application of a machine used to maccerate flax, in an adjusted mode to maccerate flax, is not a proper subject matter for a patent.
Morgan v Seaward (1836) Web Pat Cas 170 (NP); (1837) Web Pat Cas 187, 2 M&W 544, 150 ER 874 (Ex)	A manufacture may be a machine when completed, or a mode of constructing a machine.	
Jupe v Pratt (1837) Web Pat Cas 145 (Ex)		A principle coupled with the mode of carrying it into effect is patentable, provided that the mode (and not merely the principle) has been invented. The patent then protects all other modes of carrying the same principle into effect.
Neilson v Harford (1841) Web Pat Cas 295 (NP); (1841) Web Pat Cas 331 (Ex)		A claim not merely for a principle, but for a very valuable machine embodying a principle, is valid. A practical mode of carrying a principle into effect is patentable, but not the principle embodied in that mode itself. There is no difference between claiming a principle to be carried into effect in any way you will, and claiming a principle.
Walton v Potter (1841) Web Pat Cas 585 (NP); (1842) Web Pat Cas 597 (CP)		A new and superior method of achieving an old end discovered by a man's own genius and discovery, and the adaptation of a substance of known properties to a particular (new) purpose, are patentable.

(*Cont'd*)

Diagram A Some cases of note regarding patentable subject matter, 1600 to 1851 (*Cont'd*)

Case	Judicial understandings of key concepts	Examples of subject matter regarded as appropriate/inappropriate for a patent
		Coltman J—A very useful application and adaptation of a substance, the properties and qualities of which for the purpose have never been known before, is patentable.
Carpenter v Smith (1841) Web Pat Cas 530 (NP); (1842) Web Pat Cas 540 (Ex)		A new and improved combination of mechanical parts already known in combination is patentable.
Gibson v Brand (1841) Web Pat Cas 627 (NP); (1842) Web Pat Cas 631 (CP)		Tindal CJ—A process that is described with great distinctness and minuteness may be patented.
		Erskine J—An improved process may be patentable if the improvement can be shown to amount to a new invention and a new manufacture.
Walton v Bateman (1841) Web Pat Cas 613 (NP)		A claim for a new principle developed, carried out, and embodied in a mode of using that principle is valid.
Crane v Price (1842) Web Pat Cas 393 (CP)		A new combination that produces a new or improved or cheaper article is an invention or manufacture, and may be patentable.
		The application of stone coal (anthracite), combined with a hot air blast, in the smelting or manufacture of iron from ironstone, mine or ore is patentable, notwithstanding that the application of a hot blast to the smelting of iron has previously been patented.

		A combination that produces a new, better or cheaper article than that produced before, even if it includes the subject matter of an existing patent, is patentable provided such subject matter is not claimed to be part of the patented combination. (A patent that claims as part of its protected invention the subject matter of a previous patent will be void for lack of newness and infringement.)
Househill Iron Co v Neilson (1842) Web Pat Cas 676 (NP); (1843) Web Pat Cas 673 (HL)		Hope CJ—An abstract philosophical principle is not patentable apart from any mode of turning it to account in the practical operations of manufacture, or the business, arts and utilities of life. The mere discovery of such a principle is not an invention in the patent law sense of the term. A principle can be the subject of a patent if it is applied, by the specification, to any special purpose so as to thereby effectuate a new and practical result and benefit, and may specify (and therefore protect) one or more modes for achieving that purpose.
Muntz's Patent (1846) 2 Web Pat Cas 113 (PC)		The application of old plates to a new and useful purpose is patentable.
Stead v Williams (1848) 5 CB 528, 136 ER 985 (CP)		A method of paving with wood might be patentable.
Pow v Taunton (1845) G 391 (QB)		The application of an old machine in a new way is not patentable, unless the means, essential for such application, are also new for the purpose.

(*Cont'd*)

Diagram A Some cases of note regarding patentable subject matter, 1600 to 1851 (*Cont'd*)

Case	Judicial understandings of key concepts	Examples of subject matter regarded as appropriate/inappropriate for a patent
Allen v Rawson (1845) 1 CB 551, 135 ER 656 (CP)		The application of soap and water in combination with rollers is patentable, notwithstanding that the use of both soap and water, and of rollers, in the relevant process is known.
Heath v Unwin (1852) 2 Web Pat Cas 236, 12 CB 522, 138 ER 1010 (Ex Ch); (1855) 5 HL Cas 505, 10 ER 997 (HL)		The use of cheaper materials in a known composition is a new discovery or invention and thus patentable.
R v Cutler (1847) Macr 124, 3 C&K 215, 175 ER 526 (NP); (1849) Macr 137, 3 C&K 232, 175 ER 534 (QB); (1849) Macr 138, 3 C&K 233, 175 ER 534 (NP)		The application of an article to a particular purpose is not patentable where the applicant has no claim either to the mode of producing the article, or to the mode of applying it for attaining the result.
Stevens v Keating (1847) 2 Web Pat Cas 176 (VC); (1847) 2 Web Pat Cas 181 (NP); (1848) 2 Web Pat Cas 192 (Ex)		The making of an old substance by a new process makes that substance a new manufacture for which a patent may be granted.

Steiner v Heald (1851) 6 Exch 697, 155 ER 686 (Ex)	The application of an old process to spent madder to obtain a dye (garancine), previously obtained by applying the same process to fresh madder, may be a new manufacture of garancine depending on whether spent madder is in its nature and properties the same as fresh madder (which is a question for the jury in a particular case).
Electric Telegraph Co v Brett (1851) 10 CB 838, 138 ER 331 (CP)	A claim not merely for an idea or principle, but for the application of that idea or principle, can support a patent.
Newton v Vaucher (1851) 6 Ex 859, 155 ER 794 (Ex)	A claim not merely for a new principle, but for a new principle embodied in a new machine with a particular purpose in limited applications, is valid.

in the specification. Arguments of invalidity moving from classification of the subject matter as a method or principle not within the scope of 'manufacture' were, he suggested, merely 'verbal'.[34] More expansive in result was Eyre CJ's understanding. In his view, 'exposition of the statute, as far as usage will expound it' had bestowed a much wider meaning on 'manufacture' than ordinary language would support.[35] While not quite conceding the equivalence of that term with the phrase 'any thing', Eyre CJ did consider that it applied not only to things made, but also to practices of making, ways of operating, and principles carried into practice in a new manner or to produce a new (including non-physical) result.[36] In his conclusion, an invention included any artificial product or human action producing beneficial results in 'any art, trade, mystery, or manual occupation'.[37] In contrast were the views of Heath and Buller JJ, who interpreted 'manufacture' in a more restrictive sense to mean a (vendible[38]) machine or substance.[39]

Related to this issue were their Lordships' views of the threshold exclusions from patentability. All agreed that principles were unsuitable for protection and, with the exception of Heath J, understood 'principle' in this context to mean an abstract notion[40] or elementary truth of the arts and sciences.[41] For Heath J, principles extended beyond this to include methods of production, which he also regarded as inherently unpatentable.[42] Buller J agreed, but on the basis of the nature of a method as a 'double use' rather than a principle per se;[43] uses thus constituting a third category of subject matter identified in *Boulton* as beyond the limits of the patent system.

The statement that a particular subject matter is inherently unpatentable suggests that the reason for the unpatentability derives from the subject matter's properties. Thus, Buller J declared in *Boulton* that the reason patents were not available for principles was evident from 'the very statement of what a principle is',[44] although in reality it emerged more clearly from the Judges' statements of what a principle was not. For Eyre CJ, an abstract principle was not 'so far embodied and connected with corporeal substances as to be in a condition to act, and to produce effects in any art, trade, mystery, or manual occupation'.[45] For Buller and Heath JJ,

[34] (1795) 126 ER 651, 659.
[35] ibid, 666.
[36] ibid.
[37] ibid, 667.
[38] ibid, 660–661.
[39] ibid, 662–663.
[40] ibid, 667 per Eyre CJ.
[41] ibid, 659 per Rooke J, 662 per Buller J.
[42] ibid, 661.
[43] ibid, 663.
[44] ibid, 662.
[45] ibid, 667.

on the other hand, it was not a (vendible) substance or machine,[46] while for Rooke J it was not capable of novel creation.[47]

The reasons given by Buller and Heath JJ for the unpatentability of methods of production were different. For Buller J, such methods were equivalent to a 'double use' of a known thing, the unpatentability of which was conceded before the Court.[48] For Heath J, a method had no physical form and could never be sufficiently described in a specification to justify the rights which its patent would confer,[49] namely, a monopoly covering all products resulting from application of its underlying principle.[50] It followed that unless a method resulted in a vendible machine or substance it could not support a patent, and if it did so result, the patent would be for the machine or substance when produced by the method, and not independently for the method as such.[51]

The premise of Heath J's finding, that the rights conferred by a patent needed to be proportionate to the benefit secured, was consistent with the contract theory of patents expressed by the Court in *Liardet v Johnson*. It was also consistent with the reasoning of Eyre CJ, who supported a requirement for fair disclosure of a patentable invention. On the other hand, Eyre CJ seems also to have supported patents for products, independent of the methods by which they were made, and notwithstanding that a disclosure in support of such a patent could not include every method protected by the patent. Thus, related to the question of the nature of the invention was the question of what constituted a proportionate patent.

2. The Requirement for an Accurate and Complete Representation of the Invention

In the years after 1800, the courts desisted from further attempts to define the statutory category of 'manners of new manufacture'. Instead, they renewed their focus on the specification, concentrating on whether the disclosure it contained was of the novel and useful discovery of the patentee.[52] This can be seen in *R v Wheeler*,[53] where Abbott CJ upheld his decision to repeal a patent on the ground

[46] (1795) 126 ER 651, 660–661 per Heath J, 662–663 per Buller J.

[47] ibid, 659.

[48] ibid, 663.

[49] ibid, 661.

[50] ibid.

[51] ibid.

[52] See *Cartwright v Eamer* (1800) G 112, LC; *Bainbridge v Wigley* (1810) 1 Carp Pat Cas 270, NP; *Manton v Parker* (1814) 1 Carp Pat Cas 274, NP; *Bovill v Moore* (1816) 1 Carp Pat Cas 320, NP; *Wood v Zimmer* (1815) 171 ER 161, NP; *Cochrane v Smethurst* (1816) 171 ER 448, NP; *Macfarlane v Price* (1816) 171 ER 446, NP; *R v Cutler* (1816) 171 ER 495, NP; *Hill v Thompson* (1817) 36 ER 239, LC; *R v Wheeler* (1819) 106 ER 392, KB; *R v Fussell* (1826) 1 Carp Pat Cas 449, KB; *Sturtz v De la Rue* (1828) 38 ER 1048, LC; *Lewis v Davis* (1829) Web Pat Cas 488, NP; *Lewis v Marling* (1829) Web Pat Cas 490, NP, verdict aff'd (1829) Web Pat Cas 493, KB.

[53] (1819) 106 ER 392, KB.

that the subject matter for which it was claimed[54] was not the same as that disclosed in the specification:[55]

> The language in which the supposed invention is described in a patent of this nature is the language of the patentee himself. He represents to the Crown, that he has invented this or that thing, and that he is the first and sole inventor thereof, &c.; and the Crown yielding to his representation, and willing to give encouragement to all arts and inventions that may be for the public good, grants to the patentee the sole liberty and privilege of using his said invention, for a certain term, under the conditions before noticed. It is obvious, therefore, that if the patentee has not invented the matter or thing of which he represents himself to be the inventor, the consideration of the Royal grant fails, and the grant consequently becomes void. And this will not be the less true, if it should happen that the patentee has invented some other matter or thing, of which, upon a due representation thereof, he might have been entitled to a grant of the exclusive use.

As was subsequently said in *Brunton v Hawkes*,[56] '[i]t is quite clear that a patent granted by the Crown cannot extend beyond the consideration of the patent,'[57] and that in order for a claim to be valid in law, it needed to be properly supported by the specification, and not exceed 'that, which being both matter of actual discovery and of useful discovery, is the only proper subject for the protection of a patent'.[58] On the nature of an actual and useful discovery, his Lordship further stated as follows:[59]

> Now the word 'manufactures' has been generally understood to denote either a thing made, which is useful for its own sake, and vendible as such, as a medicine, a stove, a telescope, and many others, or to mean an engine or instrument, or some part of an engine or instrument, to be [so] employed, either in the making of some previously known article, or in some other useful purpose, as a stocking-frame, or a steam-engine for raising water from mines. Or it may perhaps extend also to a new process to be carried on by known implements, or elements, acting upon known substances, and ultimately producing some other known substance, but producing it in a cheaper or more expeditious manner, or of a better and more useful kind.

This conception of the invention as a method of acting on the physical world to produce either a new or improved product, or an existing product in a new or improved way, was consistent with the judgments of Heath and Buller JJ in *Boulton*. However, it was not uniformly supported by the courts of this era. Indeed, the

[54] By 'claimed' here I do not mean claimed in the post-1883 statutory sense, but rather in the sense of applying for a patent on the basis of a representation as to an alleged invention. Consistent with this, before the introduction of the statutory requirement for patent claims in the Patents, Designs, and Trade Marks Act 1883, s 5(2), the invention 'claimed' by an applicant was discerned from the patent title and/or specification in general.

[55] (1819) 106 ER 392, 393, 394–395.

[56] (1821) 106 ER 1034, KB.

[57] ibid, 1038 per Abbott CJ.

[58] ibid, 1039 per Best J (quoting *Hill v Thompson* (1817) 171 ER 367, NP, verdict rev'd in part (1817) 36 ER 239, LC).

[59] (1819) 106 ER 392, 394–395.

more common approach was to assume that any subject matter of practical utility could support a patent, and that the nature and scope of protection conferred by a patent were matters for careful negotiation by the patentee when drafting the specification.[60] This assumption resulted in the courts' implicit concession after 1800 of the inherent patentability of a variety of previously untested or unpatentable subject matter such as methods,[61] uses,[62] and conjunctions of known objects,[63] and their acceptance that an invention for patentability purposes could be either physical or abstract in nature. On its face, the effect of that acceptance was to resolve the issue raised in *Arkwright* and argued in *Boulton* concerning non-physical subject matter such as processes and improvements, and to refute the conclusion of Heath J in the second of those cases that an inability to list exhaustively all the manifestations of a new method would preclude the level of disclosure required to support a patent. Further, that this attitude to inherent patentability was borne of pragmatism, rather than a principled reading of section 6 of the Statute of Monopolies or the requirement for an invention per se, receives some support from the ambivalence of those cases in which inherent patentability was considered directly. An example is *R v Wheeler* itself, where Lord Abbott CJ queried the scope of 'manufacture' beyond vendible contrivances before embracing a test of patentability that depended exclusively on the identification of a novel and useful discovery accurately described in the specification.[64]

D. 1830 to 1851: The (Patentable) Invention as a Mechanically Inventive Subject Matter

1. The Invention as a Principle 'Turned to Account'

After 1830, the courts' suspicion of patents relaxed further, resulting in a less rigorous construction of the specification and a decreased tendency to strike

[60] cf *Forsyth v Manton* (1816) 1829 Parl Rep 198, LC (method of applying chemical combustibles to the discharge of firearms not 'a subject for a patent', albeit for want of novelty).

[61] See, eg *Cochrane v Smethurst* (1816) 171 ER 448, NP (method of lighting cities, towns, and villages presumed inherently patentable).

[62] See, eg *Hall v Boot* (1822) 1 Carp Pat Cas 423, NP (use of a gas flame to burn fibres from lace or net inherently patentable); *Hill v Thompson* (1817) 36 ER 239, LC (methods of applying known materials capable of supporting a patent); *Webster v Uther* (1824) Web Pat Cas 128n, NP (method of using a washer and screw in place of a bolt inherently patentable). cf *Boulton v Bull* (1795) 126 ER 651, 663 per Buller J; *Forsyth v Manton* (1846) 3 Hay Pat Cas 705, LC.

[63] See, eg *Huddart v Grimshaw* (1803) 1 Carp Pat Cas 200, NP (combination of known materials to produce cordage presumed inherently patentable); *Cochrane v Smethurst* (1816) 171 ER 448, NP (method of improving a street lamp by a new combination of known parts presumed inherently patentable); *Hill v Thompson* (1817) 36 ER 239, LC (new combination of materials previously used for the same purpose patentable if properly described); *Lewis v Davis* (1829) Web Pat Cas 488, NP (combinations of known things not previously combined inherently patentable).

[64] (1819) 106 ER 392, 395–396.

down patents.[65] That relaxation was also reflected in the ascendancy of the view from *Arkwright* that inventions needed to be construed with reference to their substance rather than their form: a view possibly encouraged by the tendency of applicants to describe their inventions in abstract terms in order to maximize the protection conferred.[66] However, that tendency worked against applicants as well, by increasing the vulnerability of their patents to attack. With respect to the substance of inventions, the courts focused on their underlying mechanical or chemical principle, distinguished from its actual manifestation.[67] This can be seen from *Jupe v Pratt*:[68]

> [Y]ou cannot take out a patent for a principle; you may take out a patent for a principle coupled with the mode of carrying the principle into effect, provided you have not only discovered the principle, but invented some mode of carrying it into effect. But . . . if you have done that, then you are entitled to protect yourself from all other modes of carrying the same principle into effect, that being treated by the jury as piracy of your original invention.

This reasoning was followed in *Househill Iron Co v Neilson*,[69] where the House of Lords approved the approach of Alderson B in relation to all abstract (including elementary) principles, subject only to their having been 'turned to account' through 'direction to the . . . business of human life'.[70] In doing so, the Lords distinguished between an abstract principle when promulgated, not being 'an invention, in the patent law sense of the term', and the same principle when connected to some 'special purpose or practical operation', which, having been 'clothed with the language of practical application', was capable of supporting a patent, independent of the means by which it was applied.[71] The result was a conception

65 See, eg *Russell v Cowley* (1834) 149 ER 1331, Ex, 1335 per Parke B, 1336 per Alderson B; *Neilson v Harford* (1841) Web Pat Cas 295, NP, 310; verdict rev'd in part (1841) Web Pat Cas 331, Ex; *Stevens v Keating* (1847) 2 Web Pat Cas 181, NP, 187–188; verdict aff'd (1848) 2 Web Pat Cas 192, Ex. See also Dutton, *The Patent System and Inventive Activity During the Industrial Revolution*, 79 (noting an increase in the common law cases decided in favour of patentees from 39% in 1750–1799 to 76% by the 1840s).

66 For examples of this emphasis prior to 1830, see *Huddart v Grimshaw* (1803) 1 Carp Pat Cas 200, NP; *Hare v Harford* (1803) 1 Carp Pat Cas 180, NP; *R v Cutler* (1816) 171 ER 495, NP; *Lewis v Marling* (1829) Web Pat Cas 493, KB per Lord Tenterden CJ; *Bramah v Hardcastle* (1789) 1 Carp Pat Cas 168, NP.

67 In the context of anticipation, see, eg *Jones v Pearce* (1832) Web Pat Cas 123, NP; *Kay v Marshall* (1839) 132 ER 1189, CP, aff'd (1841) 8 ER 96, HL; *Neilson v Harford* (1841) Web Pat Cas 331, Ex; *Walton v Bateman* (1841) Web Pat Cas 613, NP; *Muntz's Patent* (1846) 3 Hay Pat Cas 705, PC (and commentary); *Newton v Vaucher* (1851) 155 ER 794, Ex. In the context of infringement, see, eg *Morgan v Seaward* (1836) Web Pat Cas 170, NP, verdict rev'd in part (1837) Web Pat Cas 187, Ex; *Jupe v Pratt* (1837) Web Pat Cas 145, Ex; *Walton v Potter* (1841) Web Pat Cas 585, NP, verdict aff'd (1842) Web Pat Cas 597, CP; *Househill Iron Co v Neilson* (1843) Web Pat Cas 673, HL.

68 (1837) Web Pat Cas 145, 146.

69 (1843) Web Pat Cas 673, HL.

70 ibid, 683–684.

71 (1843) Web Pat Cas 673, HL, 684. See also *Electric Telegraph Co v Brett* (1851) 138 ER 331, CP.

of the invention even wider than that of Eyre CJ in *Boulton*, namely, as any human action in the field of the practical arts.

Among other things, this view supported the patentability of alterations to existing subject matter.[72] One type of alteration consisted in the substitution of a mechanism used in a particular process or device for a different mechanism with beneficial result.[73] After 1830 the number of patents for such subject matter increased, and began to be categorized by the courts as applications of a known mechanism rather than alterations or improvements per se.[74] While the inherent patentability of subject matter of this type had previously been accepted, their proliferation in the case law after 1830 encouraged a reconsideration of their suitability for protection.

2. The Exclusion of Manifestly Non-Inventive (Non-Ingenious) Subject Matter: Analogous Uses and *Losh v Hague*

The important decision in this context was *Losh v Hague*,[75] involving a patent for 'certain improvements in the construction of wheels for carriages to be used on railways'.[76] The problem faced by the patentee was that the wheels had previously been made according to the same principle, putting the novelty of the improvements in doubt.[77] Facing a possible verdict of invalidity,[78] the patentee argued that it was his application of the improved wheel to a railway to which it had never before been applied that entitled him to a patent.[79] Lord Abinger CB dealt with that argument before the jury with reference to a distinction between the

[72] See, eg *Bainbridge v Wigley* (1810) 1 Carp Pat Cas 270, NP (improvements in the manufacture of flutes); *Manton v Parker* (1814) 1 Carp Pat Cas 274, NP (improvements in the construction of locks); *Cochrane v Smethurst* (1816) 171 ER 448, NP (improvements in the manufacture of street lamps); *Macfarlane v Price* (1816) 171 ER 446, NP (improvements in the manufacture of umbrellas and parasols); *R v Cutler* (1816) 171 ER 495, NP (improvements in the manufacture of fire grates); *Hill v Thompson* (1817) 36 ER 239, LC (improvements in the smelting and working of iron); *Sturtz v De la Rue* (1828) 38 ER 1048, LC (improvements in copper and other plate printing).

[73] See, eg *Webster v Uther* (1824) Web Pat Cas 128n(b), NP.

[74] See, eg *Jones v Pearce* (1832) Web Pat Cas 123, NP (improvements in the manufacture of wheels for carriages construed as the use of suspension rods in place of spokes in a known method); *Saunders v Aston* (1832) 110 ER 324, KB (improvements in the manufacture of buttons construed as the application of a shank to buttons in a known method); *Walton v Bateman* (1841) Web Pat Cas 613, NP (and *Walton v Potter* (1841) Web Pat Cas 585, NP; verdict aff'd (1842) Web Pat Cas 597, CP) (improvements in the manufacture of cards for carding fibres construed as the use of India rubber in place of leather in a known method); *Muntz's Patent* (1846) 3 Hay Pat Cas 705, PC (improvements in the manufacture of plates for sheathing the bottom of ships construed as the use of pure copper and zinc in a known method); *R v Cutler* (1849) 175 ER 534, NP (improvements in the construction of tubular flues of boilers construed as the use of iron or steel tubes in a known method).

[75] (1838) Web Pat Cas 202, NP.

[76] The specification is extracted at (1838) Web Pat Cas 197.

[77] (1838) Web Pat Cas 202, 204–207.

[78] ibid.

[79] ibid.

(unpatentable) application of the wheel as old contrivance to the railway as new object, and the (patentable) application in *Hall v Boot*[80] of the gas flame as new contrivance to the task of burning the ends of fabric as old object. While the description of the flame as a 'new contrivance' is ambiguous, the point being made by his Lordship was not that the flame itself was unknown, but that its application to the task at hand was not suggested by any of the uses to which it had previously been put. That point, which depended on an expanded conception of novelty as requiring that the subject matter be unknown as well as unsuggested by what was previously known, accounted for the verdict reached in *Losh* that the use of the wheel was anticipated by its previous uses so as to be lacking in novelty.[81] Hence the principle of that case, that analogous uses of a known object were inherently unpatentable by reason of their manifest inability to advance the mechanical arts sufficiently to justify the grant of a patent. The case is important, for it suggests a conception of the invention as a mechanical (or chemical) art, in the sense of an ingenious method of working materials directed to advancing the industrial arts.

From the outset, the courts distinguished analogous uses from combinations of two or more objects, and affirmed the inherent patentability of the latter even without considering whether they were sufficiently novel (ingenious) in the *Losh* sense.[82] An example is *Crane v Price*,[83] where a patent for the combined use of anthracite and a hot air blast in the manufacture of iron[84] was held to be patentable as an invention or manufacture within the meaning of section 6.[85] In the view of the Court, while anthracite and a hot air blast had previously been used independently in the manufacture of iron, their use in combination for that purpose was new 'as to the public use thereof in England',[86] and was therefore patentable on the ground that it produced 'either a new article, or a better article, or a cheaper article to the public, than that produced before by the old method'.[87] Other considerations such as the ingenuity of the combination were regarded as irrelevant; the novelty and utility of the combination were determinative of its entitlement to a patent.[88]

80 (1822) 1 Carp Pat Cas 423, NP.

81 (1838) Web Pat Cas 202, 207–209.

82 See, eg *Carpenter v Smith* (1841) Web Pat Cas 530, NP, verdict aff'd (1842) Web Pat Cas 540, Ex. See also *Foxwell v Bostock* (1864) 46 ER 934, LC, 939 ('The term "combination of machinery", [which] has of late been a favourite form of words with patentees, is nothing but an extended expression of the word "machine"'). cf *Brunton v Hawkes* (1821) 106 ER 1034, KB.

83 (1842) Web Pat Cas 393, CP. But see *Murray v Clayton* (1869) LR 7 Ch 570, CA; *Rushton v Crawley* (1870) LR 10 Eq 522, VC.

84 The specification is extracted at (1842) Web Pat Cas 375.

85 (1842) Web Pat Cas 393, 409.

86 ibid, 410.

87 ibid.

88 ibid, 409–410.

E. 1852 to 1882: The Basis of *Losh v Hague* in the Requirement for an 'Invention' the Patenting of which was not 'Generally Inconvenient'

The recognition in *Losh* of a threshold exclusion from patentability in respect of analogous uses of known objects on the basis of their manifest lack of inventiveness gave rise to a practice among litigants of challenging patents for relating to a use the protection of which would burden the public.[89] This was particularly so after 1852, when new patent legislation introduced a definition of the 'invention' that referred expressly to section 6 of the Statute of Monopolies, focusing attention on the language of that section, including its prohibition against 'generally inconvenient' patents.[90] Inevitably, the distinction between uses and other categories of inherently patentable subject matter, including particularly between uses and combinations, became important;[91] and in *Harwood v Great Northern Railway Co*[92] the House of Lords considered the basis for that distinction.[93]

[89] Further pre-1852 cases in which a patent was invalidated for claiming a mere use include *R v Cutler* (1847) 175 ER 526, NP, retrial ordered 175 ER 534, QB, retrial (1849) 175 ER 534, NP; *Bush v Fox* (1852) Macr 140, NP, verdict aff'd (1856) 10 ER 1080, HL.

[90] See Patent Law Amendment Act 1852, s LV, reproduced at page 2.

[91] See, eg *Brook v Aston* (1857) 120 ER 178, QB (improvements in a method for finishing cotton and linen, an analogous use of a machine previously used in the finishing of yarns of wool or hair), aff'd (1857) 28 LJ QB 175, Ex Ch; *The Patent Bottle Envelope Company v Seymour* (1858) 141 ER 65, CP (improvements in the manufacture of cases or envelopes for covering bottles, an analogous use of a mandril to produce model bottles out of certain materials); *Ralston v Smith* (1865) 11 ER 1318, HL (improvements in embossing and finishing woven fabrics, and in the machinery used therefore, an analogous use of a roller engraved with a particular pattern to do two things simultaneously that had previously been done separately using a roller with different patterns); *Spencer v Jack* (1863) 11 LT 242, CA (improvements in producing steam-engines, a combination); *Thompson v James* (1863) 55 ER 224, MR (improvements in hoops for use in the manufacture of petticoats, an analogous use (or substitution) of steel springs where previously elastic materials had been used); *Newall v Elliot* (1864) 10 Jur 954, Ex (improvements in apparatuses for laying down submarine telegraph wires, a combination); *Curtis v Platt* (1866) 35 LJ Ch 852, HL (improvements in machines for spinning or doubling cotton and other fibrous substances, a combination); *Jordan v Moore* (1866) LR 1 CP 624, CP (improvements in the construction of ships and other vessels navigating on water, an analogous application of wooden planking to the iron frame of a vessel where previously applied to the wooden frame); *Parkes v Stevens* (1869) LR 8 Eq 359, VC (improvements in the manufacture of lamps involving the use of a sliding door, a combination), aff'd (1869) LR 5 Ch 36, LC; *Tatham v Dania* (1869) 1 Grif Pat Cas 213, CP (improvements in machinery for teazing and opening textile materials, an analogous use of rollers); *Murray v Clayton* (1869) LR 7 Ch 570, CA (improvements in the manufacture of bricks, a combination); *Cannington v Nuttall* (1871) LR 5 HL 205, HL (improvements in the manufacture of glass, a combination); *Bamlett v Picksley* (1875) 1 Grif Pat Cas 40, Ex Ch (improvements in reaping and mowing machines, an analogous use of a known wheel to a new machine). cf *Penn v Bibby* (1866) LR 2 Ch 127, VC (improvements in the bearings and bushes for the shafts of screw and submerged propellers, where bearings of a different description had previously been used in grindstones and waterwheels, a non-analogous application of known objects to a new purpose); *Hayward v Hamilton* (1881) 1 Grif Pat Cas 115, CA (improvements in the manufacture of pavement lights, a non-analogous use of a known glazing).

[92] (1860) 121 ER 1044, QB.

[93] Overruled (1862) 121 ER 1056, Ex Ch, aff'd (1865) 9 ER 1488, HL.

Diagram B Some cases of note regarding patentable subject matter, 1852 to 1882

Case	Patent Title and Claimed Invention	Précis of Judicial Reasoning and Comments of Relevance
Bush v Fox (1852) Macr 140 (NP); (1856) Macr 179, 5 HL Cas 707, 10 ER 1080 (HL)	Improvements in the means of and in the apparatus for building and working under water. A mode of constructing the interior of a caisson in such a manner that the workpeople may be supplied with compressed air, and be able to raise the materials excavated, and to make or construct foundations and buildings.	Pollock CB—An invention must be a production of something that can be used or sold, or a method resulting in such a thing. An application of an old invention to a new purpose is not a manufacture, and so if nothing new results from such application, no patent can be granted. HL—A mere application of an old machine is not a new manufacture, and is therefore not the subject of a patent.
Tetley v Easton (1852) Macr 33 (NP); (1853) Macr 78 (NP); (1853) Macr 82, 2 E&B 956, 118 ER 1024 (QB)	Improvements in machinery for raising and impelling water and other liquids.	The use of an old machine in an old manner is not an invention or application for which a patent may be granted.
Crossley v Potter (1853) Macr 239 (NP)	Improvements in weaving figured fabrics.	A patent must be for a manufacture, and not a principle. While patents are frequently obtained for so-called 'processes' and 'methods', the real object of such a patent, the end of the patent secured by the statute, is a new manufacture—that is, the result of the process; the thing that is manufactured or the process by which it is produced.
Re Martin and Hyam's Patent (1855) 3 WR 433 (LC)	Improvements in goloshes or overshoes. The application of gutta percha to make clogs.	An application that is useful and successful, and is patentable.
Bewley v Hancock (1856) 43 ER 1285 (LC)	Improvements in machinery for giving shape and configuration to plastic substances.	The distinction between a patent for a product and the machine by which it is produced does not apply for applications, where the distinction is between the fact of applicability and the mode of application; only the latter being a real invention.

Booth v Kennard (1856) 156 ER 1311 (Ex Ch)	Improvements in the manufacture of gas. The discovery that the first step in the process of making gas from seeds (separating the oils from the seeds) might be dispensed with.	The improvement consisted in the omission of a step, which was a useful invention and therefore patentable.
Bovill v Pimm (1856) 156 ER 1019 (Ex)	Improvements in manufacturing wheat and other grain into meal and flour. The application of ventilating screws or vanes for blowing and sustaining the necessary blast of air between the grinding surfaces of millstones.	A patent for a method of doing a known thing (eg cooling substances undergoing grinding) only protects the particular method.
Bovill v Keyworth (1857) 7 E&B 725, 119 ER 1415 (QB)	Improvements in manufacturing wheat and other grain into flour. The combination of a blast of air between the millstones used in grinding corn with an exhausting apparatus attached to the chamber in which the stones were enclosed.	Failure to specify how a particular result is achieved (eg exhausting air from cases of millstones) in a patent for that object combined with another result will invalidate the patent as being for a mere principle.
Brook v Aston (1857) 120 ER 178 (QB), 28 LJ QB 175 (Ex Ch)	The application of a machine previously used in the finishing of yarns of wool or hair to finish the threads of cotton and linen.	QB—The application of an old invention to a new purpose is patentable if novel in the sense of involving some discovery or invention. The application of an old machine in an old manner to an analogous substance is not patentable, because the results and process are the same. Ex Ch—The effects, machinery and process are the same; the material to which the process is applied is so similar that the process cannot support two patents differentiated only by that material. The application of a well-known tool to work previously untried materials or to produce new forms is not patentable unless (possibly) it produces an entirely different result.

(*Cont'd*)

Diagram B Some cases of note regarding patentable subject matter, 1852 to 1882 (*Cont'd*)

Case	Patent Title and Claimed Invention	Précis of Judicial Reasoning and Comments of Relevance
Lister v Leather (1855–1858) G 283, 120 ER 373 (QB); G 283, 120 ER 384 (Ex Ch)	Improvements in combing wool. A new combination of machinery, including the replacement of a lashing process with a nipping process combining 5 movements of the wool to be made in a certain order.	A patent for a combination of machinery protects subordinate combinations that are new and material to the entire combination.
Higgs v Goodwin (1858) 120 ER 607 (QB)	A method of treating chemically the collected contents of sewers and drains in cities, towns and villages.	The intention to produce a profitable matter is of the essence of a patent.
The Patent Bottle Envelope Company v Seymour (1858) 141 ER 65 (CP)	Improvements in the manufacture of cases or envelopes for covering bottles. The use of a mandril to produce model bottles out of certain materials.	The application of a mandril in the manufacture of bottle envelopes is not patentable where its use to produce given forms was well known. The only merit/novelty/discovery was in the material to which the mandril was applied and the form of utensil produced by it, which is not a new 'manufacture'. To allow a patent for the application of a well known tool to work previously untried materials or to produce new forms might produce oppressive monopolies.
Ralston v Smith (1860) 142 ER 45 (CP); (1861) 142 ER 880 (Ex Ch); (1865) 11 ER 1318 (HL)	Improvements in embossing and finishing woven fabrics, and in the machinery or apparatus employed therein. The use of a roller engraved with a particular pattern to do two things simultaneously that had previously been done separately using a roller with different patterns; the 'discovery' being the particular pattern.	CP—The possibility of combining inequality of speed with a patterned roller must have been well known, but was not used because it tore material. Query whether one particular (albeit new and beneficial) use of a known machine is patentable, as it virtually deprives the owner from an existing right over his own property. HL—The patentee discovered that a particular pattern might be engraved upon the roller to impress and gloss the cloth. Had the patentee claimed as his invention the construction of a machine or combination of machinery

		(the roller engraved with that pattern) that would do both those things, the patent would be good, but instead he claimed a new process, and there is no improvement in the process of embossing or finishing woven fabrics amounting to a new manufacture, because the mode of embossing and the mode of finishing were each well known. To be patentable a discovery must be more than useful; it must fall within a fair extension of the words 'a new manufacture'. The discovery that a particular pattern when impressed on a roller will not tear fabric is very useful, but it would be strange to say it was a new manufacture, and that the owner of a roller is therefore to be deprived of the most useful way of using his roller. There is nothing new in the invention, except that it teaches what was not known before; that by a particular use of it, a result shall be obtained which it was not known before could be obtained. All the patentee can claim is to have discovered that by giving a differential motion to different parts of an old machine, a power existing in it might be developed and brought into action; and such a discovery is not patentable.
Harwood v Great Northern Railway Co (1860) 121 ER 1044 (QB); (1862) 121 ER 1056 (Ex Ch); 9 ER 1488 (HL)	Improvements in fishes and fish joints for connecting the rails of railways. The application of a mechanical contrivance used for bridges, to railways.	QB—The question in every case of a patent for a use is whether the amount of affinity or similarity between the two purposes is such that they are substantially the same; and that determines whether the invention is sufficiently meritorious to be deserving of a patent. A patent for a new combination of old things is good if the result is beneficial, and the application of an old principle to what is practically the same object is not patentable, but between these two extremes it is difficult to draw the line and say where the invention commences.

(*Cont'd*)

Diagram B Some cases of note regarding patentable subject matter, 1852 to 1882 (*Cont'd*)

Case	Patent Title and Claimed Invention	Précis of Judicial Reasoning and Comments of Relevance
		Ex Ch—The patent was for a new application of an old mechanical contrivance. The contrivance was complete in itself when first invented and, though not immediately applied, immediately applicable to all forms of pieces of iron used for holding together other materials by the aid of bolts having a bearing upon the pieces of iron. It required no new invention, but merely an application of the mechanical contrivance already invented and used. Thus it was a mere application of an old contrivance, in the old way, to an analogous subject, without any novelty or invention in the mode of application. And an application such as this does not make a valid subject matter for a patent. HL (minority)—To bring subject matter within the Statute of Monopolies there must be invention applied to produce a practical result. A mere application of an old contrivance in the old way to an analogous subject, without any novelty or invention in the mode of application is not patentable. The question is whether the purpose to which the contrivance was before applied and the new purpose are so analogous that there is no discovery or invention in the new application—whether it is a mere application or not. If there is invention or discovery producing a practical benefit (eg *Crane v Price*) it is the valid subject of a patent; if there is any real invention, though a slight one, producing a practical beneficial result, the patent is good—but the question is whether there is such an amount of cognateness in the purposes that there was no real invention or discovery. We differ with the Ex Ch not as lawyers, but as mechanics and engineers. The essence of the invention lay in the thought, that inasmuch as the fish was intended to resist a strain in its own plane, the metal in the centre of the fish, which was comparatively inert for the purpose of resisting such a strain, might with advantage be partially removed. The article thus produced was an improved and useful article, and also new.

		HL (majority)—There can be no invention in taking that which was a fish for a bridge, and applying it as a fish to a railway, for there would be no end to the interference with trade and with the liberty of adopting any mechanical contrivance if every slight difference in the application of a well-known thing should be held to constitute ground for a patent. You cannot have a patent for a well-known mechanical contrivance merely when it is applied in a manner or to a purpose, which is not quite the same, but is analogous to the manner of the purpose in or to which it has hitherto been used.
Calvert v Ashburn (1862) JPM 84 (Ex Ch)	The application of caustic alkalies previously used to dissolve gluten in flour used in the manufacture of size, to dissolve gluten in the manufacture of starch.	The distinction between inventive and non-inventive mechanical applications applies with regard to applications of the chemical properties of matter.
Horton v Mabon (1862) 142 ER 1213 (CP)	Improvements in the construction of gas-holders. The application of angle-iron to construct gas-holders.	The alleged invention is nothing more than the application of a known instrument to purposes analogous to which it had previously been applied. It is not a new article, or an improved article, but only a cheaper way of using the known materials. A new combination of old machinery or instruments, whereby a new and useful result is attained, may be the subject of a patent; but there must be some invention and there is none here. The merit is due to the person who first produced the article.
Mackelean v Rennie (1862) 143 ER 21 (CP)	Improvements in floating-docks. The application of iron to form air-tight and water-tight chambers.	Query whether this application is patentable.

(Cont'd)

Diagram B Some cases of note regarding patentable subject matter, 1852 to 1882 (*Cont'd*)

Case	Patent Title and Claimed Invention	Précis of Judicial Reasoning and Comments of Relevance
Ormsom v Clarke (1862) 143 ER 134 (CP); (1863) 143 ER 531 (Ex Ch)	An improvement in the manufacture of cast tubular boilers. The casting in one piece of what was previously cast in two pieces.	QB—Casting in one piece what had previously been cast in two pieces is not patentable, because there is no claim for any novelty in the process of casting. A mere improvement in performing an operation already well known and long practised cannot be the subject of a patent—to hold otherwise would extend the protection granted by letters patent to a most inconvenient degree. Ex Ch—The patent might have been good had it claimed an improved mode of casting tubular boilers in one piece; but all it claims is 'the casting of a boiler, such as is above described, in one piece', making no claim to the boiler in several pieces.
Spencer v Jack (1863) 11 LT 242 (CA)	Improvements in steam-engines and in the apparatus connected therewith.	A combination of several known things of such use and advantage to the public that the combination itself is an improvement, is patentable. This invention is a new and useful thing, and therefore patentable.
Thompson v James (1863) 55 ER 224 (MR)	The use of hoops made of steel watch springs (in place of whalebone, cane and other substances) in the manufacture of petticoats.	There is no claim for the number of hoops or the fact of using hoops for this purpose, but only for the use of steel springs in combination with suspending tapes or bands to make this form of petticoat. The mere substitution of steel springs in the place of previously used elastic materials is not patentable, as there is no invention and nothing that can properly be called an invention in that. What is needed is some real novelty in the invention, either by a new combination of existing materials, or by the discovery of something that did not previously exist.
Morton v Middleton (1863) 1 CS 718 (CS)	Improvements in the construction of fences, and the posts or pillars for the same. The construction of the foot of a pillar to resist pressure.	The construction of a new pillar to achieve an old object is novel and patentable if it attains the object better.
Willis v Davidson (1863) 1 NR 234 (QB)	Improvements in the construction of organs. The application to pneumatic levers of an improved escape valve previously applied to organ bellows.	The application of the valve to the levers is not novel because of its previous application to organ bellows.

Young v Fernie (1864) 66 ER 836 (LC)	Improvements in the treatment of certain bituminous mineral substances and matters, or products therefrom.	The discovery (by laborious experimentation) of the thing required to enable known chemical substances to be produced in merchantable/commercial quantities is of great value and is patentable. A distinction exists between the discoveries of the merely scientific chemist and those of the practical manufacturer, who invents the means of producing in abundance, suitable for economical and commercial purposes, that which previously existed as a beautiful item in the cabinets of men of science. The law looks to the inventor and discoverer who finds out and introduces a manufacture which supplies the market for useful and economical purposes with an article which was previously little more than the ornament of a museum. Inventions in mechanics are widely different from inventions in chemistry, as the laws and operations of mechanical forces differ from the laws of chemical affinities, and due also to the results of analysis and experiment in the comparatively infant science of chemistry, with its boundless field of undiscovered laws and substances.
Newall v Elliot (1864) 10 Jur 954 (Ex)	Improvements in apparatus for laying down submarine telegraph wires.	A patent is good if it is for a new and useful apparatus. A patent for a combination does not prevent the use of anything previously in use. The fact that something is not prepared and vendible does not prevent it from being patentable.
Curtis v Platt (1866) 35 LJ Ch 852 (HL)	Improvements in certain machines for spinning or doubling cotton and other fibrous substances.	The similarity of combinations is to be determined with reference to their effects. A new mode of accomplishing an old object is patentable.
Dangerfield v Jones (1866) 13 LT 142 (VC)	An improved mode of, and apparatus for, bending wood for the handles of walking sticks, umbrella and parasol sticks, etc. Channelling heat through a mandril instead of directly on to the wood.	The application of general principles of mechanics to a new manufacture is patentable provided the invention is new, desirable and of benefit to the public; but a mere trifling matter or thing of no value will not do, as the whole theory of the patent law is based upon the assumption that there is something of real value. You must show that you have invented something useful—a new and useful improvement in manufacture.

(*Cont'd*)

Diagram B Some cases of note regarding patentable subject matter, 1852 to 1882 (*Cont'd*)

Case	Patent Title and Claimed Invention	Précis of Judicial Reasoning and Comments of Relevance
		The application of heat through the medium of a hollow mandril, instead of directly to wood, is a useful and valuable invention and therefore patentable.
Fox v Dellestable (1866) G 198 (VC)	Improvements in umbrellas.	The validity of the patent depends on whether the alleged invention is novel and important. Here it is not; the only invention claimed is to open one side of a tube where previously it was closed in the making of umbrellas.
Jordan v Moore (1866) LR 1 CP 624 (CP)	Certain improvements in the construction of ships and other vessels navigating on water. The application of wooden planking to the iron frame of a vessel (where previously applied to a wooden frame).	The application is not patentable, as it is merely the substitution of one well-known and analogous material for another to effect the same purpose, as distinct from the method of planking itself.
Penn v Bibby (1866) LR 2 Ch 127 (VC)	An improvement in the bearings and bushes for the shafts of screw and submerged propellers. (Bearings of a different description had previously been used in grindstones and waterwheels.)	The application of an old invention to a new purpose is patentable because there will necessarily be novelty in the application. In each case the main consideration is whether the new application lies so much out of the track of the former use as not naturally to suggest itself to a person turning his mind to the subject, but to require some application of thought and study. Here the patented invention is not merely a different application, but something in itself essentially different.
Hinks v Safety Lighting Company (1876) LR 4 Ch D 608 (MR)	Improvements in lamps for burning paraffin oil and other volatile liquid hydrocarbons. The replacement of a flat wick with a round wick.	Improvements in a lamp consisting (solely) in the replacement of a flat wick for a round wick are patentable, because the patent is for a new combination of known parts. However the combination must still have merit, and how such merit is to be determined I do not know—the authorities suggest it depends on the result produced. Thus a slight alteration in a combination that produces a very useful and important improvement is patentable.

White v Toms (1867) 37 LJ Ch 204 (VC)	Improvements in ladies' mourning bonnet and hat falls. Sewing the fold on both sides, whereas previously it had been sewn on one side only.	The improvement of a method of manufacturing an old article of great commercial merit and novelty is not patentable because there is no invention and nothing that can be called an invention. The inclination of modern times is to restrict rather than to enlarge patent law, and the patent right has been so much abused that it is necessary to restrain the inconvenience suffered by the public and the trade by the number of existing patents, so that the patent right shall only be granted to those who really invent something that is for the public benefit. The principle of *Harwood* is that a mere improvement in the mode of manufacture, however meritorious as an improvement, is not patentable.
Parkes v Stevens (1869) LR 8 Eq 359 (VC); (1869) LR 5 Ch 36 (LC)	Improvements in the manufacture of railway-station and other lamps.	The use of a sliding door on a spherical lamp (where previously it had been applied to a cylindrical lamp) is not patentable.
Tatham v Dania (1869) 1 Grif Pat Cas 213 (CP)	Improvements in machinery for teazing and opening textile materials.	To succeed, a patentee must show newness resulting from the production of a thing which requires some exertion of mind that could properly be called invention; the application of an old tool to a new material, even where it produces a greatly improved effect, can never support a patent because the purpose is analogous to that for which the tool was previously known to be applicable. Here that which has been taken is not patentable, because rollers have previously been used in the same way in respect of similar materials for the same purpose, and thus the use claimed is of an old process which was applied to another state of things different only in the circumstances and not in the species.

(*Cont'd*)

Diagram B Some cases of note regarding patentable subject matter, 1852 to 1882 (*Cont'd*)

Case	Patent Title and Claimed Invention	Précis of Judicial Reasoning and Comments of Relevance
Murray v Clayton (1869) LR 7 Ch 570 (CA)	Improvements in machinery for making bricks. A combination of machinery that enables the cutting of clay into bricks by one turn of the wrist.	Bacon VC—The Statute of Monopolies is declaratory of common law, which has long authorized patents to any man who, by his own wit and invention, did bring any new trade into the realm, or any engine tending to the furtherance of a trade that was never used before'; a principle that has never been departed from or lost sight of. The reasons and principle of patent law are obvious and plain: a petitioner alleges to the Crown that he has made a new and useful discovery and, assuming that allegation to be true, receives as a just reward for such discovery, the exclusive use of his invention. The test of the truth of the allegation depends on the invention's novelty. The mere arrangement of common elementary mechanical materials to construct a machine that produces no new result cannot support a patent, otherwise there would be no protection of the public or earlier patents against the ingenuity of any artisan who might have the skill to arrange the old mechanism in a new shape, and thereby to appropriate to himself the fruits of previous inventors in the proper sense of that term, so that the privilege and reward which the law only concedes to art and wit and invention might be bestowed upon mere skill in handicraft. There must be such novelty in the arrangement as would give to what is called his invention the character of an art (*Foxwell v Bostock* applied). Sir WM James LJ (overturning the VC)—The patent claims the combination of machinery that enables the cutting of clay into bricks by one turn of the wrist. The machine is very simple, and yet that is the case with many noted inventions—when the thing is once hit it seems a marvel that it was not hit before. The V-C's statement regarding the insufficiency of skill in handicraft is irreconcilable with *Crane v Price*. That is a decision I have never been satisfied with, but only because I could not see how the introduction of a particular type of fuel into a machine that had been patented for the use of every kind of fuel in the making of iron could be a 'combination'. Its principles (as per Tindal CJ) have not been questioned. On those principles the combination here is valid, for it produces a better thing in a more expeditious and economical manner.

Rushton v Crawley (1870) LR 10 Eq 522 (VC)	Manufacture of artificial hair for ladies' head-dresses and frizettes. The application of Russian tops (wool) to make chignons that had previously been made by other wools.	Patent rights must be preserved and protected for those who introduce really valuable inventions and improvements, but the public should not be hampered by persons taking out patents for frivolous articles, eg for the employment of a different material to produce the same result. The use or application of wool to make chignons and like articles is not patentable, for it would preclude the use of wool for purposes for which it has long been used. It is a gross violation of the patent privilege to take out a patent for a known article that has been used for years upon discovery that it can be produced cheaper or by a new material, or that such use can support a patent. The validity of *Crane v Price* is open to doubt.
Wright v Hitchcock (1870) LR 5 Ex 37 (Ex)	Improvements in the manufacture of frills or ruffles, and in the machinery or apparatus employed therein.	An improved method of manufacturing frills or ruffles using a sewing machine to do what had previously been done by hand produced an old result more cheaply and conveniently, and was therefore patentable. The patent protected both the machinery and the articles manufactured by it, even though such articles had formerly been made by hand or some other process, for otherwise the patent would be useless. The situation may be different in the case of ordinary products such as flour.
Arnold v Bradbury (1871) LR 6 Ch 706 (LC)	An improved ruffle and sewing-machine for producing the same, applicable also to ordinary stitching.	When you have purchased the plaintiff's machine you have purchased the right of using it with a sewing-machine, or in any other way that you may think fit, but you must not use his machine with a sewing-machine unless you have purchased it.
Cannington v Nuttall (1871) LR 5 HL 205 (HL)	Improvements in the manufacture of glass. The introduction of air all round the tank to cool the substance in the tank and to produce congelation in the case of the fluid exuding.	The use of an old process to effect the same object with a different material is never patentable, but that is not the case here, where the invention is for an effective method of removing sieges and pots from the manufacture of glass by use of basic principles of temperature. Those principles were not new, but so it is with every new invention—the skill and ingenuity of the inventor are

(*Cont'd*)

Diagram B Some cases of note regarding patentable subject matter, 1852 to 1882 (*Cont'd*)

Case	Patent Title and Claimed Invention	Précis of Judicial Reasoning and Comments of Relevance
		shown in the application of well-known principles to achieve a practical result not yet achieved. Thus the test of novelty is: is the article produced by the apparatus for which the patent is claimed effectively obtained by means of that apparatus, whereas it had never before been effectively obtained by any of the apparatus's separate portions? The invention is the introduction of air all round the tank to cool the substance in the tank and to produce congelation in the case of the fluid exuding (cf the cooling of hot substances), and the patent for that invention is infringed by any process that achieves this same result. The patent is for the combination of ingredients comprised in the apparatus. The only new discovery was the application of the external air to the sides of the tank—which was not something for which a patent could independently have been obtained. I may construct an apparatus the merit and benefit of which depends on the application of some known natural force or property; but the invention consists in the construction of the apparatus in such a manner as to bring the natural agency or power to bear upon and effect the object which I desire to effect, and that I do by means of any apparatus constructed so as to bring into action that natural power. My invention consists in the arrangement of the apparatus in order to receive that ordinary and well-known dynamic agent, and make it a fit instrument for effecting a new result.
Binney v Feldtman (1871–5) 1 Grif Pat Cas 49 (HL)	Improvements in the manufacture of packing for joints of steam engines. A method of mixing two substances during the manufacture of one of those substances (cf combined use of two substances).	The joint use of fibrous materials with a mineral lubricant to achieve an old object would not be patentable where such materials and lubricant had been independently used to achieve the same object. Here the patent did not claim such joint use, but rather a method of manufacturing braided packing by which the powder was worked into the fibrous material, and that patent was good, protecting only the particular method. HL—The patent is not infringed by a method of impregnating and saturating fibrous strands with mineral lubricant, etc, where that method produces a useful and improved article more cheaply.

Clark v Adie (1873) LR 10 Ch 667 (CA); (1877) LR 2 App C 315 (HL)	A combination of several improvements to instruments for clipping horses.	The principle underlying *Lister v Leather* is that a patent for a new combination or arrangement is entitled to the same protection and on the same principles as every other patent. In fact, every or almost every patent is a patent for a new combination. The patent is for the entire combination, but there is, or may be, an essence or substance of the invention underlying the mere accident of form; and that invention, like every other invention, may be infringed. *Lister v Leather* does not apply here, where the patent is for several distinct improvements that are a combination only in the sense that they combine with each other to make the machine a better one. HL—A patent protects the subordinate invention residing in the larger invention, if that subordinate invention is claimed. For a combination to be patentable there must be novelty of invention and production of a new and beneficial result. There is an established principle that old things used in a combination to produce different or new results may support a patent (*Hill v Thompson*), eg *Crane v Price*, which has gone as near the extreme end as any decided case has gone, and which shows that an old combination, if producing a new and beneficial result, may be the subject of a patent. Whether it is so or not must depend upon the facts of the particular case.
Johnson v Rylands (1873) 1 Grif Pat Cas 138 (Ex Ch)	Improvements in testing the strength of wire for telegraphic and other purposes.	The patent covers a method of causing wire to expand by the motion of wheels at different speed, which does not restrict use of a method of causing wheels to move at different speeds by expanding wire.

(*Cont'd*)

Diagram B Some cases of note regarding patentable subject matter, 1852 to 1882 (*Cont'd*)

Case	Patent Title and Claimed Invention	Précis of Judicial Reasoning and Comments of Relevance
Bamlett v Picksley (1875) 1 Grif Pat Cas 40 (Ex Ch)	Improvements in reaping and mowing machines.	The application of a method of causing wheels to move in a particular way to a new purpose (reaping and mowing machines) is not patentable regardless of whether there was any merit in the construction of the machine, for it is merely the use of an old broom to sweep a new carpet. This case is to be distinguished from those in which an old thing is applied to something so totally different that there is an obvious invention in the application—eg a man who uses the method by which flour is spun into macaroni pipes to spin red-hot iron into pipes—where the product and result are completely new and different even though the method was the same. The difficulty here is not the law but its application. No terms can be used which will so absolutely exclude any possible misconception as to lay down an abstract definition of what is not the subject matter of a patent. The question is whether the purpose to which this is applied is a cognate purpose. It is something very much nearer than a cognate purpose—it is mechanically identical. *Crane v Price* can only be supported on the ground that there a new product was produced of a different class and character. Whether it is supportable upon that ground or not some judges have doubted. The invention was called a combined reaping and mowing machine; but it is no more a combination than a lock that can be put on a window and door; a 'combination' at law must be a combination producing a new and useful result. This is not a new combination as the wheel, its working, and the mode are all old. All that was required was a different velocity for the new purpose.
Patterson v The Gas Light and Coke Company (1875) G 362, LR 2 Ch D 812 (CA); (1876) G 362, 3 App Cas 239 (HL)	Improvements in the purification of coal gas.	The patent claims a chemical truth (that pure sulphides of calcium will absorb the sulphur compounds), and while the plaintiff believed he had discovered that truth, in fact it was well known. The patent claims the employment of sulphides of calcium for the removal of sulphur, without limitation to any particular process etc, and that is clearly not patentable. The patent also claims a method of employing lime purifiers, but does not claim any new apparatus or process. The patentee claims to have discovered that if the carbonic acid is not wholly taken up by the lime at the beginning it will have devastating effects; and while this is a direction

		and instruction of the greatest possible value and utility, it is not patentable. How could an infringement of such a patent be predicated? Could the court say, we restrain you from working your lime-purifying process in any such way as will not allow the carbonic acid to enter the last purifier in sufficient quantity to do substantial mischief, or in less quantity on an average than it used to do in former times on an average? No one has a right to prevent a workman from using care to keep his tools in the most efficient state. No one has a right to prevent a manufacturer from cleansing his vessels and throwing away the useless contents whenever he likes, or to ask him his motives or intentions in doing so. HL—The consideration for a patent is the communication to the public of a new process. As Hindmarch said, if the public once become possessed of an invention by any means whatever, no subsequent patent can be granted for it, either to the true or first inventor himself, or to any other person, for the public cannot be deprived of the right to use the invention, and a patentee of the invention could not give any consideration to the public for the grant, the public already possessing everything that he could give. The mere discovery of the theory and reason of that which had before been done empirically does not justify a patent.
Bailey v Roberton (1878) G 26, LR 3 App Cas 1055 (HL)	Improvements in preserving animal substances.	The patent did not claim the chemical application of bisulphite of lime to animal substances, such as meat and fish, but the mode of applying a coating or film to the outside of animal substances, which is not patentable.
Sykes v Howarth (1878) LR 12 Ch D 826 (Ch)	Improvements in fancy rollers of machines for carding wool and other fibres.	An improvement consisting in the introduction of wide spaces between cards to produce an exhaust current of air and make the roller operate better in performing the old function is patentable (*Murray v Clayton* applied).
Eades v Starbuck Wagon Co (1881) WN 160 (Ch)	Improvements in the construction of tramcars.	Lengthening a truck and adding two segmental circles is not such an invention as to entitle a person to a patent.

(*Cont'd*)

Diagram B Some cases of note regarding patentable subject matter, 1852 to 1882 (*Cont'd*)

Case	Patent Title and Claimed Invention	Précis of Judicial Reasoning and Comments of Relevance
Hayward v Hamilton (1879–81) 1 Grif Pat Cas 115 (CA)	Improvements in pavement lights.	The plaintiff is an inventor, having found out and manufactured a thing, an article (a light directing pavement light), despite the lack of any very great quantity of invention in it. It is not as though a man had sat himself down to consider how he could make a sewing machine, or a grain-pressing machine; it is a much humbler piece of invention than that, and it may possibly be nothing more than that the idea struck him, and immediately the idea struck him he could apply it or carry it into execution by a very obvious apparatus. But still it is an invention, and it is not the less an invention because it required but small inventive powers to enable him to do it; its patentability does not depend on the quality of invention. Nor is it in this case the patenting of an idea. He does not say, 'I thought of such and such a thing and I will show you how to carry it into execution'; nor is it open to attack for comprising old constituent parts, because the combination of them is new. Thus the plaintiff really is an inventor—he has found out something, and made an article that was not made before. This particular case may be upon the verge, but one cannot help making this remark, that it is very strange if it is no invention that it has never been done before. Why has it never been done before? Why, because nobody else found it out, which I take to be equivalent to inventing. In all previous cases it has been taken for granted that if a thing were new and useful there must have been an invention in order to arrive at a thing that can be so described, and I should say that in nine hundred and ninety-nine cases out of a thousand that must be so. I say if the thing is new and useful it is impossible to suppose there is not sufficient to make an invention; but I do not think as a matter of law that could be predicated as an absolute rule of law, because it is possible that there may be no invention in a new and useful thing. But in this case the thing claimed, taking it to be a combination, apparatus, commercial article, or whatever you want to call it, involved sufficient invention to be an invention and is therefore patentable. It was contended that the claim was for inventing the prism, which would not be sufficient, but it is really for an improvement in pavement lights by substituting for the old glazing, a glazing that was not simply a prism, but a prism so constructed as to direct the light in any required direction.

Hill v Tombs (1881) JPM 82 (MR)	The use of a guide in a frilling machine.	The use of a guide in a frilling machine to keep down the work is not patentable because that was the use of known means for an analogous purpose, guides of similar character having been employed in many other machines.
Otto v Linford (1882) 46 LT 35 (CA)	The use of a known machine to carry out an idea or principle.	A mode of carrying out a principle or idea is patentable, even though no invention is required once the principle or idea itself is discovered. The patent does not claim a mere principle or idea; it claims the carrying out of the principle or idea in one or the other machine, and that is the subject of a patent. If you have a new principle, or a new idea, as regards any art or manufacture, and then show a mode of carrying that into practice, you may patent that, though you could not patent the idea alone, and very likely you could not patent the machine alone, because the machine alone would not be new. One of the strongest illustrations I know of is the patent for the hot blast in the iron manufacture, where there was nothing new at all except the idea that the application of hot air instead of cold air to the mixture of iron ore and fuel would produce most remarkable results in the shape of economy in the manufacture of iron. This case here is a much stronger illustration of the validity of a patent. It is true that, given the state of knowledge as regards mechanics, no invention was required to carry the idea out (ie to choose the machine and apply it) once that idea had been formulated. In the case of the hot blast the man did not pretend to invent anything; he said a machine of any shape in which you can heat air is sufficient. Overall, we have a principle and mode of carrying it out, and this is a good subject matter for a patent.

Harwood involved a patent for 'improvements in fishes and fish joints for connecting the rails of railways'.[94] The improvement consisted in the use of grooved plates of iron to secure longitudinally placed slabs of timber for use in the construction of railways;[95] the same plates having previously been used to secure vertically and horizontally placed slabs of timber for use in the construction of bridges.[96] In the Queen's Bench,[97] Lord Cockburn CJ upheld the patent as being for the application of a known mechanical contrivance (the plates) to a useful purpose (securing timber) on the ground, consistent with *Losh*, that that purpose was sufficiently distinct from the purpose for which the contrivance had previously been applied to entitle the applicant to a patent.[98] The other members of the Court agreed, albeit adopting different approaches. For Hill J, patentability depended on a new discovery producing results of benefit to the public, consistent with the decision in *Crane v Price*.[99] For Blackburn J, there existed in patent law a spectrum of subject matter differentiated by the novelty of their mechanical principle—'the degree to which the thing has been done in a mode known before, as applied to a similar purpose'[100]—with uses and combinations at opposite ends.[101]

On appeal to the Court of Exchequer Chamber,[102] the decision of the Queen's Bench was reversed. Agreeing with Lord Cockburn CJ's categorization of the subject matter as the application of the grooved plates as a known mechanical contrivance,[103] the Court disagreed with his finding of fact that the purpose of that application was not analogous to the purpose of previously known applications of the same plates.[104] Thus, the patented application consisted in the mere use of a known object, which could not support a patent on *Losh* grounds.[105] This decision was upheld by a majority of the House of Lords.[106]

At the time, *Harwood* was the leading authority on the unpatentability of analogous uses, and their distinction from (patentable) combinations.[107] In the cases

[94] The specification is extracted at (1860) 121 ER 1044, 1045.

[95] See ibid.

[96] Before the patent, railway rails had been secured by fishes constructed from flat (cf grooved) plates or iron. On the difference between the two fishes, see (1865) 9 ER 1488, 1494.

[97] (1860) 121 ER 1044, QB.

[98] ibid, 1050.

[99] ibid, 1051. Hill J also relied on *Hill v Thompson* (1817) 36 ER 239, LC; *R v Wheeler* (1819) 106 ER 392, KB; *Brunton v Hawkes* (1821) 106 ER 1034, KB.

[100] (1860) 121 ER 1044, 1052.

[101] ibid.

[102] (1862) 121 ER 1056, Ex Ch.

[103] ibid, 1059.

[104] ibid, 1058–1059.

[105] ibid, 1059.

[106] (1865) 9 ER 1488, HL.

[107] See, eg *Horton v Mabon* (1862) 142 ER 1213, CP, 1219; *Ormsom v Clarke* (1862) 143 ER 134, CP, 155; *Penn v Bibby* (1866) LR 2 Ch 127, VC, 136; *Jordan v Moore* (1866) LR 1 CP 624, 635–636; *White v Toms* (1867) 37 LJ Ch 204, VC, 207; *Curtis v Platt* (1866) 35 LJ Ch 852, HL, 869; *Parkes v Stevens* (1869) LR 8 Eq 359, VC, 365; *Hinks v Safety Lighting Co* (1876) LR 4 Ch D 608, MR, 615–616; *Hayward v Hamilton* (1881) 1 Grif Pat Cas 115, 123.

that followed, the basis for that distinction in the requirement for a subject matter of mechanical or chemical ingenuity was similarly confirmed,[108] and supported with further reference to the 'general inconvenience' that would result from allowing patents for non-inventive uses,[109] and the statutory requirement for an invention itself.[110] The first of these references in particular underlined the basis of early patents for inventions in a limited exception to a statutory prohibition against monopoly grants.

Before concluding, a final point should be made regarding the 'general inconvenience' proviso of that exception. In *Patterson v The Gas Light and Coke Co*,[111] in a decision upheld by the House of Lords, James LJ found a direction and instruction regarding a means for improving a method of purifying coal gas to be incapable of supporting a patent, despite its admitted utility and value, on the ground that '[n]o one has a right to prevent a workman from using care to keep his tools in the most efficient state . . . [or] to prevent a manufacturer from cleansing his vessels and throwing away the useless contents whenever he likes, or to ask him his motives or intentions in doing so.'[112] This decision was important, for it supported the pre-1623 understanding of 'general inconvenience' as providing a basis for denying protection for subject matter the establishment or protection of which would interfere with existing industrial practice.[113]

[108] See, eg *Brook v Aston* (1857) 120 ER 178, QB, aff'd (1857) 28 LJ QB 175, Ex Ch; *The Patent Bottle Envelope Co v Seymour* (1858) 141 ER 65, CP, 69; *Harwood* (1865) 9 ER 1488, HL; *Calvert v Ashburn* (1862) JPM 84, Ex Ch; *Horton v Mabon* (1862) 142 ER 1213, CP; *Thompson v James* (1863) 55 ER 224, MR; *Willis v Davison* (1863) 1 NR 234, QB; *Ralston v Smith* (1865) 11 ER 1318, HL; *Jordan v Moore* (1866) LR 1 CP 624, CP; *Parkes v Stevens* (1869) LR 5 Ch 36, LC; *Tatham v Dania* (1869) 1 Grif Pat Cas 213, CP; *Rushton v Crawley* (1870) LR 10 Eq 522, VC; *Bamlett v Picksley* (1875) 1 Grif Pat Cas 40, Ex Ch; *Hill v Tombs* (1881) JPM 82, MR. See also *Ormsom v Clarke* (1863) 143 ER 134, CP, aff'd (1863) 143 ER 531, Ex Ch; *Clark v Adie* (1877) LR 2 App Cas 315, HL, 329, 334.

[109] See, eg *Ormsom v Clarke* (1862) 143 ER 134, 135 per Byles J; *Rushton v Crawley* (1870) LR 10 Eq 522, VC; *The Patent Bottle Envelope Co v Seymour* (1858) 141 ER 65, 69.

[110] See, eg *The Patent Bottle Envelope Co v Seymour* (1858) 141 ER 65, 69; *Ralston v Smith* (1865) 11 ER 1318, 1327, 1329. See also *Spencer v Jack* (1863) 11 LT 242, CA, 243; HF Moulton, *The Present Law and Practice Relating to Letters Patent for Inventions* (London: Butterworth, 1913) 15–16. See also *Patterson v The Gas Light and Coke Co* (1875) LR 2 Ch D 812, CA, 833–834, aff'd (1876) 3 App Cas 239, HL, 244; *White v Toms* (1867) 37 LJ Ch 204, VC, 207; *Young v Fernie* (1864) 66 ER 836, LC, 850–851; *Otto v Linford* (1882) 46 LT 35, CA, 38–39.

[111] (1875) LR 2 Ch D 812, CA, aff'd (1876) 3 App Cas 239, HL.

[112] (1875) LR 2 Ch D 812, 834.

[113] See further *Liardet v Johnson* (1778) 1 Carp Pat Cas 35, NP, in Hulme, 'On the History of the Patent Law in the Seventeenth and Eighteenth Centuries', 287; *Morgan v Seaward* (1837) Web Pat Cas 187, 197. cf *Arkwright v Nightingale* (1785) 1 Carp Pat Cas 38, 47 (rejecting the relevance of public policy when determining utility, on the ground that '[the court] must never decide private rights upon any idea of public benefit'); *Walton v Bateman* (1841) Web Pat Cas 613, 623 (stating, in directions to the jury in respect of a plea that a patent for improvements in cards for carding fibrous substances be invalidated as being of no public use or benefit and thus injurious to the public: 'I really do not understand to what that [plea] applies—I do not understand how it is raised').

F. Conclusion

The history of patentability from 1623 to 1883 is a history of the courts' adaptation of the early requirements for novelty and utility to accommodate the various derivative forms in which the mechanical and chemical subject matter of the era were patented. It was in the process of that adaptation that the understanding of patentable inventions as mechanically or chemically inventive subject matter emerged, along with the related exclusions of analogous uses and mere collocations of two or more objects. It was also in this period that the question of the need for production of an artefact arose, and the related question of the focus of the system between the practical, industrial, and manufacturing arts. Finally, it was in this period that the need for fair disclosure of inventions was expressed, and the role of the specification acknowledged as the vehicle by which the standards of patentability were expressed in terms of actual protective value, and the balance of rights secured by a patent consequently mediated by the courts.[114] As with their conception of the inventions itself, the courts' mediation of this balance was substantially affected during the nineteenth century by the reality of the patents in existence.[115] Dissatisfaction regarding that reality came to a head in 1865, when concern over the 'multiplicity of monopolies' for (among other things) 'alleged inventions of a trivial character' and their '[obstruction of] the progress and improvement of arts and manufactures' prompted the first extensive review of the patent system.[116] In the report that was published following that review, the UK Commissioners stated as follows:[117]

> The majority of witnesses . . . decidedly affirm the existence of practical inconvenience from the multiplicity of Patents. It is clear that Patents are granted for matters which can hardly be considered as coming within the definition in the Statute of Monopolies, of 'a new manufacture.' It is in evidence that the existence of these monopolies embarrasses the trade of a considerable class of persons, artisans, small tradesmen, and others, who cannot afford to face the expense of litigation, however weak the case against them may seem to be; and a still stronger case is made out as to the existence of what may be called obstructive Patents,

[114] Criticisms were frequently levelled against the pre-1883 specification on account of its failure generally to ensure the equality of the bargain thus struck. See United Kingdom, *Report of the Commissioners appointed to inquire into the Working of the Law relating to Letters Patent for Inventions* (C (1st series) 5974, 1865) ix, xv; United Kingdom, *Report from the Select Committee on Letters Patent* (C (2nd series) 193, 1872) iii.

[115] The reasons suggested for the liberal granting practices of the Law Officers in the pre-1883 inquiries into the UK patent system included the absence of any right of repeal by patentees against the refusal of a patent and the connection between administrators' salaries and the number of patents granted. See United Kingdom, *Report of the Commissioners appointed to inquire into the Working of the Law relating to Letters Patent for Inventions*, vii; United Kingdom, *Report from the Select Committee on Letters Patent*, iv.

[116] United Kingdom, *Report of the Commissioners appointed to inquire into the Working of the Law relating to Letters Patent for Inventions*, v.

[117] ibid, ix.

and as to the inconvenience caused thereby to manufacturers directly, and through them to the public.

One type of patent of particular concern was that granted to manufacturers of chemical products. As recounted in the parliamentary debates of the day, German manufacturers were in the practice of obtaining UK patents for chemical products and refusing either to work the products or to licence their patents to local firms.[118] In 1883 the Government responded with a compulsory licensing system. That system was aimed at preventing the unreasonable refusal of patent licences, and balancing the interests in rewarding first inventors with protection of the public against obstructive claims.[119] The system was included in the Patents, Designs, and Trade Marks Act 1883, aimed at modernizing the patent system after the findings of the 1865 report above, and their repetition in the report of the Select Committee on Letters Patent published during the following decade.[120]

[118] See Parl Debs (series 4) vol 278, cols 349–394 (16 April 1883).

[119] ibid: 'In this and similar matters the patentee was only the first discoverer. Others were working on the same lines, and it was only a question of time which would arrive first at a satisfactory result. It was all very well to reward the first inventor; but it was not necessary nor just to give to the first inventor an absolute right of monopoly, which might be used for purposes of extortion, or to the injury of the country which granted these reward for invention'. For the provisions, see Patents, Designs, and Trade Marks Act 1883, s 22.

[120] United Kingdom, *Report from the Select Committee on Letters Patent*, iii.

3

The Requirement for an Invention in the Modern Era of 'New' Technologies: 1883 to 1976

A. Introduction

During the nineteenth century, the courts refrained from attempting to define the 'invention', presumably to avoid the historical problems associated with doing so. After the introduction of the Patents, Designs, and Trade Marks Act 1883,[1] however, this began to change. That Act retained the use of the term 'invention' to denote the subject matter for which a patent could be granted. However, it extended the Patent Law Amendment Act 1852 definition of that term[2] to include 'alleged invention[s]',[3] and expressly distinguished the requirement for an invention from the requirement for novelty.[4] Equivalent distinctions between an invention on one hand and inventiveness and utility on the other were introduced in the second and third quarters of the twentieth century, when each was given statutory expression as a ground for opposing a patent application or seeking revocation of a patent grant.[5] The effect of that expression was to require decision-makers from the mid-twentieth century (and earlier in the case of novelty) to consider inherent patentability as a discrete requirement in order to distinguish the grounds for statutory action.

This change in the structure of patent law coincided with an expansion in the types of subject matter for which patents were sought, from the mechanical and chemical subject matter of the nineteenth century to new informational and biotechnological methods and products. In combination, the effect was to test the courts' conception of inventions, and the threshold limits of the patent system.

[1] 46 & 47 Vict c 57.

[2] See Patent Law Amendment Act 1852 (15 & 16 Vict c 83), s LV, reproduced at page 2.

[3] Patents, Designs, and Trade Marks Act 1883, s 46. See also Patents and Designs Act 1907, s 93; Patents Act 1949, s 101.

[4] Patents, Designs, and Trade Marks Act 1883, s 11.

[5] See Patents and Designs Act 1907, ss 25, 26; Patents and Designs Act 1932, s 3(d), amending the Patents and Designs Act 1907, s 25; Patents Act 1949, ss 7, 14.

B. 1883 to 1959: The Invention as a Manner of *New* Manufacture, Including an *Alleged* Invention

1. Literal Interpretations of the Statutory Definition of the 'Invention'

The cases of the late nineteenth and early twentieth centuries reflect the courts' increased focus on the statutory language and concern to rationalize the nineteenth century principles of inherent patentability with reference thereto. Of particular significance in this regard was the word 'invention'. In *Lane-Fox v The Kensington and Knightsbridge Electric Lighting Co (Ltd)*,[6] for example, the Court of Appeal explained the exclusion of analogous uses of known objects with reference to the distinction between an 'invention' on one hand and a 'discovery' on the other.[7] According to the Court, while an invention involved an act undertaken to a useful practical end, a discovery involved the formulation of abstract knowledge.[8] A similar distinction between practical and abstract subject matter was reflected in the reasons offered immediately after 1883 for the exclusion of principles, and insubstantial improvements to known products and methods; the latter of which were also referred to as 'working directions'.[9]

The other language relied on by decision-makers immediately after 1883 to rationalize existing exclusions from patentability was taken directly from section 6 itself, namely, 'manner of new manufacture'[10] and 'mischievous to the State,

[6] (1892) 9 RPC 413, CA.

[7] See also *Ralston v Smith* (1865) 11 ER 1318, HL, 1327, 1329. On the distinction between inventions and discoveries, see also *Partington v The Hartlepools Pulp and Paper Co Ltd* (1896) 12 RPC 295, HC (working direction regarding the conduct of a known process); *Acetylene Illuminating Co Ltd v United Alkali Co Ltd* (1904) 22 RPC 145, HL ('discovery' that the passing of an arc through materials produced no injurious electrolysis); *Birch v Harrap* (1896) 13 RPC 615, VC (distinguishing between an invention on one hand and an arrangement of known elements to adopt a known mode of using an apparatus involving 'the ordinary adaptation of ordinary means of actuation' on the other).

[8] (1892) 9 RPC 413, 416. See also *Hickton's Patent Syndicate v Patents and Machine Improvements Co Ltd* (1909) 26 RPC 339, CA (supporting a conception of the invention as a combination of idea and practical manifestation the 'merit' of which need not derive from the manifestation itself).

[9] See, eg *Young v Rosenthal* (1884) 1 RPC 29, QB (describing ideas and mathematical principles as mere abstract discoveries that cannot be regarded as a manufacture or invention having previously unknown utility, and as consequently incapable of supporting a patent); *Morgan v Windover* (1890) 7 RPC 131, HL per Lord Herschell (describing the use of a known device as involving the mere extension of a known idea); *Gadd v The Mayor, &c, of Manchester* (1892) 9 RPC 516, CA (gasometers improved by application of known principles); *Wood v Raphael* (1896) 13 RPC 730, Ch (improved eye-glasses having a pivoted placquet); *Partington v The Hartlepools Pulp and Paper Co Ltd* (1896) 12 RPC 295, HC (method of treating sulphite pulp used in the manufacture of paper and the like from wood described by the Court as a working direction relating to the cleaning of machinery); *Elias v Grovesend Tinplate Co* (1890) 7 RPC 455, QB (method of manufacturing a pot to improve its performance in a known operation).

[10] See, eg *Young v Rosenthal* (1884) 1 RPC 29, 31. See also *Hickton's Patent Syndicate v Patents and Machine Improvements Co Ltd* (1909) 26 RPC 339, 348 per Buckley LJ (emphasizing, consistent with the reasoning of the other judges, that the statutory requirement for a new manufacture does not mean that the invention itself is the 'manufacture' as distinct from its underlying idea).

by raising prices of commodities at home, or hurt of trade, or generally inconvenient'.[11] The effect of this language was to focus decision-makers' attention on the original purpose of the section in creating a limited exclusion from the Statute's general prohibition in respect of certain mechanical and chemical subject matter directed to advancing the industrial arts. It was not long before that purpose was tested in the courts by a variety of non-mechanical and non-chemical subject matter;[12] an early example being *Cooper's Application*.[13]

2. The Requirement for 'Invention with Reference to a Manufacture': *Cooper's Application*

Cooper involved a patent application in respect of an improved form of newspaper page featuring a blank space along which the page could be folded without disrupting the line of its text.[14] Allowing an appeal from a decision of the Comptroller-General, Sir Robert Finlay AG affirmed the patentability of the subject matter on the ground that it involved 'invention with reference to a manufacture'[15] resulting in a material artificial product. In so doing he approved the Comptroller's direction that[16]

> [a] Patent may be properly refused in any case in which no material product of a substantial character is realised or effected by the alleged invention, or in which the only material product is a printed sheet, or its equivalent, and the only alleged invention an arrangement of words, or the like, upon such sheet

with the following remarks:[17]

> The first branch of that proposition . . . seems tolerably obvious. You cannot have a Patent for a mere scheme or plan – a plan for becoming rich; a plan for the better Government of

[11] See, eg *Elias v Grovesend Tinplate Co* (1890) 7 RPC 455, CA, 467 per Lindley LJ ('It is very singular that the Statute of James says nothing whatever about utility, but utility has been engrafted into it because of the words to which I have called attention, that is to say, it has been found by experience to be mischievous to the State to grant patents which are not useful as well as new. I call attention to that to show the extreme importance of those words'; and discussing *Morgan v Windover* (1890) 7 RPC 131, HL, 'if such a patent as this were granted it would be mischievous to the State; that is to say, that a monopoly for making this trifling alteration, excluding all the world from doing the same, would be a mischief and not a benefit to the public. That appears to me to be what underlies the whole thing'); *Gadd v The Mayor, &c, of Manchester* (1892) 9 RPC 516, 524 (describing patents for the adaptation of an old contrivance to a new purpose that presents no difficulties to persons skilled in the art as 'intolerable nuisances' and as 'mischievous to the State' in the sense of s 6 of the Statute of Monopolies). See also *Cropper v Smith* (1884) 1 RPC 81, CA, 90 per Bowen LJ; *Thierry v Riekmann* (1895) 12 RPC 412, CA, 550; *Wood v Raphael* (1896) 13 RPC 730, 735; *Waterhouse's Patent* (1906) 23 RPC 470, CA.

[12] See, eg *Lane-Fox* (1892) 9 RPC 413, CA (improvements in obtaining light by electricity and in conveying, distributing, measuring, and regulating the electric current etc).

[13] (1901) 19 RPC 53, AG. On the case's legal weight, see (1901) 19 RPC 53, 55; *T's Application* (1919) 37 RPC 109, SG.

[14] (1901) 19 RPC 53, 54.

[15] ibid.

[16] ibid. See also *Fishburn's Application* (1938) 57 RPC 245, PAT, 246–247.

[17] (1901) 19 RPC 53, 54–55.

a State; a plan for the efficient conduct of business. The subject with reference to which you must apply for a Patent must be one which results in a material product of some substantial character . . . The second part of the proposition must be taken as a whole . . . If the case is one in which the only material product is a printed sheet, or its equivalent, and the only alleged invention an arrangement of words or the like upon such sheet . . . the Patent might be properly refused . . . [Such alleged invention] would not be a manufacture; in substance it would be a scheme, and the mere arrangement of printed matter in connection with such scheme would not prevent you from looking at the real object of the Applicant. To my mind the present application is of a different kind . . . The present Applicant in no way proposes to arrange printed matter for its more convenient use from a literary point of view. What he proposes is a particular way of manufacturing a newspaper; and the alleged utility of his supposed invention is purely mechanical.

By this reasoning, the need for 'invention with reference to a manufacture' was treated as requiring a method for producing a material product of a substantial character the 'alleged utility' of which was mechanical. In substance, this affirmed the conception of the invention supported in *Losh v Hague*[18] albeit restricted to subject matter of mechanical or chemical utility directed to advancing the *manufacturing* arts. The implication that business schemes and other plans were inherently unpatentable was confirmed in the cases to 1959.[19] Also confirmed in those cases was the further implication regarding the inherent unpatentability of arrangements of literary or artistic content.[20]

[18] (1838) Web Pat Cas 202, NP.

[19] See, eg *Johnson's Application* (1901) 19 RPC 56, SG, 56 ('systems of business' inherently unpatentable, even where their implementation requires the use of a printed envelope with a particular arrangement of works thereon); *Ward's Application* (1911) 29 RPC 79, SG, 79 ('ideas' or 'schemes' or 'mere methods' inherently unpatentable); *W's Application* (1914) 31 RPC 141, SG, 142–143 ('mere systems, more convenient methods of arranging well-known objects so that you may have them handy for reference' inherently unpatentable); *C's Application* (1919) 37 RPC 247, SG, 247 ('system of writing such as a system of shorthand' inherently unpatentable); *F's Application* (1920) 37 RPC 112, SG, 113 ('scheme whereby those in authority and in command of destroyers and submarines or both are to have some control exercised over the powers which they possess, and which they ought to use under, and which must be left to, their own judgment' inherently unpatentable); *A E W's Application* (1924) 41 RPC 529, SG, 529–530 ('colouring or . . . other method of indication on an odometer wheel' effecting a 'plan or scheme for reminding the driver that attention is required' inherently unpatentable); *D A and K's Application* (1925) 43 RPC 154, SG, 158 (plan for using confectioners' ordinary stock in trade inherently unpatentable); *Johnson's Application* (1930) 47 RPC 361, SG, 364 ('distinctive colour mark of origin of . . . goods, having a trade mark value' inherently unpatentable); *Littlewoods' Application* (1954) 71 RPC 185, SE, 189 (plans for the conduct of business inherently unpatentable). cf *Fishburn's Application* (1938) 57 RPC 245, 248; *Cobianchi's Application* (1953) 70 RPC 199, PAT, 200.

[20] See, eg *Ward's Application* (1911) 29 RPC 79, SG (arrangement of information to create indexing system); *C's Application* (1920) 37 RPC 247, SG (method of musical notation consisting in distinguishing between the sharps, flats, and naturals by differences in the colours, shapes, or designs of the characters on the printed sheet); *M's Application* (1924) 41 RPC 159, SG (arrangement of musical notes for instructional purposes); *P's Application* (1924) 41 RPC 201, SG (arrangement of printed matter on sheets to enable the sheets to be used with a copying apparatus); *E S P's Application* (1944) 62 RPC 87, SE (architectural plans); *Kelvin's Application* (1953) 71 RPC 103, PAT (arrangement of information on a chart for instructional purposes); *American Optical Co's Application* [1958] RPC 40, PAT (method of distorting the images on a film). cf *Fishburn's Application* (1938) 57 RPC 245,

The requirement for subject matter of alleged mechanical or chemical utility was justified in the early twentieth century with reference to the need for something more than an idea or abstract contribution to knowledge. After the 1950s, however, the prevailing tendency was to treat it as implicit in the requirement for a 'manner of new manufacture'.[21] Thus, while a newspaper or book might be a manner of manufacture, considered as such it was unlikely to differ from existing newspapers or books, and thus unlikely to be a manner of *new* manufacture.[22] Similarly, if manifestly lacking mechanical utility, it could not be regarded as directed to advancing the manufacturing arts, and for that reason could not be regarded as an 'alleged invention'.[23] Aside from *Cooper* itself, this principle emerged from the High Court of Australia's decision in *Griffin v Isaacs*.[24] The implication of Dixon J's finding in that case was that if a subject matter when conceived as an object of mechanical or chemical utility—ie, *qua* invention—was indistinguishable in its constitutive aspects from existing such objects, it could not claim to advance the manufacturing arts, and would thus be excluded from (inherent) patentability.[25]

PAT (method of arranging a cinema ticket so that it could be torn in half either way, with all necessary information appearing on either half); *Alderton's Application* (1941) 59 RPC 56, PAT (method of arranging time charts on a time recording card to improve a known system of recording workmen's work times). On the need for mechanical or chemical utility and inventiveness generally, see further *Klaber's Patent* (1906) 23 RPC 461, HL; *Hill v Thomas* (1907) 24 RPC 415, CA; *Thermos Ltd v Isola Ltd* [1910] RPC 388, HC; *M's Application* (1924) 41 RPC 159, SG; *P's Application* (1924) 41 RPC 201, SG; *Loth's Application* (1924) 41 RPC 273, SG; *W R's Application* (1924) 41 RPC 216, SG; *Compagnies' Application* (1930) 48 RPC 185, SG; *L & G's Application* (1940) 58 RPC 21, PAT; *F M M's Application* (1940) 58 RPC 115, PAT; *Alderton's Application* (1941) 59 RPC 56, PAT; *E S P's Application* (1944) 62 RPC 87, SE; *Kelvin's Application* (1953) 71 RPC 103, PAT; *F's Application* (1954) 72 RPC 127, SE; *American Optical Co's Application* [1958] RPC 40, PAT; *Battig's Application* (1929) 49 RPC 415, AC; *Johnson's Application* (1930) 47 RPC 361, SG.

21 See, eg *L'Oreal's Application* [1970] RPC 565, PAT; *Jacques Kessler's Application* [1973] RPC 413, PAT.

22 See *Edgar Rhodes' Application* [1973] RPC 243, PAT, 247, discussed at page 103.

23 See, eg *Johnson* (1901) 19 RPC 56; *Ward's Application* (1911) 29 RPC 79, SG, 79 (referring to the requirement that the novelty manifested by a subject matter 'consist . . . in the manufacture'); *W's Application* (1914) 31 RPC 141, SG; *B A's Application* (1915) 32 RPC 348, SG; *Johnson's Application* (1930) 47 RPC 361, 364 (finding that 'the words "new" and "manner of manufacture" must not be considered separately but together' for the purpose of determining the existence of an invention). See further, *Compagnies' Application* (1930) 48 RPC 185, 188–189 (rationalizing the approach adopted in these cases with the Patents and Designs Act 1907, and finding that applications not directed to an 'invention' within the meaning of that Act must be refused, and that the question 'what is an invention?' involved 'two distinct questions . . . first, "Is there a manner of manufacture?"; second, "Is the manufacture a new manufacture?"' According to the Solicitor-General, the first question was to be determined from the face of the specification, and the second as one of inventive step, depending on the relevant prior art); *Commissioner of Patents v Microcell Ltd* (1959) 102 CLR 232, HCA (adopting the approach in *Compagnies' Application*).

24 (1938) 12 ALJ 169, HCA.

25 ibid, 169, 170. See also *Monier Industries Ltd v Hume Pipe Co (Australia) Ltd* (1943) 13 AOJP 575, HCA, 577–578 per Dixon J; *Gum v Stevens* (1923) 33 CLR 267, HCA.

Aside from abstract schemes and ideas, arrangements of aesthetic and informational content, analogous uses, and collocations of known objects,[26] three additional categories of subject matter were opposed for their alleged failure to constitute an invention during the late nineteenth and early twentieth centuries. They were methods not resulting in a product, products as such, and 'immoral' or otherwise socially detrimental subject matter.

(a) The Exclusion of Subject Matter not Resulting in a (Vendible) Product

As has been seen, the exclusion from patentability of abstract schemes was rationalized in *Cooper* with reference to a general exclusion from patentability of methods not involving the production of an artefact. After 1901, recognition of such an exclusion led to the denial of patents for methods of colouration,[27] a method of navigation,[28] a method of transferring an object unchanged from place to place,[29] and other methods and applications of known compounds which, though possibly inventive and of practical utility, did not involve the manufacture of a product.[30] Then in *G E C's Application*,[31] Morton J upheld a challenge to a patent for a method of extinguishing fires using a known chemical substance for its failure to result in the production, improvement, restoration, or preservation of some 'vendible product'. This was consistent with the earlier decision of the High Court of Australia in *Rogers v The Commissioner of Patents*,[32] where a method of burning timber by causing a self-feeding slow fire to act continuously against the side of a tree was denied a patent on the ground that, while ingenious and useful, it was merely 'a direction how best to use materials in everyday use to achieve an everyday object',[33] and involved neither the production of any vendible article nor the inventive application of any mechanical contrivance or chemical substance.[34]

[26] See, eg *Sharp v Brauer* (1886) 3 RPC 193, VC; *Morgan v Windover* (1890) 7 RPC 131, HL; *Teckelpenny v Army and Navy Co-operative Society (Ltd)* (1888) 5 RPC 405, CA; *Williams v Nye* (1890) 7 RPC 62, CA; *Gadd v The Mayor, &c, of Manchester* (1892) 9 RPC 516, CA; *West v James Keeves & Sons Ltd* (1911) 28 RPC 474, Ch; *Wilson v Wilson Bros Bobbin Co Ltd* (1911) 28 RPC 612, Ch; *A F's Application* (1913) 31 RPC 58, SG; *B A's Application* (1915) 32 RPC 348, SG; *D Co's Application* (1921) 38 RPC 897, SG; *C G R's Application* (1924) 42 RPC 321, SG; *Simon-Carves Ltd's Application* (1928) 45 RPC 407, SG; *Johnson's Application* (1930) 47 RPC 178, SG; *Esau's Application* (1932) 44 RPC 85, SG; *Axien's Application* (1935) 53 RPC 127, PAT; *Naamloose Vennootschap de Bataafsche Petroleum Maatschappij's Application* (1939) 57 RPC 341, PAT; *L & G's Application* (1940) 58 RPC 21, PAT; *G E C's Application* (1942) 60 RPC 1, PAT; *Virginia-Carolina Chemical Corporation's Application* [1958] RPC 38, PAT.

[27] See, eg *T's Application* (1919) 37 RPC 109, SG; *C's Application* (1920) 37 RPC 247, SG; *A E W's Application* (1924) 41 RPC 529, SG; *Johnson's Application* (1930) 47 RPC 361, SG.

[28] *A L's Application* (1924) 41 RPC 615, SG.

[29] *R H F's Application* (1944) 61 RPC 49, PAT.

[30] See, eg *D A and K's Application* (1925) 43 RPC 154, SG.

[31] (1942) 60 RPC 1, PAT.

[32] (1910) 10 CLR 701, HCA.

[33] ibid, 709 per Griffiths CJ.

[34] ibid, 712 per O'Connor J.

In a strong dissent, reminiscent of Eyre CJ's judgment in *Boulton*, Isaacs J rejected the majority's conception of inherent patentability as overly narrow and argued that any subject matter could support a patent if it involved a new and useful combination of ideas and *modus operandi*.[35]

The effect of the decision in *G E C* was to restrict even further the conception of inherent patentability articulated in *Cooper*. Nevertheless, some relaxation of the 'vendible product' requirement was effected almost immediately by Evershed J in *The Cementation Co Ltd's Applications*.[36] In that case, his Lordship held that such a requirement should not restrict the ordinary meaning of 'manner of new manufacture'[37] (and through it the statutory definition of 'invention' per se).[38] The focus of Evershed J in this regard was on the term 'product', which he held could only be reconciled with 'manner of new manufacture' if it were understood as having been used liberally by Morton J, namely as[39]

> a convenient and compendious term to indicate the article or material resulting from the activity, and [not as] intending to limit with reference to what may be the common acceptation to-day of the word 'product' that which results from a manufacture.

The effect was a conception of the invention as a method resulting in an 'article or material'. This led, in 1945, to the acceptance for grant by Evershed J of a method of treating subsoils to render them impervious to subterranean fire,[40] and in 1946, of a method of producing electrical oscillations.[41] It did not, however, save from opposition a method of fumigating enclosed spaces, which was held by Evershed J not to involve such 'alteration in the structure of the enclosing walls of the space or of the articles within it; so that it may be said, within the phrasing of the well known *G E C* case, that it improves or protects from deterioration . . . some "product".'[42]

[35] ibid, 718.

[36] (1945) 62 RPC 151, PAT.

[37] See ibid, 153 (defining 'manufacture' to include '(a) the action or process of making by hand (b) the making of articles or material . . . by physical labour or mechanical power and (c) a branch of productive industry').

[38] ibid, 153–154.

[39] ibid, 154. According to Evershed J, this reading of 'product' was in any case supported by the ordinary meaning of that term as (among other things) 'anything produced by nature or a natural process . . .; fruit; and . . . that which is produced by any action, operation or work'.

[40] *The Cementation Co's Application* (1945) 62 RPC 151, PAT.

[41] *Rantzen's Application* (1946) 64 RPC 63, PAT. See also *Philips Electrical Industries Ltd's Application* [1959] RPC 341, HC (treating visible light as a 'product' within the meaning of *G E C*).

[42] *Bovingdon's Application* (1946) 64 RPC 20, PAT, 21. See also *The Dow Chemical Co's Application* [1956] RPC 247, SE (method of treating seed-infected soils distinguished from the method in *The Cementation Co's Application* and held to be inherently unpatentable on the ground that the treatment had no effect on the physical structure of the soils itself); *American Chemical Paint Co's Application* [1958] RPC 47, PAT (method of treating a growing cotton plant); *British Petroleum Coy Ltd's Application* [1958] RPC 253, PAT (method of testing).

(i) Methods of Medical Treatment

In addition to clarifying the meaning of 'product', Evershed J during the 1940s elucidated the more restrictive aspect of Morton J's rules, which required that the product of a patentable process be vendible. While that requirement was not unknown to patent law,[43] its doctrinal basis was not fully explored until the twentieth century, when it first emerged as a ground for the exclusion from patentability of methods of treating the human body. Thus, in *C & W's Application*,[44] a method of extracting lead from bodies was held ineligible for patent protection because of its lack of association with the manufacture or sale of a 'commercial product'. In reaching his conclusion, the Solicitor-General relied on the original statutory context of the phrase 'manner of new manufacture'[45] to support a conception of the invention as 'a thing newly made, or a substance which, if made before, is improved in its manufacture', or 'a machine or a process that can be used in making something that is, or may be, of commercial value'.[46] While a human being could be considered as a working organism that would be improved by the method in question, the Solicitor-General considered that to be insufficient to make the method one of relevance to 'manufacture or trade'.[47] Put differently, a method of medical treatment was not directed to advancing the industrial arts.

The general principle of *C & W* was subsequently upheld by an equivocal High Court of Australia in *Maeder v Busch*,[48] the only other case prior to 1959 to consider the patentability of medical methods. That case involved a patent application for a method of forming permanent waves in human hair. While refusing conclusively to decide the matter, two of the four judges cast doubt on the correctness of *C & W*.[49] In the most detailed judgment of the Court, Dixon J expressed the problem as follows:[50]

> To be patentable an invention must relate to an art. Perhaps the widest statement is one of the earliest [referring to the statement of Eyre CJ in *Boulton*]. But the ultimate end in view is the production or treatment of, or effect upon, some entity. Applications of old things to a new use, accompanied by the exercise of inventive power, are often patentable though there be no production of a new thing. But in every case the invention must refer to and be applicable to a tangible thing.
>
> In the present case there is nothing to be affected but the hair. The chemical compounds already exist. The use of them, the application of heat and the method of treatment constitute nothing but method, procedure, treatment or process. Can the hair growing upon the human head be regarded as satisfying the condition that the process shall in some way relate to the productive arts?

[43] See, eg *Boulton v Bull* (1795) 126 ER 651, CP, 660–661 per Heath J.
[44] (1914) 31 RPC 235, SG.
[45] Especially the 'general inconvenience' proviso; see ibid, 235.
[46] ibid, 235–236.
[47] ibid, 236.
[48] (1938) 59 CLR 684, HCA.
[49] ibid, 706–707 per Dixon J, 707 per Evatt J. cf ibid, 699 per Latham CJ, 708 per McTiernan J.
[50] ibid, 706.

His Honour was ambivalent in answering this question. On one hand, a method of treating hair was not, he suggested, aimed at producing or helping to produce any article of commerce or contribution to the productive arts.[51] On the other hand, the method could be considered to be 'embodied in a manual art or craft' on the ground that its purpose was to treat an object of purely mechanical utility.[52] Implicitly, while the method was directed to advancing the *practical* arts, it was not directed to advancing the *manufacturing* arts. It followed in Dixon J's view that the patentability of the method depended on whether an invention involved the production or treatment of a 'merchantable article', or merely the production or treatment of some object of mechanical utility so as to establish the method as a manual art or craft.[53] According to Dixon J, even the latter conception would not save methods of treating or manipulating a vital part of the human body from unpatentability, on the ground that such methods would always lack the mechanical utility of a practical art.

(ii) Ephemeral and Biological Subject Matter

As has been seen, the question raised by Dixon J concerning the need for an invention to result in a vendible product was answered affirmatively by Morton J in *G E C*.[54] In *Rantzen's Application*,[55] Evershed J held that a 'vendible product' in this sense was not confined to something that could be passed from one person to another upon a transaction of purchase or sale, but encompassed anything that might 'fairly be regarded as the outcome of a process of manufacture'.[56] However, this comparatively liberal interpretation was not favoured by other judges,[57] who relied on Morton J's rules to justify the exclusion from patentability of two main types of process from 1942 to 1959. The first involved the production or treatment of ephemeral matter,[58] and the second involved the treatment of soil to improve its

[51] ibid.

[52] ibid.

[53] ibid. For Dixon J, the commercial nature of the process was irrelevant. In his judgment (ibid), '[i]t is difficult to base any legal distinction on the motive or purpose of the operator or manipulator or on the vocation he pursues. It can hardly matter whether he acts in the exercise of a profession or art or trade or business. The purpose of the patentee and those intended to employ the process may be entirely commercial. The process may be intended for use in ordinary trade or business such as that of hairdressing, manicure, pedicure. The purpose, on the other hand, may be the relief of suffering by surgical or manipulative means.'

[54] (1942) 60 RPC 1, PAT.

[55] (1946) 64 RPC 63, PAT.

[56] ibid, 66.

[57] cf Lloyd-Jacob J (see *Lenard's Application* (1954) 71 RPC 190, PAT: *Elton and Leda Chemical Ltd's Application* [1957] RPC 267, PAT).

[58] See, eg *C M's Application* (1944) 61 RPC 63, PAT (method of making a cinematographic photograph); *F's Application* (1954) 72 RPC 127, SE (process for producing an image by cinematographic projection); *Huber's Application* [1956] RPC 50, SE (process for producing an image by cinematographic projection); *Philips Electrical Industries Ltd's Application* [1959] RPC 341, HC (process for producing visible light); *British Petroleum Coy Ltd's Application* [1958] RPC 253, PAT (process for producing information).

crop-bearing utility;[59] the latter on the ground that the only vendible products resulting were the crops nourished by the soil, which were not affected by the treatment itself.

A third type of process, involving the 'modification of the natural conditions under which living phenomena pursue their course',[60] was also excluded from patentability during this period for its failure to relate to the manufacturing arts.[61] These methods were treated by decision-makers as being purely biological rather than mechanical in nature, and as incapable of supporting a patent for that reason. This distinction between so-called biological and mechanical processes was also adopted in the first case to consider expressly the inherent patentability of a horticultural method, which was *Rau GmbH's Applications*.[62] That case involved a method for producing, by selective cultivation, improved stocks of seeds, in which the applicant was refused a patent on the following ground:[63]

> Selective breeding of animals and cultivation of plants for the obtainment of improved stocks by the rigorous selection of and breeding from the few individuals which are nearest the ideal has . . . been practised from the earliest times as a part of agricultural or horticultural development . . . and the exercise of art or skill in these directions has not been regarded as coming within the term 'manufacture'. I think that the subject of the present applications is clearly distinguishable from the invention considered in *Commercial Solvents Corporation v Synthetic Products Co Ld*; . . . the process in [that] case is analogous to a chemical manufacturing process or operation, the cultured bacteria playing a part in the conversion of starchy bodies into acetone and alcohols similar to that of chemical reagents or catalysts.

The invention in *Commercial Solvents v Synthetic Products*[64] consisted of 'improvements in the bacterial fermentation of carbohydrates and in bacterial cultures for the same', described as 'the employment of [a butyl producing bacillus] to produce large yields of acetone and butyl alcohol under aerobic or anaerobic conditions'.[65] The question of inherent patentability was not addressed in detail by the Court, which simply declared that '[i]n such an invention there is . . . patentable subject matter'.[66]

59 See, eg *Standard Oil Development Co's Application* (1951) 68 RPC 114, PAT; *The Dow Chemical Co's Application* [1956] RPC 247, SE; *American Chemical Paint Co's Application* [1958] RPC 47, PAT. See also *N V Philips's Application* (1954) 71 RPC 192, PAT (improved method of producing poinsettia); *Rau GmbH's Applications* (1935) 52 RPC 362, PAT (method of producing seeds having certain desired properties by selective cultivation).

60 The phrase is from *N V Philips's Application* (1954) 71 RPC 192, 194.

61 See, eg *Lenard's Application* (1954) 71 RPC 190, PAT; *N V Philips's Application* (1954) 71 RPC 192, PAT; *Goldhaft's Application* [1957] RPC 276, SE; *Virginia-Carolina Chemical Corporation's Application* [1958] RPC 38, PAT.

62 (1935) 52 RPC 362, PAT.

63 ibid, 363.

64 (1926) 43 RPC 185.

65 ibid, 225.

66 ibid.

The exclusion from patentability of processes for modifying natural conditions was confirmed in all subsequent cases to consider the issue prior to 1959, with two exceptions. The first was *Szuec's Application*,[67] where the Superintending Examiner drew a distinction between methods of treating or producing 'agricultural or horticultural products' on one hand, and lower forms of life such as micro-organisms like 'yeasts, bacteria and tissues' on the other. He concluded that, while the former were inherently unpatentable, the latter were not, on the ground that while the production of micro-organisms was 'of course a biological process', it could not unequivocally be described as[68]

> a 'natural phenomenon' of the kind to which Lloyd-Jacob J was referring when, in [*N V Philips' Application*], he decided that an invention, which is a modification of the conditions under which natural phenomena pursue their inevitable course does not disclose a manner of manufacture.

According to the Superintending Examiner, this conclusion was supported by the fact that the process in question 'comprise[d] as necessary steps . . . two sub-processes carried out in the factory' using apparatus from a chemical plant.[69]

The decision in *Szuec* was approved in *Goldhaft's Application* in the following year on the ground that the process involved did not result in the production of an agricultural or horticultural product, or 'one of the usual products of a . . . farm'.[70] This confirmed the existence by 1959 of a distinction for inherent patentability purposes between methods involving the treatment of higher and lower forms of life respectively. Pursuant to that distinction, methods involving the production or treatment of micro-organisms were sufficiently artificial by analogy to chemical methods to be regarded as inventions, while methods involving the production or treatment of living plants, animals, and humans were not.[71]

[67] [1956] RPC 25, SE.

[68] ibid, 26.

[69] ibid.

[70] [1957] RPC 276, 277.

[71] Note, however, that even the status of higher-life-form methods was equivocal. In relation to methods of treating plants, see, eg *The Cementation Co's Application* (1945) 62 RPC 151, PAT, discussed at page 77. In relation to methods of treating humans, see, eg *Maeder v Busch* (1938) 59 CLR 684, discussed at pages 78–79. In relation to methods of treating animals, see, eg *C & W's Application* (1914) 31 RPC 235, SG (leaving open the possibility that a method of treating the body of an animal to make it a more marketable product might be an invention); *Canterbury Agricultural College for L P's Application* [1958] RPC 85, SE (affirming the Patent Office practice of refusing patents for methods of treating animals on the authority of *C & W* while also noting the various considerations against that practice, and concluding that methods of treating the body of an animal were inventions if they resulted in the creation of a vendible product).

3. The Uncertain Position of (Chemical) Products

(a) 1883 to 1918: Products as Manners of New Manufacture

As seen in chapter 2, nineteenth century concerns over the granting of obstructive patents to foreign manufacturers of chemical products was a central motivation for the statutory reforms that eventually led to the Patents, Designs, and Trade Marks Act 1883. An important feature of those reforms was the introduction of a compulsory licensing system to prevent the unreasonable refusal of patent licences, and thereby balance the interests of rewarding first inventors with the protection of the public against obstructive grants. In effect, this system sought to resolve the tension between an expansive conception of inherent patentability, assumed by the Legislature to cover products independent of their associated methods of manufacture, and concerns to confine the scope of protection conferred by a patent to the inventor's specific contribution, which contribution was a teaching or practical art.

After 1883, the inherent patentability of products was assumed by the courts, albeit without consideration of patent scope. In *Thomson v American Braided Wire Co*,[72] for example, the House of Lords held a patent for pillows and cushions to be lacking in novelty but otherwise valid. In *Vorwerk v Evans*,[73] a majority of the Court of Appeal interpreted a claim to 'the waistband of the formation substantially as set forth and indicated hereinbefore' as being for the waistband-by-process rather than the waistband itself.[74] And in *Adhesive Dry Mounting Co Ltd v Trapp & Co*,[75] Parker J construed a claim to be for a pellicle rather than a pellicle-by-process, as had been argued, and to be invalid as such for anticipation.[76] In each case, the inherent patentability of mechanical products seems to have been accepted. To similar effect in the context of chemical products was the dictum of Lord Davey in *Acetylene Illuminating Co v United Alkali Co Ltd*, as follows:[77]

> I need scarcely say that if the Patentee had discovered a new material of the character which he mentions [a compound of calcium and carbon, namely, a calcium carbide], a material having the very valuable commercial properties which he ascribes to it, that would have been a good subject matter for a Patent, and there could have been no question, supposing he had done so, that his Patent was a good one. He would be bound, of course, to state the means by which he produced that material; but the novelty of the means or the process by which the material was produced would have been immaterial, because the merit and the novelty of the invention would consist in the substance produced itself.

[72] (1889) 6 RPC 518, HL.
[73] (1890) 7 RPC 265, CA.
[74] ibid, 272 per Cotton LJ, 274 per Bowen LJ. cf ibid, 274 per Fry LJ.
[75] (1910) 27 RPC 341, Ch.
[76] ibid, 353.
[77] (1904) 22 RPC 145, HL, 153.

By contrast is the decision in *Kopp v Rosenwald Bros.*[78] There, Buckley LJ construed the phrase 'manner of new manufacture' as importing a process element, with the result that what was argued to be a product claim was rather a product-by-process claim, or perhaps even (he suggested) a selection claim.[79] The implication was that there was no protection for products per se. Aside from its being supported by the statutory language, this view reflected the concerns of the day regarding obstructive (chemical and other) patents. Those concerns extended to expansive patent drafting practices, as reflected in *Thomson v American Braided Wire Co* itself:[80]

> The Appellants maintain that the Specification of 1885 is framed in terms so comprehensive as to include in the invention all bustles and dress-improvers composed of one or more sections of tubular braided wire. Had the claim of the inventor been of that extent, I should have been of opinion that the patent was void.

Similarly in *The Badische Anilin und Soda Fabrik v Levinstein*,[81] where a patent for naphthylamine was held to be invalid by reason of the patentee's failure to distinguish the product's known isomeric forms. According to the Court of Appeal, the specification operated as 'a drag-net, sweeping into its wide ambit all the isomers of the sulpho-acid of oxy-azo-naphthylamine, without regard to whether they were good, bad, or indifferent,' and leaving for subsequent investigation and experiment the resolution of those issues and choice between the specified methods of production.[82] And in 1915, in *Natural Colour Kinematograph Co Ld v Bioscemes Ld*,[83] Lord Loreburn made the following comments regarding patent drafting practices:[84]

> Some of those who draft specifications and claims are apt to treat this industry as a trial of skill, in which the object is to make the claim very wide upon one interpretation of it, in order to prevent as many people as possible from competing with the patentee's business, and then to rely upon carefully prepared sentences in the specification which, it is hoped, will be just enough to limit the claim within safe dimensions if it is attacked in Court.

It seems reasonable to assume that decisions such as these encouraged suspicion of product patents. One person who harboured such suspicions was the

[78] (1900) 19 RPC 205, Ch.

[79] In addition to the judgment (see ibid, 211–212), this can be seen in the exchange between Buckley LJ and Terrell QC (see ibid, 209).

[80] (1889) 6 RPC 518, 525.

[81] (1885) 2 RPC 73, CA.

[82] ibid, 116.

[83] (1915) 32 RPC 256, HL.

[84] ibid, 266 per Earl Loreburn; see also per Lord Parmoor.

Comptroller-General of Patents, Mr Temple Franks, who believed that product claims were 'in the majority of instances' obstructive ones:[85]

Another point to be noticed [concerns the] use of what are called 'product claims'. These claims are claims to any new product per se irrespective of the *process* by which it is made and are in the form e.g. 'as a new product the dyestuffs made as above or by any other process'. The consequence of such claims especially in chemical manufacture is that the inventor of a process producing a new chemical product is enabled to attack as infringements products produced not only by the process discovered by him but by any other method. These are, in my opinion, in the majority of instances, obstructive and injurious claims, and they very largely aid the establishment of a monopoly in the case of chemical manufacture as they prevent research and invention on analogous lines by other persons.

It can therefore be said that in the early decades of the 1883 Act, there existed a view among some lawmakers that patents were unsuitable for products on the ground that the monopoly they conferred exceeded the benefits which they secured. Ironically, this was the same reason given by Heath J in 1795 for opposing patents for methods of production, and accounts for the conflicting views later expressed by reformers,[86] academics,[87] and courts[88] regarding the inherent patentability of products in this era.

*(b) 1919 to 1949: The Statutory Exclusion of Chemical Products and Fair Basis/*Mullard Radio *Solutions to the Problem of Obstructive Claims*

Dissatisfaction with the 1883 compulsory licensing scheme as a means of resolving the impact of chemical product patents was aired in the House of Commons in 1898, but action was neither suggested nor taken.[89] In the view of the President of the Board of Trade, that which industry perceived as a weakness of the system was on the contrary one of its strengths: the disregard of the position under foreign law, which ensured non-discrimination of foreign inventors.[90] Then, in 1916, the

[85] *Memorandum on German Use of Our Patent Law* (1916), quoted in *Lundbeck A/S v Generics Ltd* [2008] EWCA Civ 311, [43] per Lord Hoffmann.

[86] cf, eg United Kingdom, *The British Patent System: Report of the Committee to Examine the Patent System and Patent Law* (Cmnd 4407, 1970) (Banks Report) [396]; United Kingdom, *Report of the Department Committee on the Patents and Designs Acts and on the Practice of the Patent Office (Patent Office Practice)* (Cmd 3829, 1930–1931) (Sargant Report) [176].

[87] cf, eg RD Satchell, 'Chemical Product Patent Practice in the United Kingdom' (1970) 1 IIC 179, 179; C Robinson, 'Patent Protection for Chemical Products in Canada, Great Britain and the United States' (1972) 2 IIC 139, 140.

[88] cf, eg *Lundbeck A/S v Generics Ltd* [2008] EWCA Civ 311, [43] per Lord Hoffmann; *Generics Ltd v Landbeck A/S* [2009] UKHL 12, [71] per Lord Neuberger.

[89] See Parl Debs (series 4) vol 63, cols 450–569 (29 July 1898). In other contexts members of the House of Commons expressed their support for chemical product patents, or at least their assumption that such patents were valid; see, eg Parl Debs (series 4) vol 180, cols 645–676 (9 August 1907) (objecting to a proposal to require that patented products be marked with the word 'patent' as infeasible in the case of chemical products).

[90] See Parl Debs (series 4) vol 63, cols 565 (29 July 1898).

Parker Committee—which included Temple Franks himself as a member—recommended a general prohibition against chemical product claims on two specific reported grounds, namely, the desirability of harmonizing English law with the law of other (European) states, and the desirability of reducing the hold of German firms on the British chemical industry.[91] The Committee's recommendation was accepted, and section 38A(1) of the Patents and Designs Act 1907 was introduced.[92]

According to Hansard, the 'principal object' of section 38A(1) was 'to prevent an inventor who had found a way of preparing a new chemical substance from claiming the substance when prepared in other ways'.[93] To that end it prohibited claims for chemical substances 'except when prepared or produced by the special methods or processes of manufacture described and claimed or by their obvious chemical equivalents'.[94] The section also created an express evidentiary presumption that had the effect of expanding the patentee's rights to the extent permitted by the prohibition in alignment of the English and German positions.

In addition, there remained the secondary patentability requirements of novelty, inventiveness, and utility. According to Lord Shaw in *British Thomson-Houston v Charlesworth*,[95] '[a] thing discovered in ordinary trade or adventure, with no novelty or invention of means employed, is not patentable any more than would the discovery by a sailor of an island cast up by the sea'.[96] In *Sharp & Dohme Inc v Boots Pure Drug Co Ltd*,[97] the Court of Appeal agreed, finding that chemical products could only support a patent if produced by a novel and inventive method.[98] The Sargant Committee of 1931 disagreed, however, criticizing the principle in *Sharp & Dohme* that chemical products needed to be produced by a patentable method.[99] In the Committee's opinion, it was the product as produced by the process that was patentable, and not the process of production itself.[100] An amendment to section 38A(1) was recommended[101] and introduced by the Legislature the following year.[102] That the amendment achieved its intended effect is apparent

[91] See Banks Report, [396].
[92] See Patents and Designs Act 1919, s 11(1).
[93] Sargant Report, [178].
[94] See Hansard HL (series 5) vol 37, cols 602–671 (4 December 1919). See also Hansard HC (series 5) vol 118, cols 1841–1868 per Sir W Pearce (28 July 1919).
[95] (1925) 42 RPC 180, HL.
[96] ibid, 207.
[97] (1928) 45 RPC 153, CA.
[98] See also *M's Application* (1922) 39 RPC 261, SG; *W's Application* (1922) 39 RPC 263, SG. cf *E M's Application* (1924) 41 RPC 590, SG.
[99] Sargant Report, [179]–[184].
[100] ibid, [185].
[101] ibid, [184]; United Kingdom, *Memorandum (Patents and Designs Bill)* (Cmd 4067, 1931–1932) 35–56.
[102] Patents and Designs Act 1932, s 8.

from *May & Baker Ltd*,[103] where Jenkins J affirmed the patentability of any 'invention consisting of the production of new substances by known methods from known materials', though emphasizing that it needed to be 'truly new, as opposed to being merely additional members of a known series (such as the homologues),' and that the product's useful qualities needed to be attributable to the inventor's discovery itself.[104]

The claim at issue in *May & Baker Ltd* was for a class of chemical compounds. According to Jenkins J, the fact that only some of the compounds covered by the claim had the chemo-therapeutic value described in the patent meant that the claim was unduly broad in scope:[105]

> The specification claims the whole range of products falling within its terms although only some of them are known to be useful for the purpose in view . . . This, I think, amounts to such a failure of consideration, or in other words, such a disparity between the inventive step and the width of the monopoly claimed, as to afford good ground for objection under Sec. 6 of the Statute [of Monopolies]. See *Mullard Radio Valve Coy. Ltd. V. Philco. Corpn.* (52 R.P.C. 261, 278; 53 R.P.C. 323, 347, 348).

Of particular note is the suggestion here that the inventive step disclosed in a patent needed to correspond to the monopoly claimed. In the case cited for that proposition—*Mullard Radio Valve Co Ltd v Philco Radio & Television Corp*[106]—the House of Lords invalidated a claim on grounds expressed by Lord Alness as follows:[107]

> [T]he invention and the claim do not equiparate, and . . . the monopoly claimed by the appellants does not conform with the consideration set out in the specification. The fundamental rule of patent law . . . has always been that the monopoly claimed must not go beyond the consideration. In my opinion, [the claim] sins against this cardinal rule. The appellants have, I think, claimed more than is necessary for the protection of their patent.

Explaining this 'fundamental rule' to the European Committee of Experts on Patents in 1953,[108] the UK cited *Esau's Application*.[109] In that case, a claim to 'an apparatus for influencing substances by means of high frequency electric energy'[110] was rejected by the Law Officer on the ground that the patentee's disclosure in the

[103] (1948) 45 RPC 255, Ch; appealed on another issue to the House of Lords. See also Committee of Experts, Observations of the United Kingdom Delegation: Industrial Character, EXP/Brev (56) 3 (3 May 1956) 1–2.

[104] (1948) 45 RPC 255, 281.

[105] ibid, 288.

[106] [1936] 2 All ER 920, HL.

[107] ibid, 934.

[108] See Committee of Experts, Reply to the Questionnaire drawn up by the Bureau of the Committee of Experts of the Council of Europe from the point of view of legislation in the United Kingdom, EXP/Brev (53) 7 (17 March 1953) 11. See also Committee of Experts, Criteria of Novelty and Patentability, CM/WP IV (51) 9 (undated) 48.

[109] (1932) 44 RPC 85, SG.

[110] ibid, 86.

specification did not correspond to the monopoly claimed in the patent. In his judgment, the Officer admonished patentees to realize[111]

> that it is not the practice of the Patent Office to allow broad and indeterminate claims of a speculative character, and that, if they put such claims into their complete specifications, they must expect to find them disallowed unless they are able to give a sufficiently detailed and full description to support them.

Similarly in the case of *Shell Development Co's Application*,[112] Wynn-Parry J approved the practice of the Patent Office of not allowing 'broad and indeterminate claims of a speculative character' on the basis that such practice was 'based on a true construction of the Act as interpreted by the authorities'. He upheld the decision of the Superintending Examiner to refuse a patent for new solvents of use in the separation of organic mixtures on the ground that the claim was excessive in scope.[113]

(c) 1949 to 1959: The (Inherent) Patentability of Chemical Products

In 1947, the Swan Committee recommended that section 38A(1) of the Patents and Designs Act 1907 be abolished, and that claims for new chemical substances be allowed so as 'to give the patentee an exclusive right to manufacture that substance' including against those who discovered new methods of making the substance.[114] Two reasons were given for that recommendation. The first was the definitional difficulties created by the provision's language, including particularly the phrases 'obvious chemical equivalents' and 'chemical processes'.[115] And the second was the permissibility of per se claims for non-chemical products, including new alloys, which were arguably produced by a chemical process.[116] The Legislature accepted the Committee's recommendation and repealed section 38A(1). In so doing, it supported the inherent patentability of chemical products independent of the methods by which they were made.[117] It also suggested that the

[111] ibid, 87.

[112] (1947) 44 RPC 151, PAT.

[113] ibid, 154.

[114] United Kingdom, Patents and Designs Acts; Final Report of the Departmental Committee (Cmnd 7206, 1947) [96].

[115] ibid, [94].

[116] ibid.

[117] See Robinson, 'Patent Protection for Chemical Products in Canada, Great Britain and the United States', 141. See further EXP/Brev (53) 7, 6; EXP/Brev (56) 3, 1–2; United Kingdom, *United Kingdom Patent Law: The Effects of the Strasbourg Convention of 1963* (Cmnd 2835, 1965) [33] (proposing alternative statutory definitions of inherent patentability, both of which covered chemical products). The same view was taken of the impact of the repeal in 1967 of the German equivalent of section 38A(1); see K Nastelski, 'Product Protection for Chemical Inventions in Germany' (1972) 3 IIC 267, 270.

compulsory licensing system[118] would be sufficient to off set any adverse social impact resulting from that change.[119]

In addition to repealing section 38A(1), the Legislature introduced a fair basis requirement. According to section 4(4) of the Patents Act 1949, '[t]he claim or claims of a complete specification [needed to] relate to a single invention, [to] be clear and succinct, and [to] be fairly based on the matter disclosed in the specification'. Section 32(1)(i) supported this by affirming the courts' power to revoke a patent on the ground 'that the scope of any claim of the complete specification is not sufficiently and clearly defined or that any claim of the complete specification is not fairly based on the matter disclosed in the specification'. Describing the effect of these provisions to the European Committee of Experts on Patents, the UK emphasized that it was 'not to limit an applicant's invention strictly to what he has described but to something which is "fairly based" on his description'. The description itself, they said, 'should give a full disclosure of the nature of the invention and of how it is to be carried into effect'.[120] This was consistent with an understanding of claims as needing to 'define with precision and without ambiguity the scope of the monopoly' sought from the patent, and of the description itself as needing to convey the information required to secure the invention for the benefit of the public, thereby establishing the applicant's entitlement to the patent.[121] The question remained as to what that involved. According to the rule in *Mullard Radio*, consistent with the principle of *Esau's Application*, it required equiparation of a claim with the specification. Still, this begged the question alluded to above, and first raised by the decision in *Boulton*, whether a claim for a product per se, covering every method of producing the same, could equiparate with a specification notwithstanding the inability ever to disclose all methods of producing a product.

4. The Exclusion of Immoral Subject Matter

A final issue in relation to inherent patentability that arose expressly for the first time after 1883 concerned the relevance of morality for inherent patentability. The legislative basis for that issue was the exclusion from section 6 of the Statute of Monopolies 1623 of subject matter (the patenting of which would be) 'contrary to law [or] mischievous to the State . . . or generally inconvenient', and the power

[118] See Patents Act 1949, s 37.

[119] See Committee of Experts on Patents, Experience of the effects of UK legislation relating to licensing of patents, with particular reference to patents for chemical products, food, medicine and surgical or curative devices, Report presented by the UK Experts, EXP/Brev (52) 10 (6 March 1952) 7–9.

[120] See Committee of Experts on Patents, Report on the form and content of patent specifications, presented by the Experts of the UK of Great Britain and Northern Ireland, EXP/Brev (52) 11 (18 March 1952) 5–8. See also EXP/Brev (53) 7; Committee of Experts on Patents, Observations of the United Kingdom Delegation, EXP/Brev (56) 2 (11 June 1956) 9.

[121] ibid.

of the Comptroller, first conferred in 1883, to 'refuse to grant a patent for an invention . . . of which the use would, in his opinion, be contrary to law or morality'.[122]

The inherent patentability of 'immoral' subject matter was considered directly in two cases prior to 1959,[123] which involved, respectively, the method of extracting lead from human bodies discussed above and a contraceptive device.[124]In the second of these cases, the Comptroller rejected the application for the patent in exercise of his legislative discretion above, and consistent with a Patent Office rule of not supporting contraception with a grant of a patent.[125] On appeal that decision was upheld, though without relying on the legislative provision in question. Instead, the Solicitor-General denied patentability on the ground that 'it would not be fitting to sell [the device] as being protected by Royal Letters Patent'. He further stated as follows:[126]

> The question . . . arises whether, quite apart from Section 75 [of the Patents and Designs Act 1907], the Crown in the exercise of its prerogative could possibly be expected to exercise its discretion to grant a patent for an article designed as an apparatus for the prevention of contraception . . . I decline to be any part to the grant of a patent for this class of article . . . Even if, as to which I express no opinion, its use as a contraceptive is consistent with morality, I am not prepared to exercise on behalf of the Crown the Crown's discretion in favour of the grant of a patent in respect of it . . . I express no opinion as to whether the use of these articles is consistent with morality, because I am not aware that the law has laid down what the exact standards of morality are. I am a Court of Law, and not a Court of Morality. All I say is I think these are not articles for which, whether the specification be amended or not, the Crown can be expected to exercise its discretion by way of granting a patent. I therefore dismiss the Appeal.

The reasons for judgment were more ambiguous in *C & W*, with the Solicitor-General in that case concluding as follows:[127]

> It has been urged, and I think quite rightly, that the question of humanity ought not to affect the decision in such a case as this. I agree. Of course, it is well known that the medical profession do all in their power to discourage members of their body from obtaining protection for any discovery that has for its object the alleviation of human suffering, and it is impossible to speak too highly of such conduct, but it cannot affect my judgment in arriving at a conclusion upon the terms of the Section of the Act of Parliament, and I have altogether excluded such consideration from my mind.

[122] See Patents, Designs, and Trade Marks Act 1883, s 86. This provision was reproduced in the Patents and Designs Act 1907, s 75, amended in 1932 (see Patents and Designs Act 1932, s 11).

[123] See also *British Thomson-Houston Co Ltd v Corona Lamp Works Ltd* (1921) 39 RPC 49, HL, 67 per Viscount Haldane (describing 'questions of public policy' as the exclusive concern of Parliament); *T S's Application* (1924) 41 RPC 530, SG.

[124] See respectively *C & W's Application* (1914) 31 RPC 235, SG, discussed at pages 78–79, and *A & H's Application* (1927) 44 RPC 298, SG.

[125] See Hansard HL (series 5) vol 379, cols 290–291 (24 January 1977).

[126] ibid, 298.

[127] (1914) 31 RPC 235, 236.

Nevertheless, the Solicitor-General's success in not allowing his views regarding the moral culpability of applicants for patents for methods of medical treatment to affect his legal judgment must be doubted, given the absence at the time of any precedent supporting either his conclusion or the statement of law from which it was derived. Read in that context, the decision in *A & H's Application* supports the impression created by *C & W*, namely, that while morality was expressly eschewed as a basis for denying patentability before 1959, it was an influential factor, at least in the context of methods of medical treatment.

5. Conclusion

By 1959, an invention for UK patent law purposes was settled to be any human action of mechanical utility directed to advancing the manufacturing arts by producing (or improving) a physical vendible product. The result was a restrictive conception of inherent patentability that was regarded as supporting the threshold exclusion of three specific categories of subject matter: (i) surgical and therapeutic methods; (ii) agricultural, horticultural, and other biotechnological subject matter; and (iii) certain informational and expressive works, including business methods and so-called schemes. By contrast, considerable uncertainty existed with respect to the inherent patentability of products as such, notwithstanding the Legislature's support of such patentability. Similarly, the relationship between morality and inherent patentability was unclear, along with the relevance, when identifying an invention, of considerations of non-industrial social policy and benefit.

C. Post-1959: The Invention as a 'Proper Subject of Letters Patent According to the Principles . . . Developed for the Application of s 6 of the Statute of Monopolies'

1. The Invention as a Human Action on the Physical World Producing an Artificial End of Practical Utility: *National Research Development Co v Commissioner of Patents*

National Research Development Co v Commissioner of Patents[128] involved a type of subject matter that had been considered in a line of early twentieth century applications for patents for horticultural methods, consisting of a method of using known chemical substances to treat soil in order to eradicate and control weeds from crop areas without affecting the crops themselves.[129] The issue for the Court was whether that method was precluded from being a manner of manufacture

128 (*NRDC*) (1959) 102 CLR 252, HCA.

129 The claims are extracted at ibid, 260–261, and described by the Court at ibid, 264–265.

(and thus supporting a patent) because of its nature as a mere use of a known substance and/or its failure to result in any new or improved vendible product, as the Deputy Commissioner of Patents had found.[130] In a unanimous decision, the High Court of Australia held that it was not so precluded on two grounds. First, the method depended, for its success, on a feature of the known substance not previously known to make that substance effective in the treatment of soil, and the discovery and exploitation of that feature reflected the mechanical ingenuity required to take the method out of the category of analogous uses.[131] And second, the method was a proper subject of letters patent according to the historically developed principles governing the application of section 6 of the Statute of Monopolies, on the ground that its purpose was to create a 'vendible product' in the sense of an artificially created end of practical utility.[132] The first of these grounds was consistent with the reason offered by the Court for the correctness of the principle, expounded by Sir Stanley Buckmaster SG, in *B A's Application*,[133] that the new use of a substance is not patentable where the methods producing that substance have already been ascertained and its characteristics and constituents well defined;[134] namely,[135]

> [b]ecause in the postulated state of knowledge the new purpose is no more than analogous to the purposes for which the utility of the substance is already known, and therefore your suggestion of the new purpose lacks the quality of inventiveness

Having rejected the Commissioner's argument that the method at issue was an analogous use, the Court went on to consider whether it was an invention within the meaning of the Act.[136] In determining this question it stated as follows:[137]

> It is of the first importance to remember always that the [Patents Act] defines the word 'invention', not by direct explication and in the language of its own day, nor yet by carrying forward the usage of the period in which the Statute of Monopolies was passed, but with reference to the established ambit of s 6 of that Statute. The inquiry which the definition demands is an inquiry into the scope of the permissible subject matter of letters patent and grants of privilege protected by the section. It is an inquiry not into the meaning of a word so much as into the breadth of the concept which the law has developed by its consideration of the text and purpose of the Statute of Monopolies. . . . The word 'manufacture' finds a place in the present Act, not as a word intended to reduce a question of patentability to a question of verbal interpretation, but simply as the general title found in the Statute of Monopolies for the whole category under which all grants of patents which may be made in accordance with the developed principles of patent law are to be subsumed. It is therefore a mistake, and a mistake likely to lead to an incorrect conclusion, to treat the question

[130] ibid, 253.
[131] ibid, 265.
[132] ibid, 276–277.
[133] (1915) 32 RPC 348, SG.
[134] ibid, 349.
[135] (1959) 102 CLR 252, 262. See also (1959) 102 CLR 252, 265.
[136] ibid, 264.
[137] ibid, 269.

whether a given process or product is within the definition as if that question could be restated in the form: 'Is this a manner (or kind) of manufacture?' It is a mistake which tends to limit one's thinking with reference to the idea of making tangible goods by hand or by machine, because 'manufacture' as a word of everyday speech generally conveys that idea. The right question is: 'Is this a proper subject of letters patent according to the principles which have been developed for the application of s 6 of the Statute of Monopolies?' It is a very different question.

Having thus articulated the proper question for determination, the Court proceeded as follows. First, it noted the comment by the Chief Justice in *Maeder v Busch*[138] that a widening conception of the 'invention' had been a central characteristic of the development of patent law,[139] and cited as examples of the accuracy of that statement the conceptions articulated by Eyre CJ in *Boulton*[140] and Tindal CJ in *Crane v Price*.[141] Second, it considered, as the central unresolved issue concerning the conception of an invention, 'whether it is enough that a process [to be patentable] produces a useful result or whether it is necessary that some physical thing is either brought into existence or so affected as the better to serve man's purposes',[142] and reviewed the early twentieth century case law in respect of the issue, including in particular *G E C*[143] and the line of cases in which its rules were subsequently applied.[144] And third, it concluded its review by stating that the rules laid down in *G E C* could only be upheld as 'wide enough to convey the broad idea which the long line of decisions on the subject has shown to be comprehended by the Statute'[145] if the word 'product' was understood 'as covering every end produced',[146] and the word 'vendible' as pointing only to the requirement of utility in practical affairs'.[147] The effect of this conclusion was to bring *NRDC's* method within the scope of those rules, as a method '[having] as its end result an artificial effect falling squarely within the true concept of what must be produced by a process if it is to be held patentable'.[148] The Court extrapolated:[149]

This view is, we think, required by a sound understanding of the lines along which patent law has developed and necessarily must develop in a modern society. The effect produced by the appellant's method exhibits the two essential qualities upon which 'product' and 'vendible' seem designed to insist. It is a 'product' because it consists in an artificially created state of affairs, discernible by observing over a period the growth of weeds and crops

[138] (1938) 59 CLR 684, HCA.
[139] (1959) 102 CLR 252, 270.
[140] (1795) 126 ER 651, CP.
[141] (1842) 134 ER 239, CP.
[142] (1959) 102 CLR 252, 270.
[143] (1942) 60 RPC 1, PAT.
[144] (1959) 102 CLR 252, 271–276.
[145] ibid, 276.
[146] ibid.
[147] ibid.
[148] ibid, 277.
[149] ibid.

respectively on sown land on which the method has been put into practice. And the significance of the product is economic; for it provides a remarkable advantage, indeed to the lay mind a sensational advantage, for one of the most elemental activities by which man has served his material needs, the cultivation of the soil for the production of its fruits.

Thus, in the view of the High Court of Australia, any human action resulting in an (objectively discernible) artificially created state of affairs of economic significance would be a proper subject of letters patent within the meaning of the Statute of Monopolies; subject matter of this type being methods producing an end of utility in practical affairs.

The immediate effect of *NRDC* was to override a series of cases in the 1950s involving methods of treating soil to improve its crop-bearing qualities.[150] However, the Court was careful to emphasize that not all horticultural methods would satisfy its test. In particular, methods involving the modification of 'the conditions under which natural phenomena pursue their inevitable course'[151] were held by the Court to be unpatentable under the (revised) test of *G E C* on the ground that the thing they produced (eg fruit) was the inevitable result of some characteristic inherent in the thing itself,[152] and could not therefore be said to have resulted from the patented method.[153] Also confirmed by the Court, however, was the distinction between such methods and processes involving the use of microorganisms which, it was held, were 'analogous to a chemical process in that, given the microorganisms and the appropriate conditions, the desired result inevitably follows from the working of the process'.[154] This was consistent with its view of the invention as a purposive human action directed to (and responsible for) a specific artificial end.

A further class of historically unpatentable subject matter which the High Court tentatively suggested might fall outside the limits of even its revised 'vendible product' test was methods of medical treatment. In respect of such methods it noted as follows:[155]

[A] process, to fall within the limits of patentability which the context of the Statute of Monopolies has supplied, must be one that offers some advantage which is material, in the sense that the process belongs to a useful art as distinct from a fine art . . . – that its value to the country is in the field of economic endeavour. (The exclusion of methods of surgery and other processes for treating the human body may well lie outside the concept of invention because the whole subject is conceived as essentially non-economic: see *Maeder v. Busch*).

150 See (1959) 102 CLR 252, 278–279 (describing the historical denial of patents to methods by reason of their 'agricultural or horticultural' nature as 'a classic illustration of thinking in terms of the everyday concept of manufacture instead of following the lines along which, over a long period, the courts have given effect to the real purpose and operation of s 6 of the Statute of Monopolies').

151 *N V Philips's Application* (1954) 71 RPC 192, 194.

152 (1959) 102 CLR 252, 279.

153 ibid.

154 ibid, supporting *Szuecs's Application* [1956] RPC 25, SE.

155 (1959) 102 CLR 252, 275 (footnotes omitted).

For present purposes, the significance of *NRDC* derived from its establishment of the following four principles. First, the question to be asked in determining the existence of an invention was 'is this a proper subject of letters patent according to the principles which have been developed for the application of section 6 of the Statute of Monopolies?' Second, this question would only be answered affirmatively for human actions that depended for their claim to patentability on their contribution to a practical art, regardless of whether it was a manufacturing art, and thus whether it resulted in the production of a physical product. Third, the subject matter would only contribute to a practical art if it produced an end of practical utility, such as an artificially created state of affairs having a certain economic significance. And fourth, the following three classes of subject matter were inherently unpatentable pursuant to this test: (a) essentially non-economic subject matter, including possibly surgical methods, which lacked practical utility and belonged to the fine (rather than the practical) arts; (b) methods consisting in the modification of conditions under which natural phenomena pursue their inevitable course, which involved insufficient human involvement to possess the artificiality required of an invention; and (c) methods consisting in the use of a known substance or device the utility of which for the relevant purpose was previously known, which, like subject matter characterized by its aesthetic or informational content, lacked the mechanical or chemical ingenuity constitutive of inventions, and could not be said to advance the practical arts. Implicit was an understanding of the invention as a legal construct defined to further the objectives of the patent system. In the Court's view, those objectives were to advance the practical arts for the economic benefit of society, rather than (as had previously been regarded) to advance the mechanical and chemical arts for the benefit of industrial manufacture.

At the time of its publication, the decision in *NRDC* represented the most authoritative statement on the nature of inventions and the scope of inherent patentability generally in Anglo-Australian law. After its publication, the case was widely hailed as a 'watershed', and not only by Barwick CJ in his now famous description in *Joos v Commissioner of Patents*.[156] For the Banks Committee, charged with reviewing the UK patent system in preparation for ratification of the Strasbourg Convention[157] in 1970, the decision in *NRDC* confirmed the capacity of the Statute of Monopolies-based model of inherent patentability to maintain flexibility in a rapidly changing technological landscape,[158] and thus warranted the retention of that model despite past recommendations that it be abolished.[159]

[156] [1973] RPC 59, HCA, 62 per Barwick CJ.

[157] Convention on the Unification of Certain Points of Substantive Law on Patents for Inventions (Strasbourg Convention) (Strasbourg, 27 November 1963, ETS 47).

[158] Banks Report, [226].

[159] See, eg United Kingdom, *Patents and Designs Acts: Final Report of the Departmental Committee* (Cmd 7206, 1947) [122]–[128]; United Kingdom, *United Kingdom Patent Law: The Effects of the*

Still, the precise implications of *NRDC* for the range of new technologies coming before the courts took some time to unravel, leaving the status of many subject matter in the two decades after *NRDC* uncertain. Principal among those subject matter were biological materials, computer programs and other information-related subject matter, and methods of medical treatment. Further, while supporting a relatively expansive conception of the invention with reference to the practical and commercial fields, *NRDC* did not consider the patentability of chemical and other kinds of products.

2. Biotechnological and Chemical Subject Matter, Including Products

In *NRDC*, the High Court expressly rejected an argument that 'agricultural' or 'horticultural' processes should prima facie be excluded from patentability, and emphasized as the relevant test in respect of such processes whether they involved more than the modification of natural conditions producing an inevitable result.[160] Methods involving the use of micro-organisms, it was suggested, would generally satisfy that test.[161]

Implicit in this treatment of agricultural and horticultural processes, and in the Court's understanding of the nature of inventions in general was a view of the inherent patentability requirement as applying consistently across all fields of technology. After *NRDC* that view was regarded as supporting the proposition that in order to constitute an invention, a subject matter needed only to manifest sufficient human involvement to take it out of the realm of nature and into the realm of (artefactual) invention without also needing to be of mechanical or chemical type.[162] The difficulty remained, however, of conceiving a subject matter *qua* invention for the purpose of determining its patentability and scope of protection.

An early case to consider this issue in the context of biotechnological subject matter was *American Cyanamid Co (Dann's) Patent.*[163] That case involved a claim for a method of producing an antibiotic from a naturally occurring strain of a known micro-organism, and for the antibiotic so produced. The issue was whether the statutory requirement for sufficient description was satisfied in the absence of any known source or means of producing the strain. In particular, the question was whether the patentee's refusal to make a sample of the strain available for public use placed it in breach of those requirements, with the answer turning on the Court's view of the invention which needed to be disclosed.

Strasbourg Convention of 1963 (Cmnd 2835, 1965) [32]. See also *Swift v Commissioner of Patents* [1960] NZLR 775, NZSC, 779 per Barrowclough CJ.

160 (1959) 102 CLR 252, 279.

161 ibid.

162 See *General Electric Co Ltd's Application (Patent)* [1961] RPC 21, SE, PAT.

163 [1971] RPC 425, HL.

According to the majority, the invention was 'the prosaic process' of isolating the strain, and not the new antibiotic as such.[164] Only Lord Diplock dissented, arguing that the invention extended to the antibiotic itself, on the basis that it was 'a new product useful to humanity which does not exist in nature'.[165] The result was to require description (by deposit) of the isolated strain, and not merely of the process of isolation itself; a conclusion his Lordship reached with reference to the technical proficiency, labour, and cost involved in 'searching for and finding hitherto unidentified strains of micro-organisms existing in the natural state from which useful new antibiotics can be prepared by what is now a well-known standard process',[166] and the following considerations of social policy:[167]

> If such research is to be encouraged in a competitive society, the monetary rewards of success must be assured to those who undertake the expense; and the means of doing so in this and in most other countries with comparable social systems is by according to the successful discoverer of the new product the controlled and limited monopoly granted for inventions under the national patent laws. . . . The concept of 'invention' as an activity which entitles the person who undertakes it to a patent monopoly of a process or its product has been continually changing by applying the social policy which underlay the Statute of Monopolies in the seventeenth century to the changing conditions resulting from advances in technology and the sciences. That policy was to provide a material inducement to persons to undertake whatever effort and expense might be needed to introduce into the United Kingdom some useful new process or new product with a view to its becoming generally available for use or manufacture by the inhabitants of the realm . . . The counterpart of the temporary monopoly was the provision of the information. . . . It is in my view consistent with this basic policy to treat the kind of research involved in the discovery of a strain of micro-organism from which a new and useful antibiotic can be prepared as an activity which entitles the person who undertakes it to a temporary monopoly under the Patents Act 1949, of the product of his success. I accept, therefore, the extension of the concept of 'invention' to include antibiotics which are discovered through this kind of research.

This analysis was important for two reasons. First, it underlined the different ways of conceiving a subject matter *qua* invention, and the importance of the conception chosen by a court for its assessment of the other patentability requirements. And second, it supported a method of conception that depended on an assessment of patent policy considerations having regard to the category of subject matter in question. This approach was consistent with the emphasis in *NRDC* on a purposive definition of the 'invention' with reference to the principles developed for application of the original patent legislation. Further, by suggesting the purpose of that legislation as a reference for testing (and thereby ensuring) the technological neutrality of the system, it gave meaning to the *NRDC* non-discriminatory approach. On the other hand, the difficulty remained of how to determine the

[164] [1971] RPC 425, HL, 448.
[165] ibid, 451.
[166] ibid.
[167] ibid, 451–452.

'material inducement' which the 'social policy' of the system justified conferring on inventors in individual fields. That difficulty does not detract from the significance of the case, however, which lies in its support for a conception of the invention as a term of legal art, defined to advance the public policy objectives of the patent system and thereby orient the system around its original purpose, and in its use of that purpose as a mechanism for ensuring the technological neutrality of the system. In this scheme, the requirement for an invention was supported by the requirement for proportionate—'equiparating'—claims.

As has been seen, by the time of the decision in *NRDC*, a requirement for fairly based claims had been enacted in sections 4(4) and 32(1) of the Patents Act 1949. Nonetheless, the common law principle of *Mullard Radio* continued to be valid. In *Pottier's Application*,[168] for example, it was applied to support the refusal of a patent for a process of treating hydrated seedlings. In *Neva Corp's Application*,[169] it was applied to support the refusal of a patent for an apparatus for inducing reduced awareness. And in *Eastman Kodak Co's Application*,[170] it was cited in support of the following principle:[171]

> Though mere width of claim is in itself no good ground of objection under section 14 [governing pre-grant opposition], it is a fundamental rule of patent law that the consideration which a patentee gives for his monopoly is the disclosure which he makes and [that] the protection which he gets cannot extend further than is necessary to protect that which, in view of his disclosure, he is fairly entitled to cover.

Of greatest interest for present purposes, however, is *Olin Mathieson Chemical Corp v Biorex Laboratories Ltd*.[172] In that case, Graham J relied on the principle in *Mullard Radio* to support a test of claim validity in respect of a class of chemical products structured with reference to the following factors:

(a) The size of the class of chemical bodies covered by the claim;
(b) The desirability of supporting research in the drug and other fields; and
(c) The need to ensure that claims are not so broad as unjustifiably to stifle research by others—while noting the existence of statutory mechanisms (including compulsory licensing provisions) for preventing such abuses.[173]

Considering the basis for this tripartite test, the Judge stated as follows:[174]

> Sir Lionel argued that [lack of fair basis under s 32(1)] was a broad objection directed against grants in which it could be shown that for one reason or another the consideration given by the patentee by the disclosure of his invention in his specification was less than he

[168] [1967] RPC 170, PAT.
[169] [1968] RPC 481, SE.
[170] [1970] RPC 548, PAT.
[171] ibid, 563.
[172] [1970] RPC 157, Ch.
[173] ibid, 192–193.
[174] ibid, 181.

should have given having regard to the width of his claims. In other words, section 32(1) covers the objection discussed at length in the case of [*Mullard Radio*] that a claim which is a 'covetous' claim, or one in which the claim does not 'equiparate' with the consideration given by the disclosure, is a bad claim. This requirement, said Sir Lionel, has always been fundamental in our patent law, and section 32(1) certainly includes it, but it is not, of course, limited to it.

By this reasoning, Graham J applied the *Mullard Radio* requirement for 'equiparation' in a claim to support the implication of Dixon LJ above, that in determining the validity of a claim for a patent (and invention), it is necessary to consider social policy considerations, and particularly the objectives of the patent system.

3. Revisiting the Exclusion of Surgical and Therapeutic Methods

In *NRDC* the High Court of Australia raised, as a possible justification for the exclusion of methods of medical treatment, their essentially non-economic nature, which, it was suggested, could be regarded as taking them out of the category of useful arts in which inventions belong. After 1959, however, an alternative basis for that exclusion was suggested in *Swift & Co's Application*,[175] deriving from the instability of methods that depend for their utility on the physiological functions of higher life forms. That principle was reflected in the treatment during the 1960s of the patentability of surgical and therapeutic methods.[176] For example, in *Neva Corporation's Application*,[177] a method of inducing a state of reduced awareness in humans (and animals) by means of sounds generated by a known apparatus was held to be an unpatentable method of medical treatment on the ground that its utility depended on the reception and interaction of the sounds with the human brain. Similarly in *London Rubber Industries Ltd's Patent*,[178] a method of controlling female ovulation through the prescribed oral administration of known hormones was refused a patent on the ground, implicitly, that the utility of the method depended upon the physiological response of the person treated and, for that reason, fell outside the realm of useful arts in which inventions existed.[179] Representing one of the very few early decisions expressly to consider the reason for the exclusion of methods of medical treatment from inherent patentability, the PAT reconciled its decision with *NRDC* by positing as the results of medical methods the human response to them, which it held lay beyond the realm of manufacture.[180]

175 [1961] RPC 129, PAT, 133, overruled [1962] RPC 37, Divisional Court with reference to *Swift v Commissioner of Patents* [1960] NZLR 775, NZSC.

176 See *Puharich and Lawrence's Application* [1965] RPC 395, PAT; cf *United States Rubber Co's Application* [1964] RPC 104, SE, 105–106.

177 [1968] RPC 481, PAT.

178 [1968] RPC 31, PAT.

179 ibid, 35.

180 ibid, 34–35.

Cracks in the façade of *London Rubber* were evident in the 1970 decision of *Palmer's Application*,[181] where a Superintending Examiner's (SE's) finding of inherent unpatentability in respect of a method of treatment that depended for its efficacy upon the physiological response of the persons treated was overruled on appeal on the basis that a method so depending could still involve the practical employment of a person's art and skill so as to constitute an invention properly conceived.[182] The effect of that recognition was to remove the primary obstacle to the inherent patentability of surgical and therapeutic methods; an effect confirmed the following year in the important case of *Schering AG's Application*,[183] the significance of which can be summarized as follows. First, it confirmed as the primary justification for the conferral of patents the rewarding of people for their contributions to the learning of a practical art through the expenditure of skill, effort, and expense; the reference to 'effort and expense' in this context recalling the emphasis of Lord Diplock in *Dann's Patent*. Second (and relatedly), it confirmed the applicability of that justification in respect of methods of medical treatment, thereby removing any first-principle policy objections to the patenting of such methods. Third, it confirmed the absence of any relevant distinction for patentability purposes between methods of medical treatment on one hand, and methods of manufacturing medical substances and devices on the other, thereby highlighting the absence of any logical basis for treating them differently. And fourth, it implied that, legislative provision aside, the only legitimate ground on which methods of medical treatment might be denied patentability was that of ethics, thereby contradicting the principle that public policy, including 'question[s] of humanity', were not relevant when determining patentability.[184]

It took only two years and two cases[185] for the import of *Schering* to be taken up at a higher judicial level. That occurred in *Joos v Commissioner of Patents*,[186] where Barwick CJ, sitting as a single judge of the High Court of Australia, held that a method of cosmetic treatment was inherently patentable by reason of its commercial significance (consistent with *NRDC*), and suggested that any general exclusion from inherent patentability of methods of 'medical' treatment[187] could only be justified at common law on 'public policy [grounds] as being, in the language of

[181] [1970] RPC 597, PAT.

[182] ibid, 601 (relying on *Wheeler* (1819) 106 ER 392, KB).

[183] [1971] RPC 337, PAT.

[184] (1914) 31 RPC 235, 236. See also *British Thomson-Houston Co Ltd v Corona Lamp Works Ltd* (1921) 39 RPC 49, HL, 67 per Viscount Haldane.

[185] See *Bio-Digital Sciences Inc's Application* [1973] RPC 668, PAT; *Calmic Engineering Co Ltd's Application* [1973] RPC 684, PAT.

[186] (*Joos*) (1972) 126 CLR 611, HCA.

[187] See ibid, 622 (defining methods of medical treatment narrowly to mean methods 'for medical treatment of human disease, malfunction, disability or incapacity of the human body or any other part of it'). See also *Lee Pharmaceuticals' Applications* [1978] RPC 51, SE (treating a method for preventing tooth decay as inappropriate subject matter for a patent as a method for the prevention of a disease).

the Statute of Monopolies, "generally inconvenient"'.[188] The implication of that suggestion was that, considerations of public policy aside, methods of medical treatment were inventions within the meaning of *NRDC* and the Statute of Monopolies respectively.[189]

Two points were central to the Chief Justice's decision in *Joos*. The first concerned the effect of *NRDC*, which was held to have established as inherently patentable any process having economic value, whether or not 'directly supplied by the nature of the activity which would utilize the process'.[190] And the second concerned the phrase 'general inconvenience' in section 6 of the Statute of Monopolies, which was suggested to provide a general policy basis for the recognition of threshold exclusions from patentability applicable in respect of subject matter otherwise comprising an invention.[191] This latter point was consistent with the nineteenth century view of 'general inconvenience' as supporting the exclusion from patentability of subject matter the monopolization of which would restrain the public in an inconvenient manner;[192] a view that was also reflected in the two (then) recent decisions of the PAT and SE in *Rolls-Royce*[193] and *Hiller's Application*[194] respectively.[195] In the first of those cases a patent was refused for a method of operating a known engine, on the basis of the burden on pilots should they be required to '[avoid] infringement of a statutory monopoly in the operation of [their] standard engine controls'.[196] Similar was the reasoning in *Hiller*, where a plan for subterranean utility distribution schemes was denied a patent with reference to the 'unfair and unreasonable' burden that the patent would impose on persons

188 (1972) 126 CLR 611, 623.

189 ibid, 622–624.

190 ibid, 624.

191 ibid, 623.

192 See, eg *Patterson v The Gas Light and Coke Co* (1875) LR 2 Ch D 812, CA, aff'd (1876) 3 App Cas 239, HL, and discussed at page 67.

193 [1963] RPC 231, PAT.

194 [1969] RPC 267, PAT.

195 Compare the decisions in *Rolls-Royce* and *Hiller* with the decision in *T S's Application* [1924] 41 RPC 530, SG, where an opposition to an application directed to a mechanism for continuous motion without external force on the basis of its alleged uselessness and the 'general inconvenience' that would result were a patent to be granted was rejected. In reaching his decision the Solicitor-General equated 'general inconvenience' with utility, and held that 'the inconvenience which might ensue by the granting of patents in frivolous cases' was 'largely mitigated by the power of the court to inquire into questions of utility' (at 538, relying on *Morgan v Seaward* (1837) Web Pat Cas 187, Ex): 'These difficult words ["generally inconvenient"] have been applied by Sir Edward Coke to an invention for fulling in a mill and not being manual labour which was held to be "inconvenient in that it would turn so many labouring men to idleness". During the development of the industrial system this ground of objection to a patent, that it would cause unemployment, has long since been abandoned, some economists taking the view that, if unemployment be caused in one occupation by an invention, employment is thereby caused in another and, in my view, whatever may be the truth of this contention, the fact is that these words in their proper meaning are now obsolete and in any event cannot have reference to an alleged invention which, on the Comptroller's assumption, is useless and cannot work and therefore could scarcely have the effect of "throwing labouring men into idleness".'

196 *Rolls-Royce Ltd's Application* [1963] RPC 231, 256.

responsible for providing gas, electricity, water, and sewage disposal services for residential properties.[197]

After *Joos*, it was clear that any historical reason in principle or logic for the denial of patent protection in respect of methods of medical treatment had been removed by the test of inherent patentability articulated in *NRDC*, but that such methods might still legitimately be regarded as beyond the scope of section 6 of the Statute of Monopolies—and thus outside the statutory definition of the invention—on public inconvenience grounds. That left only the question of whether they should be so regarded.

4. Revisiting the Exclusion of (Other) Non-Manufacturing Methods

(a) Information-Related Subject Matter, Including Computer Programs

From 1960 to 1970, methods of preparing, generating, and presenting information (information-related subject matter) were excluded from inherent patentability on the basis of their manifest failure to advance the practical arts.[198] The evidence of that failure did not lie in their absence of economic value or lack of mechanical utility,[199] but rather in their absence of the physical form[200] and exclusively practical significance[201] required of an invention under earlier law. The result was the denial of patents during the 1960s for a method of operating a known engine to improve its performance,[202] a method of recording by apparatus the journeys of passengers in a transport system to calculate their fare,[203] a method of preparing, tabulating and codifying data for mechanical computation,[204] and a method of operating a known computer apparatus.[205] The only case to consider the doctrinal basis for that denial was *Rolls-Royce Ltd's Application*,[206] where reliance was placed on the literal terms of the legislative definition of 'invention', the (purported)

197 [1969] RPC 267, 268.

198 See, eg *Rolls-Royce Ltd's Application* [1963] RPC 231, PAT; *Stahl & Larsson's Application* [1965] RPC 596, PAT; *Slee & Harris's Application* [1966] RPC 194, SE; *Badger Co Inc's Application* [1970] RPC 36, PAT; *Gevers' Application* [1970] RPC 91, PAT.

199 See *Rolls-Royce Ltd's Application* [1963] RPC 231, 255 (rejecting the vendible product test of Morton LJ on the ground, among others, that 'the beneficial economic exploitation of new ideas extends beyond mere buying and selling').

200 See, eg *Stahl & Larsson's Application* [1965] RPC 596, 600; *Slee & Harris's Application* [1966] RPC 194, 196–197; *Burroughs Corporation (Perkins') Application* [1974] RPC 147, PAT, 152. See also *Badger Co Inc's Application* [1970] RPC 36, 40 (finding a method of preparing, tabulating, and codifying data to be inherently unpatentable because of the conceptual nature of the raw material, and its failure to contribute to the fashioning of a product).

201 See, eg *Slee & Harris's Application* [1966] RPC 194, 197.

202 *Rolls-Royce Ltd's Application* [1963] RPC 231, PAT.

203 *Stahl & Larsson's Application* [1965] RPC 596, PAT.

204 *Badger Co Inc's Application* [1970] RPC 36, PAT.

205 *Slee & Harris's Application* [1966] RPC 194, SE.

206 [1963] RPC 231, PAT.

exclusion of so-called 'schemes and plans',[207] and the general inconvenience that would result to the public were a patent to be granted for the methods at issue.[208]

After 1970, the view of inherent patentability as requiring that a method produce a physical object of exclusively practical significance lest it be treated as a mere scheme or plan was revised (again) in the context of computer-related inventions. Thus in *Burroughs Corporation (Perkins') Application*,[209] the PAT held that provided a method involved an alteration in the operability of a physical object, the test of *NRDC* would be satisfied and the method inherently patentable.[210] This was followed by the same PAT four years later in *IBM's Application*,[211] where it was held that a method of operating or controlling a conventional computer to process financial data was inherently patentable, as was the computer program by which the method was effected. The reason, it was suggested in both cases, was that the alteration of a physical object's function, and the embodiment of a computer program 'in such physical forms as punched cards or tape',[212] were sufficient to 'clothe' the method 'with a practical garment in the shape of apparatus' so as to make it more than a mere idea or 'naked conception'.[213] The argument in *IBM* that a known computer programmed in a new way differed in intellectual content only and was thus inherently lacking in novelty in the *Griffin v Isaacs* sense was expressly rejected, the Court finding that '[t]here must be different holes in the card [embodying the program] or different magnetic patterns on the tape, or some other automatic control imposed to ensure that the computer carries out the particular operation required'.[214] It was thus the physical manifestation of the differences, and their mechanical interaction with the relevant computer apparatus, that precluded treatment of the program and programmed computer as being mechanically equivalent to previous programs and computers respectively.

The effect of the decisions in *Burroughs*, *IBM*, and *Rolls-Royce*, was to extend the legal conception of the invention to include methods 'result[ing] in a new machine or process or an old machine giving a new and improved result'.[215] In adopting this approach, the PAT removed the 1960s restrictions on the patenting of methods involving the preparation of information for computation by a conventional device. In addition, while supporting the need for a subject matter of mechanical utility, the PAT suggested that, in the case of computer programs, such utility

207 [1963] RPC 231, PAT 255 (citing *D A and K's Application* (1925) 43 RPC 154, SG).

208 ibid.

209 [1974] RPC 147, PAT.

210 ibid, 158 (the subject matter at issue in *Burroughs* was a method of transmitting data over a communication link between computers).

211 [1980] FSR 564, PAT.

212 ibid, 571.

213 [1974] RPC 147, 158.

214 [1980] FSR 564, 569.

215 [1974] RPC 147, 158. See also ibid, 572; *Badger Co Inc's Application* [1970] RPC 36, 40.

would be implicit in the programs' mechanical interaction with their implementing device. Thus it held that while computer programs were inherently unpatentable in isolation (as an abstract scheme, or idea or conception),[216] in conjunction with their implementing device they were inventions. The effect was to affirm the inherent patentability of automated methods as combinations of hardware and software, and thereby obviate the implication that they should be regarded as an instance of analogous use of a 'known' (ie a conventional) computer device.

(b) Methods of Presentation

Consistent with post-*NRDC* developments in respect of other subject matter, there was some indication in the 1970s that previous restrictions on the inherent patentability of methods of indication or presentation were on the wane as a result of an increasingly liberal approach by decision-makers to the question of when the claimed contribution of a subject matter could be said to have the mechanical utility required of an invention. The first case to reflect that transition was *Edgar Rhodes' Application*,[217] where a method of displaying the impact speed of a moving car by means of a speedometer on the car's dash was held to be inherently patentable, in apparent conflict with the decision in *A E W's Application*,[218] decided forty-nine years earlier. In justifying its finding, the PAT emphasized the alleged novelty and utility of the method of presentation which, it held, made the subject matter comparable to the first printed book in braille.[219] The decision in *Edgar Rhodes* was relied on by the Patents Court five years later in *ITS Rubber Ltd's Application*[220] to justify the inherent patentability of a method of colouration. In that case, Whitford J overruled a decision of the SE that a squash ball, distinguished by its colour, differed from existing inventions in aesthetic or intellectual content only and was for that reason incapable of constituting a manner of new manufacture.[221] According to his Lordship, the colour improved the operation of the ball with the result that its function could be described as mechanical and not purely intellectual or aesthetic.[222] That conclusion was reconciled with the early twentieth century decisions regarding methods of indication by distinguishing the subject matter with which they were concerned. Thus it was said that while a method of colouring a specific article to enhance its functionality could support a patent, a mere 'scheme for colouring a vast variety of articles'[223] could not.[224]

[216] [1974] RPC 147, 158, 160; [1980] FSR 564, 568.
[217] [1973] RPC 243, PAT.
[218] (1924) 41 RPC 529, SG.
[219] [1973] RPC 243, 247.
[220] [1979] RPC 318, PAT, 321.
[221] ibid.
[222] ibid, 324–325.
[223] ibid, 324.
[224] See also *T's Application* (1919) 37 RPC 109, SG.

The decisions in *Edgar Rhodes* and *ITS* reflected the recognition following *NRDC* that the need for mechanical inventiveness or 'invention relating to some manufacture' did not, as had been held in *Cooper*, require the production of a physical artificial product, but rather some human action on the physical world directed to advancing the practical arts. They also confirmed that abstract ideas aside, the only subject matter likely to fall foul of that requirement were those lacking ingenuity in the *Griffin v Isaacs* sense. Examples of such subject matter excluded from patentability during the 1960s and 1970s[225] were various methods of packaging known medicinal products,[226] and a medium carrying an instructional message.[227]

(c) 'Schemes', Including Business Methods

Following *NRDC*, the inherent patentability of 'mere schemes or plans' in the sense of *Cooper*—methods resulting in an artificially created state of affairs having practical or commercial significance but producing an end with no necessary connection to the physical world[228]—was considered by the UK courts in one case only, in the context of a transactional (ie a business) method.[229] That case was *IBM*, where the PAT, after affirming the inherent patentability of the programmed computer at issue, emphasized that had the claim 'on its true construction [protected the patentee's] business scheme however carried out', it would have been invalid, thereby confirming the view of early decision-makers that business was not among the practical arts suitable for protection by a patent, notwithstanding its undoubted economic significance.[230] This contrasted with the principles of *NRDC*, and with the view taken by the House of Lords in *Househill Iron Co v Neilson*,[231] that abstract (including elementary) principles could support a patent, subject only to their having been 'turned to account' through 'direction to . . . human life'.[232]

D. Conclusion

A central theme to emerge from the analysis above is the broad consistency in UK conceptions of the invention from the introduction of the specification in the

[225] cf *Pitman's Application* [1969] RPC 646, PAT; *Jacques Kessler's Application* [1973] RPC 413, PAT.

[226] See, eg *L'Oreal's Application* [1970] RPC 565, PAT, 571–572; *Organon Laboratories Ltd's Application* [1970] RPC 574, PAT; *Ciba-Geigy AG (Durr's) Applications* [1977] RPC 83, PAT, CA.

[227] See, eg *Nelson's Application* [1980] RPC 173, HC.

[228] See also *Dixon's Application* [1978] RPC 687, PAT (method of speech instruction involving conditioning the diaphragm by following instructions printed on a sheet inherently unpatentable); cf *Pitman's Application* [1969] RPC 646, PAT.

[229] See also *Jacques Kessler's Application* [1973] RPC 413, PAT.

[230] *IBM's Application* [1980] FSR 564, 568.

[231] (1843) Web Pat Cas 673, HL.

[232] ibid, 683–684.

eighteenth century until the *NRDC* decision in 1959. According to that conception, an invention was any purposive human action of mechanical or chemical utility directed to advancing an industrial art, and not manifestly failing in that regard. The only two significant issues of uncertainty were whether the action needed to result in a physical (and vendible) product, such that the art advanced would be a manufacturing art, and whether a patent could protect such a product itself, independent of the method by which it was made. In the twentieth century the first of these issues was resolved in favour of a requirement for production of a product. That position changed, however, following the High Court of Australia's decision in *NRDC*, according to which an invention was any subject matter properly regarded as such according to the principles developed for application of section 6 of the Statute of Monopolies. According to the Court, those principles supported the treatment as inventions of any practical art of commercial significance.

Implicit in conceptions of the invention as an art is a view of the same as not extending to products. That view is consistent with the statutory definition of the 'invention' in pre-1977 law as a 'manner' (ie a method) 'of new manufacture'. Further, while inconsistent with a widespread assumption to the contrary, when that assumption was tested in the courts following the introduction of the 1883 Act, doubts were raised with respect to its validity. In 1899, for example, Buckley LJ in *Kopp v Rosenwald* made clear his view that the law did not support the patenting of products independent of the method by which they were made. Similarly in 1908, Viscount Haldane in *British Thomson-Houston v Charlesworth*[233] described as 'wrong' the proposition of Lord Davey in *Acetylene* that 'if a new material had been discovered by no new means and by no new process of production, that material itself would have been a patentable article'.[234] These decisions were important in suggesting that the Comptroller-General of Patents was not alone in his concerns about chemical product patents. They also accounted for the view of the Sargant Committee that the weight of authority was against such patents, even in the absence of a 'reported decision to that effect'.

In 1919, the Legislature stepped in, enacting section 38A(1) of the Patents and Designs Act 1907 to limit the protection for chemical products to protection for the product-by-process alone. Even after that section was repealed, however, and the inherent patentability of products apparently affirmed, Graham J in *Olin Mathieson* drew on the 'fundamental' common law rule that a claim must equiparate with the specification in support of a test of claim validity that focused on the policy considerations of the system, assessed with respect to the subject matter in question. In so doing he supported the approach of Lord Diplock in *Dann's Patent*, and a conception of the requirement for an invention as a mechanism for orienting the system around its social policy objectives. He also resolved a longstanding

233 (1925) 42 RPC 180, HL.
234 ibid, 207.

tension between two traditions: on one hand, the practice of granting patents for chemical and other products, and on the other, the common law requirement that the monopoly of a patent equiparate with the disclosure in the specification as the consideration for which the monopoly is granted. And finally, he drew attention to the central question in law raised by the restriction of protection to patentable inventions: what constitutes a subject matter *qua* invention?

From 1623 to 1977, UK patent law reflected the culmination of judicial and administrative decision-making within the statutory context of section 6 of the Statute of Monopolies. However, this changed with the introduction of the Patents Act 1977, which substituted the Statute of Monopolies-based system with one derived from the European Patent Convention 1973.[235] At the heart of that system was a model of inherent patentability that differed substantially from the 'manner of manufacture' model of pre-1977 law, largely because of its omission of any positive definition of 'invention' in favour of a non-exhaustive list of non-inventions. That list is contained in Article 52(2) and (3) of the EPC, which provide, with Article 52(1), as follows:

> (1) European patents shall be granted for any inventions, in all fields of technology, provided that they are new, involve an inventive step and are susceptible of industrial application.
> (2) The following in particular shall not be regarded as inventions within the meaning of paragraph 1:
>   (a) discoveries, scientific theories and mathematical methods;
>   (b) aesthetic creations;
>   (c) schemes, rules and methods for performing mental acts, playing games or doing business, and programs for computers;
>   (d) presentations of information.
> (3) Paragraph 2 shall exclude patentability of the subject matter or activities referred to in that provision only to the extent to which a European patent application or European patent relates to such subject matter or activities as such.

On its face, the effect of this provision was to substitute the philosophical and policy-based conception of the invention supported under pre-1977 law, focused on the relationship between people and the material world, and that which had historically been regarded as appropriate for a patent, with a formalistic conception, focused on the existence of a subject matter not falling within any of the categories of subject matter listed. In Part II, the courts' response to this change is considered, along with the post-EPC requirement for an invention in general.

[235] Convention on the Grant of European Patents (European Patent Convention, as amended) (Munich, 5 October 1973, 13 ILM 268).

PART II

THE REQUIREMENT FOR AN INVENTION IN THE ERA OF THE EUROPEAN PATENT CONVENTION

4

The EPC Requirement for an Invention in its Interpretive Context

A. Introduction

In a well-known literary expression of the social values of his time, Mark Twain in *A Connecticut Yankee in King Arthur's Court*[1] sends Hank Morgan back more than 1300 years from the industrial age of late nineteenth century America to take up office in the royal Court of King Arthur. Determined to lift Arthurian society out of its Dark Age and push it along the path to Enlightenment, one of Morgan's first tasks in office is to establish a patent system in the belief that 'a country without a patent office and good patent laws was just a crab, and couldn't travel any way but sideways or backwards'.[2]

One hundred years after its attribution, Morgan's expression of unquestioned faith in the capacity of patents to ensure that most vaunted of nineteenth century ideals, Progress, continues to ring more of truism than hyperbole. Indeed, if there is an unreality in Morgan's belief it sounds less in the patent–Progress connection on which it rests than in the suggestion that a patent system might plausibly be fashioned in a single day.

On both counts the twentieth century history of European patent reform is a case in point. For example, it took barely four months from the creation of the Council of Europe in May 1949 for a proposal for the establishment of a European patent system to be considered.[3] While the proposal was rejected, belief in the need for a European patent office and 'good patent laws' to support the early stages of post-war integration was overwhelming, and led to the submission of a string of further proposals to the same end during the 1950s.[4] While each was similarly rejected, in combination they helped set the agenda for the future creation of a unitary European patent.

[1] M Twain, *A Connecticut Yankee in King Arthur's Court* (New York: Signet Series, 1963).

[2] ibid, 58.

[3] See Consultative Assembly of the Council of Europe, First Ordinary Session, Sixteenth sitting, 6 September 1949, Doc 75 ('Creation of a European Patents Office: Report proposed by M Longchambon on behalf of the Committee on Economic Questions').

[4] By 1952 seven proposals for a European patent system had been considered by the Committee, all involving a separation of functions between a European and national Patent Offices.

Over the last 50 years that agenda has all but defined European patent reform, and while its goal of a unitary patent remains elusive, significant achievements have nonetheless been made. The most important of these is the European Patent Convention (EPC),[5] which was signed by an initial group of countries in 1973 and today has a membership close to that of the European Union (EU).

The central provisions of the EPC are Articles 52 to 57, which define the essential criteria of patentability, including in Article 52(2) and (3) the criterion of inherent patentability itself.[6] The focus of this chapter and the following two chapters is on the meaning of these latter paragraphs. That meaning must be derived from the relevant decisions of the European Patent Office (EPO) and national courts charged with interpreting and applying the EPC, and it is therefore primarily with those decisions that the chapters are concerned. The uncertainty and controversy surrounding them, however, make it appropriate to consider also the legislative background of the Articles, including the reasons for their adoption and the original intention regarding their application. The primary evidence of that background is the drafters' *travaux préparatoires*. In addition to providing the starting point and critical context for constructions of Article 52(2) and (3), the *travaux* have direct legal relevance as interpretive aids under Article 32 of the Vienna Convention on the Law of Treaties.[7] This is so notwithstanding that the European Patent Organisation is not a party to that Convention, due to the finding of the Enlarged Board of Appeal in *EISAI/Second Medical Indication*[8] that interpretations of the EPC should nonetheless conform to Vienna Convention principles.[9] Indeed, even the UK courts have supported the applicability of those principles in the context of inherent patentability, following their decision to treat Article 52(2) and (3) as the direct source of national law in place of section 1(2) of the Patents Act 1977 (PA) in which they are transposed.[10] Hence the direct legal relevance of the *travaux* under both EPC and UK law, deriving from Article 32 of the Vienna Convention, which provides as follows:

Recourse may be had to supplementary means of interpretation, including the preparatory work of the treaty and the circumstances of its conclusion, in order to confirm the meaning

[5] Convention on the Grant of European Patents (European Patent Convention, as amended) (Munich, 5 October 1973, 13 ILM 268).

[6] Reproduced at page 106.

[7] Vienna Convention on the Law of Treaties (Vienna Convention) (Vienna, 23 May 1969, UNTS 1155, 321).

[8] G_0005/83 (*EISAI/Second Medical Indication*) [1979–85] EPOR B241.

[9] ibid, B244.

[10] See *Aerotel Ltd v Telco Holdings Ltd* [2006] EWCA Civ 1371, [10]; See also *CFPH's Application* [2005] EWHC 1589, Pat, [26]–27] per Mr Prescott QC (supporting a teleological interpretation of EPC Art 52(2) and (3) in recognition of their basis in an international treaty).

resulting from the application of article 31, or to determine the meaning when the interpretation according to article 31:
(a) leaves the meaning ambiguous or obscure;
(b) leads to a result which is manifestly absurd or unreasonable.

Having said this, it is important to note the refrain expressed regarding use of *travaux* by the Technical Board of Appeal in *AKZO/Automatic Revocation*:[11]

[T]he importance of preliminary drafts and working papers leading to the [EPC] should not be over-emphasised when considering how [it] should be interpreted. . . . [The approach of Articles 31 and 32 of the Vienna Convention] obviously accords with common sense, in that the text of a treaty is the primary source of law, and that is what tells practitioners and the public what the law is. The working papers are not widely available. If the Convention itself can be sensibly interpreted so as to lead to a clear, fair and practically workable result which is in accordance with its object and purpose, recourse to the working papers is normally unnecessary. In any event, it is to be recognised that passages in the working papers leading up to a treaty may sometimes be misleading in relation to the proper interpretation of that treaty as finally agreed. It is to be noted that the European Court of Justice has very rarely had recourse in its judgments to the working papers leading to the various treaties and conventions which it has to interpret. In accordance with the generally accepted international practice referred to above, judgments of the European Court of Justice commonly interpret treaties such as the Treaty of Rome in light of its object and purpose. In the Board's view, this approach should normally be followed when interpreting the EPC.

While not disputing these comments, there is nonetheless a circularity about them. The reason is that it is precisely to discover the 'object and purpose' of a treaty and its provisions that recourse to the *travaux* is both necessary and authorized by the Vienna Convention;[12] a point reflected in the circumstances in which the *travaux* have been used in the context of Article 52(2) and (3) themselves.[13] Indeed, it is submitted that the more apposite point in the *AKZO* remarks concerns the limited availability of the EPC *travaux*. However, rather than being a reason to ignore the *travaux* in construing the EPC, this is a reason for increasing their accessibility.

It is in the context of these considerations that I turn to the *travaux* for Article 52(2) and (3), after an initial discussion of the international and European statutory context in which they exist.

[11] T_0026/88 (*AKZO/Automatic Revocation*) [1990] OJ EPO 4, [3.6].

[12] Art 32 authorizes recourse to the *travaux* to clarify or confirm an interpretation resulting from an application of Art 31, namely, from a reading of the terms 'in light of [the treaty's] object and purpose' (see Vienna Convention, Arts 31 and 32).

[13] See, eg *Aerotel Ltd v Telco Holdings Ltd* [2006] EWCA Civ 1371, [10]; T_154/04 (*DUNS LICENSING ASSOCIATES/Estimating Sales Activity*) [2008] OJ EPO 2/46, [6].

B. The EPC Requirement for an Invention in its International and European Statutory Context

1. International and European Patent Law-Making: An Overview

Most international patent agreements to date have been concluded under the aegis of the World Intellectual Property Organization[14] (WIPO) or its predecessor, the Bureaux Internationaux Réunis pour la Protection de la Propriété Intellectuelle (BIRPI), and continue to be administered by WIPO itself. These Conventions, which include the first international intellectual property agreement, the Paris

Diagram A Overview of multilateral patent agreements by organization

WIPO Conventions

The Bureaux Internationaux Réunis pour la Protection de la Propriété Intellectuelle was established in the nineteenth century to provide countries with an administrative framework and permanent forum for discussion of intellectual property-related issues, and to administer the multinational intellectual property agreements resulting therefrom. Its successor, the World Intellectual Property Organization, was established in 1967 and at the time of writing has 184 members.

WIPO Conventions divide into three categories (see at <http://www.wipo.int/treaties/en/>):

(i) Intellectual property protection conventions, including (for patents) the Paris Convention 1883[1] and Patent Law Treaty 2000.[2]
(ii) Global protection conventions, including (for patents) the Patent Cooperation Treaty (PCT).[3]
(iii) Classification conventions including (for patents) the Patent Convention (SPC).

Non-WIPO Conventions

The other major institutional actors in the international IP arena and the instruments in the area of patents for which they are each responsible are:

(i) The World Trade Organization (WTO), responsible for the TRIPS Agreement;[4]
(ii) The Council of Europe, responsible for the SPC;

[1] Paris Convention for the Protection of Industrial Property (Paris Convention, as amended) (Paris, 20 March 1883, 13 UST 1).

[2] Patent Law Treaty (Cm 6779, 2000).

[3] Patent Cooperation Treaty (Washington, 19 June 1970, 9 ILM 978).

[4] Agreement on Trade-Related Aspects of Intellectual Property Rights (TRIPS Agreement) Marrakesh Agreement Establishing the World Trade Organization, Annex 1C, Legal Instruments—Results of the Uruguay Round, vol 31 (15 April 1994, 33 ILM 81).

[14] See Convention Establishing the World Intellectual Property Organization (Stockholm, 14 July 1967, 6 ILM 782).

(iii) The European Union, responsible for the Community Patent Convention 1975,[5] the Community Plant Variety Rights Regulation 1994,[6] The Biotech Directive 1998,[7] and the Draft Agreement on the European and Community Patents Court and Draft Statute 2009,[8] all aimed, among other things, at creating a Community patent system. Since the coming into force of the Lisbon Treaty[9] in December 2009, pursuit of that aim will be facilitated by the strengthened role of the European Parliament in EU IP-related legislative matters, as a result of changes to the EU's co-decision procedure effected by the Treaty.

(iv) The United Nations, responsible for the Universal Declaration of Human Rights 1948.[10]

Other Conventions, such as the International Convention for the Protection of New Varieties of Plants (UPOV Convention) 1961–1991[11] (establishing a *sui generis* system for new plant varieties) and the European Patent Convention (EPC) 1973 (establishing a centralized system for the granting of patents), have been created by specially convened diplomatic conferences. The London Agreement 2000[12] and Draft Agreement on the establishment of a European patent litigation system 2005—aimed at creating a European patent judiciary—are initiatives of the European Patent Organisation.

[5] Convention for the European Patent for the Common Market (Community Patent Convention) (Luxembourg, 15 December 1975, 15 ILM 5).

[6] Council Regulation (EC) 2100/94 of 27 July 1994 on Community plant variety rights [1994] OJ L227/1.

[7] Directive 98/44 of the European Parliament and of the Council of 6 July 1998 on the legal protection of biotechnological inventions (Biotech Directive) [1998] OJ L213/13.

[8] Council of the European Union, Working Document ('Draft Agreement on the European and Community Patents Court and Draft Statute') 5072/09 (8 January 2009).

[9] The Treaty of Lisbon amending the Treaty on European Union and the Treaty establishing the European Community (Lisbon, 13 December 2007).

[10] United Nations (UN) General Assembly resolution 217 A (III), Universal Declaration of Human Rights (Paris, 10 December 1948).

[11] International Convention for the Protection of New Varieties of Plants (UPOV Convention) (Paris, 2 December 1961).

[12] Agreement on the application of EPC, Art 65 [2001] OJ EPO 549.

Convention of 1883,[15] rest on the two fundamental principles of national (non-discriminatory) treatment and minimum standards of protection. According to the first of these, every country must accord citizens of other countries the same rights as it accords its own. According to the second, every country must recognize certain minimum standards of substantive protection in the subject area of the agreement. This latter principle in particular is intended to ensure that agreements among states represent a minimum intrusion on national sovereignty. While the standards of the Paris Convention have been revised upwards since the nineteenth century, it is nonetheless true that states today retain considerable discretion under them to fashion their own intellectual property policies and laws. The same can be said for non-WIPO agreements since the mid-twentieth century, such as the 1963

[15] Paris Convention for the Protection of Industrial Property (Paris Convention, as amended) (Paris, 20 March 1883, 13 UST 1).

Strasbourg Convention (SPC),[16] and to a lesser extent the TRIPS Agreement of 1994,[17] both of which are built on the 'minimum standards' model of harmonization.

Since the 1970s, international law-making outside WIPO has increased, particularly in Europe, where the goal of eradicating trade barriers between EU Member States has driven a reform agenda based in part on a closer harmonization of national laws than could ever be achieved through the likes of the Paris Convention. The main instrument used to that end has been the EU Directive. Importantly, it and other EU instruments do not depend on the agreement of Contracting States to take effect, and can be enforced by the Court of First Instance and the European Court of Justice (ECJ). To take effect in UK law, Directives need to be implemented legislatively, unlike Regulations, which take effect without national legislative action.[18] Failure to implement a Directive in an adequate manner may result in ECJ proceedings being issued by the Commission against the relevant state. Such factors, combined with the number of EU Directives and other instruments in the intellectual property field, make the EU the most important international legislative authority in this area.

In 1994, the goal of facilitating trade at the international level also led to the creation of the TRIPS Agreement as part of the Uruguay Round of the GATT. The TRIPS Agreement incorporated many of the provisions of the Paris Convention, as well as a host of other substantive protection provisions covering the spectrum of intellectual property rights. Its explicitly trade context and inclusion within the WTO enforcement and dispute resolution machinery differentiate it from the WIPO agreements, and even from the various EU Directives. While its impact on UK law has been less than that of the EU measures, it has been relied on directly by the UK courts to support certain principles of national jurisprudence.[19] Further, it is the source of the statement in Article 52(1) of the EPC that patents shall be available for inventions 'in all fields of technology'.[20]

[16] Convention on the Unification of Certain Points of Substantive Law on Patents for Inventions (Strasbourg Convention) (Strasbourg, 27 November 1963, ETS 47).

[17] Agreement on Trade-Related Aspects of Intellectual Property Rights (TRIPS Agreement) Marrakesh Agreement Establishing the World Trade Organization, Annex 1C, Legal Instruments—Results of the Uruguay Round, vol 31 (15 April 1994, 33 ILM 81).

[18] See Treaty Establishing the European Community [2002] OJ C325, Art 249.

[19] See, eg *Designers Guild Ltd v Russell Williams (Textiles) Ltd* [2000] UKHL 58, [23] per Lord Hoffmann; *Nova Productions Limited v Mazooma Games* [2007] EWCA Civ 219, [37]–[38], [68] per Jacob LJ.

[20] While PA, s 1 has not been amended to reflect this principle, the principle almost certainly applies in UK law given PA, s 130(7) and the courts' practice of deferring to the EPC in the case of differences between the PA and EPC texts, as in the context of EPC, Art 52(2) and (3) themselves. See, eg *Symbian v Comptroller General of Patents* [2008] EWCA Civ 1066, [6], discussed at pages 318–21.

2. The Paris Convention 1883 (WIPO): A Mechanism for Creating 'Special Agreements for the Protection of Industrial Property'

The starting point for more detailed consideration of patent law is the world's first multilateral intellectual property agreement, the Paris Convention 1883. Article 1 of that Convention creates a Union of Signatory States that at the time of writing comprises 173 countries, including all Members of the EU. Within that Union, the Convention establishes the principle of national treatment, a 12-month right of priority, and certain minimum standards of protection for 'industrial property'; a concept it defines 'in the broadest sense', as applying[21]

> not only to industry and commerce proper, but likewise to agricultural and extractive industries and to all manufactured or natural products, for example, wines, grain, tobacco leaf, fruit, cattle, minerals, mineral waters, beer, flowers, and flour.

Significantly for present purposes, the Convention also provides for the making between Union States of 'special agreements for the protection of industrial property, in so far as these agreements do not contravene [its] provisions'.[22] Further, 'the protection of industrial property' is expressly defined to include 'patents',[23] explaining why the EPC and its three main precursors—the Strasbourg Convention, the Community Patent Convention, and the Patent Cooperation Treaty[24]—were concluded as 'special agreements' within the framework of Paris.

3. The Strasbourg Convention 1963 (CofE): A Convention to Facilitate the Harmonization of Substantive Patent Law

The conclusion of the Strasbourg Convention in 1963 under the aegis of the Council of Europe (CofE) marked the first concrete achievement in the pursuit of a European patent system. The aim of that agreement, reflected in the terms of its Preamble, was to facilitate the harmonization of certain points of substantive patent law for the purposes of assisting European industry and contributing to the future creation of an international patent system. To that end, its Member States are required, among other things, to grant patents for (and only for) 'inventions which are susceptible of industrial application, which are new and which involve an inventive step'.[25] No definition of the 'invention' is included.

By the time of the Strasbourg Convention's adoption, work had also begun on a separate treaty with the more ambitious aim of establishing a Common Market patent for the six Member States of the then recently established European

[21] Paris Convention, Art 1(3).
[22] ibid, Art 19.
[23] ibid, Art 1(2).
[24] Patent Cooperation Treaty (Washington, 19 June 1970, 9 ILM 978).
[25] SPC, Art 1.

Economic Community (EEC).[26] The foundations of this system had been laid by the earlier work of the Council of Europe's Committee of Experts on Patents, and had the central purpose of abolishing the territorial nature of European patent law in order to facilitate the removal of trade barriers within the EEC, consistent with the obligations imposed by the EEC Treaty.[27] From 1959 the work was overseen by a dedicated Committee of Coordination, and from 1960 it became the task of a Patents Working Party headed by the then President of the German Patent Office, Dr Haertel. Within a year, Haertel's Working Party had produced a First Preliminary Draft Convention Relating to a European Patent Law, the first 'unofficial' English translation of which was published in 1963, accompanied by the following laudatory introduction:[28]

The Draft Convention relating to a European Patent Law is a *novum* in the history of patent legislation. One is amazed at its daring conception, original ideas, and breadth of field. The authors have attemp0ted to create:

1. a supranational patent system in conformity with the Paris Union,
2. a federal patent system coexisting with national systems which have been in force for at least half a century,
3. a patent system adapted to enforce the economic structure of the E.E.C,
4. a supranational patent office in a region where at least 10 languages prevail, all with historical backgrounds,
5. a compromise between the systems in which patents are granted without search and those which provide an examination as to patentability, and
6. a patent system holding a fair and equitable balance between the interests of the inventor and those of the public.

In its original form, the Common Market patent treaty sought to create a truly international patent system with harmonized granting procedures from which would emerge a patent valid in all EEC states,[29] and adopted to that end the definition of 'patentable invention' contained in the Strasbourg Convention.[30] In 1965, however, work on the treaty was shelved indefinitely for political reasons, including the rejection of the UK's application for EEC membership four years earlier.[31]

[26] The countries were France, the Federal Republic of Germany, Belgium, Italy, Holland, and Luxembourg.

[27] M van Empel, *The Granting of European Patents: Introduction to the Convention on the Grant of European Patents, Munich, October 5, 1973* (Leyden: AW Sjthoff, 1975) 14–16; Treaty Establishing the European Economic Community (EEC Treaty) (Rome, 25 March 1957, 37 ILM 56) Arts 2 and 3.

[28] G Oudemans, *The Draft European Patent Convention: A Commentary with English and French Texts* (London: Stevens & Sons, 1963) xi.

[29] See Committee of Experts, Article published in Europe's Marché commune (No 560 of 24 November 1959): 'The Six EEC Countries Plan to Establish a Uniform European Law on Industrial Property' EXP/Brev (59) 1 (3 December 1959).

[30] See Luxembourg Conference, Minutes of the meeting of Working Party I (Luxembourg 8–11 July 1969) BR/7/69 (29 July 1969) (1st Working Party I Meeting) [21].

[31] See generally DW Urwin, *The Community of Europe—A History of European Integration Since 1945* (2nd edn, London: Longman, 1995) 104ff.

It took another four years, and the impetus of separate preparations for the EPC's third precursor agreement, the Patent Cooperation Treaty, for 'the dormant plan for a European patent' to be 'awakened from its slumber'.[32]

4. The Patent Cooperation Treaty 1970 (WIPO): A Record of Patentability Exclusions

The Patent Cooperation Treaty (PCT) was a US initiative aimed at simplifying patent granting procedures worldwide, and resulted from a 1966 invitation to the Director of BIRPI to[33]

> undertake urgently a study on solutions tending to reduce the duplication of effort both for applicants and national Patent Offices, in consultation with outside experts to be invited by him and giving due regard to the efforts of other international organizations and groups of States to solve similar problems, with a view to making specific recommendations for further action, including the conclusion of special agreements within the framework of the Paris Union.

Concluded in 1970, the PCT provides a basis for the international harmonization and integration of national patent granting procedures, including procedures for the application, search, and examination stages of a grant.[34] It thus establishes an international system for the filing and examination of patent applications and the conduct of prior art searches that is administered by a network of national and regional patent offices acting as Receiving Offices, International Searching Authorities, and/or International Preliminary Examining Authorities. Its specific purpose is to help inventors obtain a patent for the same invention in two or more (PCT) countries by allowing them to assert their claim internationally and have it subjected to preliminary assessment before committing to one or more local applications. In this way it reduces the risk of wasted expenditure by enabling an applicant to postpone the decision to pursue an application to grant until confident of its success. By creating a series of pre-grant procedures with recognized validity in PCT countries, it also provides a basis for the harmonization of patentability standards.

The PCT has, at the time of writing, 139 members, and can be used by any national of a Member State seeking a patent in one or more PCT countries or

[32] Washington Diplomatic Conference on the Patent Cooperation Treaty (Washington, 25 May–19 June 1970), Verbatim and Summary Minutes, Verbatim Minutes of the First Plenary Meeting (25 May 1970) (First Plenary Meeting of the Washington Diplomatic Conference) [64.1]; see also [3.9], [95.4], [123.4]–[123.5] (reprinted in *Records of the Washington Diplomatic Conference on the Patent Cooperation Treaty 1970* (Geneva: WIPO, 1972)).

[33] Washington Diplomatic Conference on the Patent Cooperation Treaty (Washington, 25 May–19 June 1970), Post-Conference Documents, History of the Patent Cooperation Treaty, PCT/PCD/1 (16 October 1970) [3].

[34] G Paterson, *The European Patent System: The Law and Practice of the European Patent Convention* (2nd edn, London: Sweet & Maxwell, 2001) [1–34].

associated regions. It requires that an otherwise local application be commenced as an international application by (among other things) being filed with a Receiving Office and designating the countries or regions in which a patent is sought. Once filed, an international application is forwarded to an International Searching Authority which searches for evidence that the invention was known or technically obvious at its date of filing or other priority date. The results of the search are incorporated into two documents which the Authority prepares: a search report, produced for the public and made available with the application 18 months after the priority date or as soon as possible thereafter; and an opinion on the invention's patentability, produced for the applicant as a non-binding assessment of the likely success of the application.

While not dealing with inherent patentability, the PCT exempts International Searching and Preliminary Examining Authorities from the obligation to process applications directed to any of six specific categories of subject matter, chosen by reason of their exclusion from the patent regime of one or more PCT Signatory State, and in recognition of the infeasibility of having the authorities prepare reports for subject matter with which they have no (search and examination) experience.[35] The relevant provisions are Rules 39.1 and 67.1, drafted in substantially identical terms. Rule 39.1 provides as follows:

> No International Searching Authority shall be required to search an international application if, and to the extent to which, its subject matter is any of the following:
> (i) scientific and mathematical theories,
> (ii) plant or animal varieties or essentially biological processes for the production of plants and animals, other than microbiological processes and the products of such processes,
> (iii) schemes, rules or methods of doing business, performing purely mental acts or playing games,
> (iv) methods for treatment of the human or animal body by surgery or therapy, as well as diagnostic methods,
> (v) mere presentations of information, [and]
> (vi) computer programs to the extent that the International Searching Authority is not equipped to search prior Art concerning such programs.

The similarity of Article 52(2) of the EPC to this provision is striking. Of the categories of subject matter excluded by Rule 39.1, only two—'plant or animal varieties or essentially biological processes for the production of plants and animals, other than microbiological processes and the products of such processes' and

[35] This explains the justification offered by the Japanese delegation at Washington for its request to add therapeutic methods to the list, namely, that the threshold exclusion of such methods from Japanese patent law would otherwise prejudice the ability of the Japanese Patent Office to act as International Searching and Preliminary Examining Authorities for PCT purposes. See Washington Diplomatic Conference on the Patent Cooperation Treaty (Washington, 25 May–19 June 1970), Summary Minutes of the 16th Meeting of Main Committee I (3 June 1970) [1174]–[1175]. See also PCT International Search and Preliminary Examination Guidelines, PCT/GL/ISPE/1, [9.02].

'methods for treatment of the human or animal body by surgery or therapy, as well as diagnostic methods'—are absent from Article 52(2), albeit still excluded from the scope of European patentability,[36] with the same number of EPC exclusions—'discoveries' in sub-paragraph (a) and 'aesthetic creations' in sub-paragraph (b)—absent from Rule 39.1. These differences notwithstanding, a cursory comparison of the provisions suggests a relationship between them, raising the question of the nature and direction of that relationship.

That Article 52(2) and(3) were themselves influenced by Rule 39.1 is undeniable, as can be seen from the analysis of the EPC *travaux préparatoires* below. That the basis of that influence was a view of inherent patentability as concerned in part with the exclusion of non-searchable or non-examinable subject matter, is less clear. On one hand, that view was the explicit premise of the 1966 proposal of the US Senate Subcommittee to exclude computer programs from US patent laws[37] to which Rule 39.1 has been attributed by at least one commentator.[38] On the other, as an explanation of the EPC drafters' reliance on Rule 39.1 in formulating Article 52(2) and (3), it is not supported by the *travaux préparatoires*, which suggest on the contrary an acknowledgement by the drafters that modelling Article 52(2) and (3) on Rule 39.1 could only be justified on political grounds, 'as an encouragement towards national harmonisation'.[39]

What, though, of the converse influence of early drafts of inherent patentability provisions on Rule 39.1?

[36] With respect to plant and animal varieties, see SPC, Art 2(b) and EPC, Art 53(b). With respect to medical and veterinarian methods, see EPC, Art 53(c) (corresponding to EPC 1973, Art 52(4)).

[37] United States Senate Subcommittee on Patents, Trademarks, and Copyrights, '"To Promote the Progress of... Useful Arts" in an Age of Exploding Technology', Report of the President's Commission on the Patent System (17 November 1966) 90th Congress 1st Session SENATE, Doc No 5, 21 (supporting the exclusion of computer programs from patentability with reference to the difficulties and costs associated with searching their prior art, which, it was suggested, made 'the patenting of programs . . . tantamount to mere registration' conferring an 'all but nonexistent' presumptive validity). See to similar effect, United Kingdom, *The British Patent System: Report of the Committee to Examine the Patent System and Patent Law* (Banks Report) (Cmnd 4407, 1970) [483] discussed at page 182–6.

[38] See J Betten, 'Patent Protection for Computer Programs in Germany and by the EPO' [1987] EIPR 10, 10.

[39] Luxembourg Conference, Minutes of the 9th Meeting of Working Party I, Inter-Governmental Conference for the Setting up of a European System for the Grant of Patents (Luxembourg, 12–22 October 1971) BR/135/71 (17 November 1971) (9th Working Party I Meeting) [95] (responding to an argument by NGO representatives that Rules 39.1/67.1 provided insufficient justification for including a list of patentability exclusions in the EPC on account of their concern with search and examination rather than inherent patentability: see Luxembourg Conference, Minutes of the 5th Plenary Meeting (Part II: 'Hearing of the non-governmental international organizations on the Second Preliminary Draft of a Convention establishing a European System for the Grant of Patents Minutes') (Luxembourg, 26 January–1 February 1972) BR/169/72 (15 March 1972) (NGO Hearing) [11]).

As has been seen, in 1970 when the PCT was signed, a year after its penultimate draft had been agreed,[40] work on the EPC had already commenced. Indeed, the Council of Europe had been studying the Europeanization of patent law since 1949, and the EEC since 1959. It is therefore not surprising that, by the time of the disbanding of the EEC Patents Working Party in 1965, a definition of 'invention' already existed. Based on a joint German–Netherlands proposal from 1964,[41] that definition was as follows:[42]

> The following in particular shall not be regarded as inventions within the meaning of paragraph 1:
> (a) scientific knowledge and theories as such;
> (b) mere discovery of substances occurring in nature;
> (c) purely aesthetic creations;
> (d) financial or accounting methods, rules for playing games or other systems, insofar as they are of a purely abstract nature;
> (e) methods of therapy, including diagnostic methods.

Again, the similarity in the terms of this provision and Rule 39.1 of the PCT can only be explained with reference to derivation. That the PCT drafters were influenced by this text is not surprising. One reason was the identity of the drafters, some of whom were also members of the EEC Working Party.[43] Another was the concern of those drafters, reflected in the terms of the BIRPI invitation itself, to ensure the PCT's consistency with the harmonization initiatives of other regional bodies.[44] A third was the purpose of Rule 39.1, which, as a non-proscriptive record

[40] See PCT/DC/4, PCT/DC/5, republished with drafting comments in PCT/DC/11, PCT/DC/12, and forming the basis of the drafters' work at the Washington Diplomatic Conference; see First Plenary Meeting of the Washington Diplomatic Conference, [37].

[41] See Joint proposal by the German and Netherlands delegations on Art 9 regarding exceptions to the concept of invention, 9663/IV/64, referred to in Patents Working Party, Proceedings of the 15th meeting of the Patents Working Party held at Brussels from 19 to 26 October 1964, 11821/IV/64–E–Final (7 December 1964) (15th Working Party Meeting) 3.

[42] Patents Working Party, Amendments to the Preliminary Draft Convention relating to a European Patent Law (Arts 1 to 185) 2335/IV/65 (22 January 1965) ch I, Art 9(2).

[43] This included Mr van Benthem and the Working Party's Chairman, Dr Haertel, whose Vice-Chairmanship of Main Committee I at the Washington Diplomatic Conference on the Patent Cooperation Treaty ensured him a central role in formulating Rules 39.1 and 67.1. That he fulfilled that role with direct regard to European developments can be inferred from his statement at the conclusion of the Washington Diplomatic Conference, as follows: 'The successful conclusion of the Patent Cooperation Treaty today in Washington will give new impetus to the preparations for a European patent system, on which 17 European states are at present working and the aim of which is the creation of a European patent. We are convinced that the PCT and the proposed European patent system are not only compatible with each other but will successfully complement each other. . . . The German Delegation will sign the Patent Cooperation Treaty here today in the hope that it will be another step towards a better understanding among nations' (First Plenary Meeting of the Washington Diplomatic Conference, [123.4]–[123.5]).

[44] See, eg First Plenary Meeting of the Washington Conference, [3.9]–[3.10], [54.2]–[54.3], [64.1], [95.4], [156.1]. This concern was in recognition of the PCT's significance as 'a first step toward far-reaching international harmonization of patent law and procedure' (First Plenary Meeting of the Washington Conference, [3.10]).

of subject matter excluded from the systems of Signatory States,[45] could only have been based on a record of existing patentability exclusions, which Article 9(2) was believed by its drafters to be.[46]

5. The Community Patent Convention 1975 (EEC): An Attempt to Create a Unitary Patent

By 1969, substantial progress on the PCT, combined with the envisaged enlargement of the EEC to include states such as the UK not directly involved in the earlier work on the Common Market patent, had had two effects.[47] First, it had created a perceived need within the EEC and Europe generally for a treaty that would realise the PCT's objective of creating a harmonized patent granting procedure and centralized examination system without overriding national laws. And second, it had led to a decision to pursue that treaty independently of the envisaged Common Market patent, and to restrict the scope of the latter accordingly.[48] The result was an agreement in 1969 to develop two conventions in parallel: the EPC, aimed at creating an international system of examination, search, and grant that would confer on an applicant a bundle of national patents the (post-grant) validity of which would be a matter for Member States' national laws; and the CPC, aimed at creating a single, supranational patent having effect in all EEC states with a view to achieving for the post-grant life of a patent what would be achieved pre-grant by the EPC. The EPC was signed in Munich on 5 October 1973 and entered into force in Belgium, Germany, France, Luxembourg, the Netherlands, Switzerland, and the UK on 7 October 1977. The CPC was signed two years later in 1975, but, having been ratified by only one Signatory State, is unlikely to take effect.[49]

[45] Rather than, eg a record of subject matter deemed by the PCT drafters to be worthy or unworthy of such exclusion per se. See PCT International Search and Preliminary Examination Guidelines, PCT/GL/ISPE/1, [9.02].

[46] The non-prescriptive nature of rr 39.1/67.1 has been advanced as a reason for the near absence of references in the historical records to their purpose and provenance. See, eg Statement by E Schumm, Head, PCT Information Services Section, PCT Legal Division (Personal email correspondence, 5 January 2006). On the nature of the list as a declaration of subject matter excluded from the patent laws of European States, cf Committee of Experts, Memorandum on the unification of legislation (Item 4 of the Agenda for the meeting of 28th November 1960) by Mr Finniss, Rapporteur-General, EXP/Brev (60) 7 (28 November 1960) 5.

[47] Paterson, *The European Patent System: The Law and Practice of the European Patent Convention*, [1–39]–[1–42].

[48] See V Vossius (tr) and K Haertel, *European Patent Convention* (Cologne: Verlag, 1980) 2–3.

[49] Report from the Select Committee on European Communities, 'The Community Patent and the Patent System in Europe' HL (1998) 26, [1]. The CPC has been regarded nonetheless as a symbol of its signatories' commitment to a unitary patent. See, eg *Gale's Application* [1991] RPC 305, CA, 322–323 per Nicholls LJ.

6. The European Patent Convention 1973 (Munich Conference): A System of Law for the Grant of European Patents

Article 1 of the EPC states that its purpose is to establish 'a system of law, common to the Contracting States, for the grant of patents for inventions', and also to establish an organization, the European Patent Organisation, to administer it. A patent granted under the EPC is called a European patent, but takes effect as a bundle of national patents under the laws of the Contracting States in which protection is sought. The conditions of the grant of such a patent are contained in Articles 52 to 57, and mirror the provisions of the SPC in requiring an invention that is novel, inventive, and industrially applicable.

The rules that govern interpretation of the EPC have already been touched on. Most importantly for common law lawyers, principles of precedent are not applicable within the EPO; and indeed, the only sources of law binding on the EPO are the text of the EPC and its Implementing Regulations.[50] Thus, there is no express obligation on individual Boards to follow the decisions of previous Boards when deciding matters of law, nor to give reasons for their decisions; a point emphasized in *DUNS LICENSING ASSOCIATES/Estimating Sales Activity*, as follows:[51]

> [T]he legal system of the European Patent Convention gives room for evolution of the jurisprudence (which is thus not 'case law' in the strict Anglo-Saxon meaning of the term) and leaves it to the discretion of the boards whether to give reasons in any decision deviating from other decisions or to refer a point of law to the Enlarged Board.

Similarly for the Examining Division and other first instance departments, for which the 'ultimate authority on practice' is the text of the EPC, followed by the interpretive decisions of the Boards of Appeal.[52] While the EPO publishes detailed guidelines for its examiners, those guidelines are of persuasive authority only.

Despite the supremacy of the EPC and the absence of any doctrine of precedent governing its application, the Boards of Appeal have been described by Paterson as observing an 'unofficial system of precedent' based on the principle of equality that like cases should be treated alike.[53] As he notes, this is consistent with the requirement in Article 20 of the Rules of Procedure of the Boards of Appeal that a Board, upon considering it necessary to deviate from an interpretation or explanation of the EPC given in an earlier decision of another Board, should state the grounds

[50] See EPC, Art 23(3); Rules of Procedure of the Boards of Appeal [2007] OJ EPO 536, Art 20(1).

[51] T_154/04 (*DUNS LICENSING ASSOCIATES/Estimating Sales Activity*) [2007] EPOR 38, [2].

[52] See T_162/82 (*SIGMA/Classifying Areas*) [1987] OJ EPO 533, and the General Introduction to the EPO Guidelines for Examiners, discussed in Paterson, *The European Patent System: The Law and Practice of the European Patent Convention*, [1–07]–[1–08].

[53] Paterson, *The European Patent System: The Law and Practice of the European Patent Convention*, [2–129].

for such deviation unless those grounds are consistent with an earlier opinion or decision of the Enlarged Board of Appeal.[54] As with all provisions of the Rules of Procedure, however, that Article is subject to the supremacy of the EPC.[55]

As has been noted, the EPC is concerned only with the pre-grant phase of a patent application; once a patent is granted it is governed entirely by the relevant national law, although post-grant opposition proceedings are available,[56] and a European patent may only be revoked on a ground contained in Article 138 of the EPC.[57] In addition, the risk of conflicting interpretations of identical EPC provisions is reduced in UK law by sections 125 and 130(7) of the Patents Act. Section 125 mirrors Article 69 of the EPC regarding the extent of the invention, and confirms the applicability of the EPC's Protocol on the Interpretation of Article 69 regarding the scope of protection conferred by a patent. Section 130(7) was transposed directly from the Resolution on the Adjustment of National Patent Law annexed to the CPC, and provides as follows:

> Whereas by a resolution made on the signature of the Community Patent Convention the governments of the Member States of the European Economic Community resolved to adjust their laws relating to patents so as (among other things) to bring those laws into conformity with the corresponding provisions of the European Patent Convention, the Community Patent Convention and the Patent Co-operation Treaty, it is hereby declared that the following provisions of this Act, that is to say, sections 1(1) to (4), 2 to 6, 14(3), (5) and (6), 37(5), 54, 60, 69, 72(1) and (2), 74(4), 82, 83, 100 and 125, are so framed as to have, as nearly as practicable, the same effects in the United Kingdom as the corresponding provisions of the European Patent Convention, the Community Patent Convention and the Patent Co-operation Treaty have in the territories to which those Conventions apply.

In 2000, the EPC was revised in an attempt to modernize European patent law to take account of developments in international law, to satisfy the needs of users, to eliminate unnecessary requirements, and to introduce flexibility into the Convention.[58] Those revisions took effect in December 2007, and entailed a variety of organizational, jurisdictional, and substantive changes, including, of greatest relevance here, amendment of Article 52(1) to specify the availability of European patents for inventions 'in all fields of technology', consistent with Article 27.1 of the TRIPS Agreement.

[54] ibid. See also Rules of Procedure of the Boards of Appeal [2007] OJ EPO 536, Art 20(2) (making similar provision for decisions that deviate from the EPC Guidelines for Examiners).

[55] Paterson, *The European Patent System: The Law and Practice of the European Patent Convention*, [2–129].

[56] EPC, Part V.

[57] See PA, s 72.

[58] See European Patent Office, Diplomatic Conference for the Revision of the EPC, available at <http://www.epo.org/patents/law/legal-texts/archive/epc2000/diplomatic-conference-nov2000.html>. See also European Patent Organisation, Notice from the President of the EPO, 24 March 2000 (2000) 4 OJ EPO 173; Act Revising the Convention on the Grant of European Patents (European Patent Convention) of 5 October 1973, last revised on 17 December 1991 (Munich, 29 November 2000).

7. The TRIPS Agreement 1994 (WTO): The Principle of Patents for Inventions in All Fields of Technology

Like the Paris Convention, the TRIPS Agreement contains national treatment provisions and sets minimum standards of substantive protection, including some from the Paris Convention. Perhaps most importantly, it contains a dispute resolution mechanism, with the result that there is now an international forum for resolving disputes concerning those and other TRIPS provisions, including Article 27.1. Aside from guaranteeing the availability of European patents 'in all fields of technology', that Article prohibits discrimination on the grounds (among others) of technological field:

> Subject to the provisions of paragraphs 2 and 3, patents shall be available for any inventions, whether products or processes, in all fields of technology, provided that they are new, involve an inventive step and are capable of industrial application. Subject to paragraph 4 of Article 65, paragraph 8 of Article 70 and paragraph 3 of this Article, patents shall be available and patent rights enjoyable without discrimination as to the place of invention, the field of technology and whether products are imported or locally produced.[59]

On its conclusion in 1994, the TRIPS Agreement was approved by a decision of the EC Council.[60]

8. Ongoing Attempts to Create a Community Patent

(a) EEC Initiatives

In recognition of the unlikelihood of the Community Patent Convention (CPC) taking effect, the EEC in 1989 introduced the 89/695/EEC Agreement relating to Community patents for the purpose of bringing the CPC and certain supporting instruments—the Patent Litigation Protocol, the Common Appeal Court Statute, and the Common Appeal Court Privileges Protocol—into force. Among other things, the Community Patent Agreement provides for the creation of a Community Patent Appeal Court to hear referrals on questions of infringement and validity from designated national courts, and to hear appeals from the Revocation Divisions and Patent Administration Division of the EPO. However, the EEC's Community Patent Agreement has been ratified by only 12 Signatory States, and is therefore also unlikely to come into force.

[59] TRIPS, Art 27.1.

[60] See Council Decision (of 22 December 1994) concerning the conclusion on behalf of the European Community, as regards matters within its competence, of the agreements reached in the Uruguay Round multilateral negotiations (1986–1994) [1994] OJ L336/1.

(b) EU (and European Patent Organisation) Initiatives

In the 1990s, the EU renewed its own efforts to introduce a Community patent, following the failure of the EEC's Community Patent Agreement. This resulted in the introduction of Commission Proposal (EU) 2000/412 for a Council Regulation on the Community Patent.[61] Under that proposal, Community patents would be granted by the EPO (following the EU's accession to the EPC) in either English, German, or French, and later published with a translation in the other two languages. A Community Intellectual Property Court would also be established with exclusive jurisdiction to determine issues of validity and infringement.

In 2003, two proposals were put forward for the creation of such a court. One[62] was to confer jurisdiction in respect of certain Community patent disputes on the ECJ. The other[63] was to establish a Community Patent Court with judges appointed by the Council to exercise the ECJ's jurisdiction on its behalf, a specialized chamber within the Court of First Instance to hear appeals from that Court, and a mechanism for review by the ECJ. In 2004, however, both proposals faltered, following the failure of the EU Competitiveness Council to agree on certain details regarding the first proposal,[64] and a decision by the ECJ criticizing the second proposal on a range of technical, substantive, and institutional grounds.[65]

Nevertheless, work towards a Community patent continues. In May 2008, for example, the Council Presidency submitted a Draft Agreement on the European and Community Patents Court to the Working Party on Intellectual Property (Patents), followed in January 2009 by an amended version.[66] This Agreement differs from the 2003 initiatives in proposing a Court of First Instance with central and local/regional divisions, a Court of Appeal, and a mechanism for limited review by the ECJ. In March 2009, the European Union also sought negotiating directives from the Council to conclude an agreement creating a Unified Patent Litigation System, and two months later EU ministers agreed to ask the ECJ to provide an opinion on the legality of the System. And as this book was going to press, on 4 December 2009, the Council adopted conclusions on an enhanced patent system in Europe, including an approach to Commission Proposal (EC) 2000/412 above.

[61] [2000] OJ C337 E/278.

[62] Proposal for a Council Decision conferring jurisdiction on the Court of Justice in disputes relating to the Community patent, COM (2003) 827 final.

[63] Proposal for a Council Decision establishing the Community Patent Court and concerning appeals before the Court of First Instance, COM (2003) 828 final.

[64] See 'Results of the Competitiveness Council of Ministers, Brussels, 11th March 2004 Internal Market, Enterprise and Consumer Protection Issues', MEMO/04/58 (12 March 2004).

[65] See Proposal for a Council Decision establishing the Community Patent Court and concerning appeals before the Court of First Instance, Opinion of the Court of Justice of the European Communities (8 November 2004) Document No 14349/04.

[66] See Council of the European Union, Working Document ('Draft Agreement on the European and Community Patents Court and Draft Statute') 5072/09 (8 January 2009).

Finally, while other important steps towards the creation of a European Patent Court have been taken by the European Patent Organisation,[67] they have recently been 'paused' in light of the EU initiatives above.[68]

9. Conclusion

As can be seen from this discussion, since the 1960s, patent law has been the focus of substantial European and international harmonizing activity. In Europe, most of that activity has been undertaken by the EU in an effort to create a unitary European patent and litigation system that would either bring the EPC regime within EU jurisdiction or render it redundant. For the moment, however, the realization of that aim seems a distance away, leaving as the principal instruments for European nations the SPC and EPC, the latter of which transposes the provisions of the EU Biotech Directive.[69]

C. The European Drafters' Intent Regarding the Requirement for an Invention

Signed in 1963 and 1973 respectively, the EPC and SPC were the result of extensive work undertaken over a 22-year period by two expert committees and two diplomatic conferences: the Council of Europe's Committee of Experts on Patents (1951 to 1969), the EEC Patents Working Party (1961 to 1965), the Inter-Governmental Conference for the Setting Up of a European System for the Grant of Patents (Luxembourg Conference) (1969 to 1972), and the Munich Diplomatic Conference for the Setting up of a European System for the Grant of Patents (Munich Conference) (1973). Thus, preparatory work on the creation of a European patent system can be divided into three (partly overlapping) stages. The first includes the work of the Committee of Experts on Patents on the SPC and other European patent conventions[70] from its earliest discussions in 1951 to its conclusion of the SPC in 1963. The second includes the work of the EEC Patents Working Party on the Common Market patent from its earliest discussions in 1961 to its disbandment in 1965. And the third includes the work of the wider

[67] See, eg the Agreement on the Application of Article 65 EPC (the London Agreement 2000) which came into force on 1 March 2008, and the Draft European Patent Litigation Agreement 2005, which, if implemented, would establish an independent (non-EU) European patent judiciary with responsibility for patent enforcement and validity, comprising regional first instance divisions and a central appellate body, the European Patent Court.

[68] See <http://www.epo.org/patents/law/legislative-initiatives/epla.html>.

[69] See Implementing Regulations to the Convention on the Grant of European Patents, ch V.

[70] See European Convention relating to the Formalities required for Patent Applications (Paris, 11 December 1953, CETS No 016); European Convention on the International Classification of Patents for Inventions (Paris, 19 December 1954, CETS No 017).

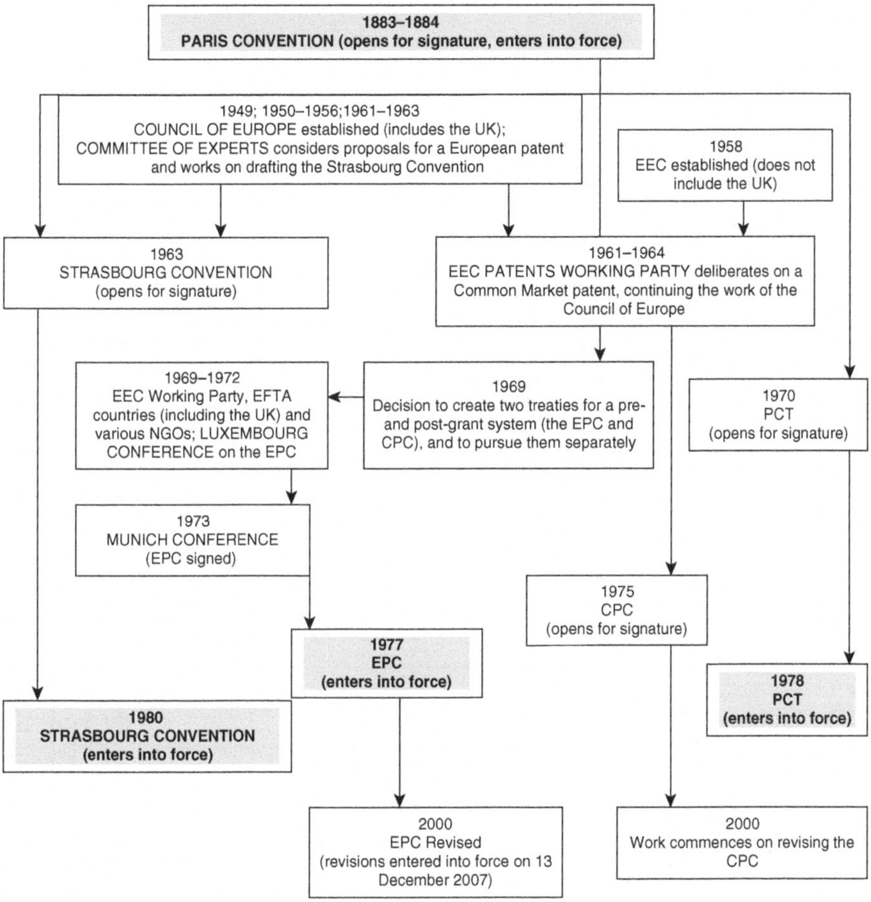

Diagram Bi. The EPC in its historical and international context

collection of states and organizations on the EPC itself from their convening at Luxembourg in 1969 to their conclusion of the EPC in 1973.[71] Strictly speaking, it is the record of deliberations from this third and final stage that is properly regarded as the EPC *travaux*. However, such is the debt of the EPC drafters to

[71] The second and third stages were referred to at the time as the 'EEC' and 'great European' phases respectively. See JB van Benthem, 'The European Patent System and European Integration' (translation of a revised version of the Romauld Singer Memorial Lecture held at the Patent Forum 1992 in Munich) <http://www.suepo.org/public/docs/2001/vanben_en.htm> [II] (quoting Haertel).

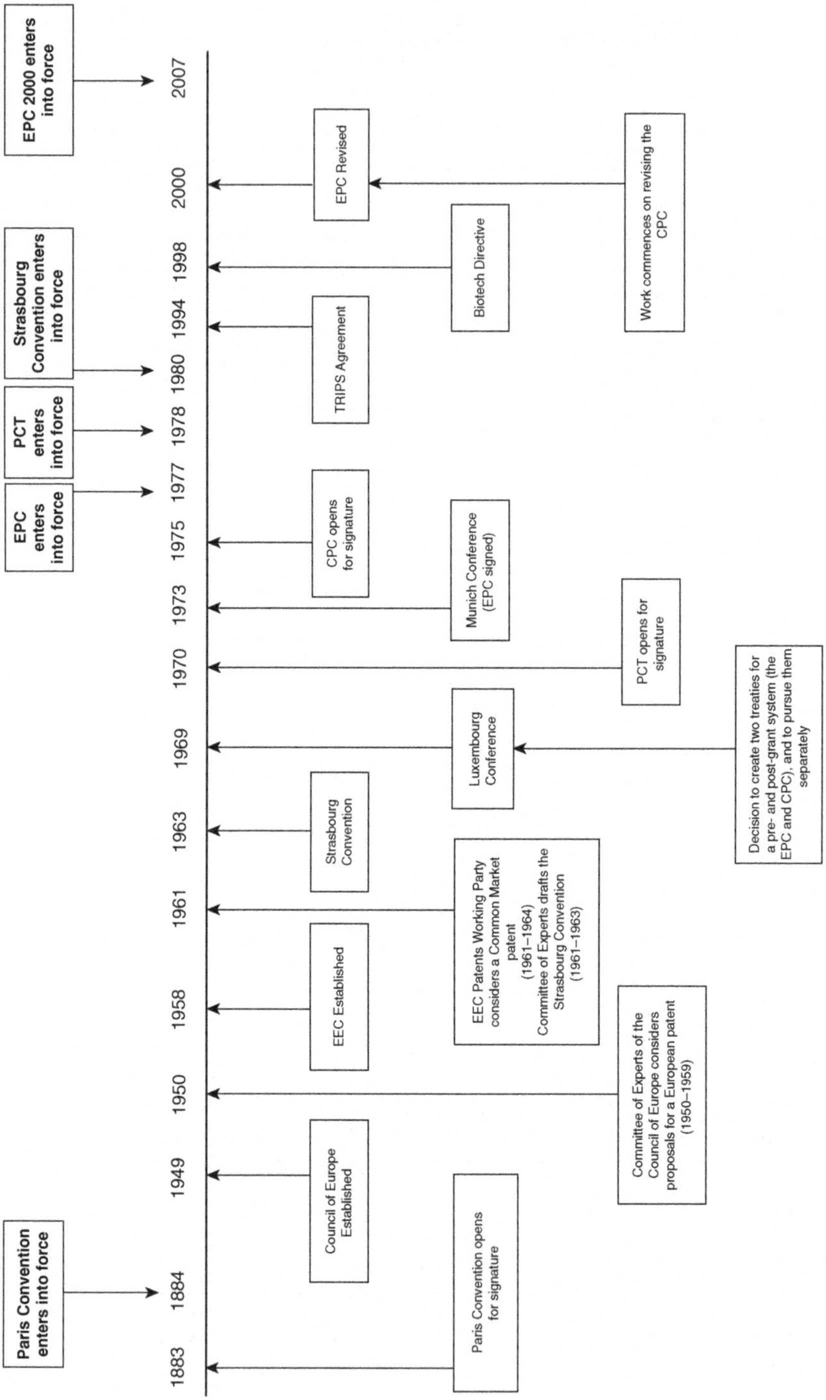

Diagram Bii. The EPC in its historical and international context

their predecessors' work[72] that it is appropriate to begin a study of their Conferences with a study of the deliberations during the first and second of the stages above.

1. The European Committee of Experts on Patents, 1951 to 1963

The original proposal to the Council of Europe for a European patent system was made by Senator Longchambon of France in 1949, and was aimed at unifying the conditions and formalities for obtaining a patent.[73] Conscious of the failure of past harmonization attempts, and concerned to protect the sovereignty of states, it eschewed unification as such, supporting instead an optional system for obtaining a European Certificate of Invention valid in all Contracting States, at least with respect to 'its form and its newness'.[74] According to the proposal, certificates would be granted by a European patent office applying criteria sufficiently wide that substantial reform would not be required.[75]

The Longchambon proposal was sceptically met.[76] For example, some believed a European patent office could not be created without unifying national patentability standards, including with respect to the requirement for inventions.[77] On the other hand, some did not believe unification would be possible, at least without a European patent jurisprudence,[78] and perhaps also a European property law.[79]

[72] See Comments on the Patents Working Party First Preliminary Draft Convention relating to a European patent law of 14 March 1961 (Arts 11 to 29) IV/2071/61, Section 14 (undated) 2.

[73] 'Creation of a European Patents Office: Report proposed by M Longchambon on behalf of the Committee on Economic Questions' (see n 3).

[74] See Consultative Assembly of the Council of Europe, First Ordinary Session, Sixteenth sitting, 6 September 1949, Doc 110 ('Creation of a European Patents Office: Recommendation to the Committee of Ministers adopted 8th September 1949 on the conclusion of the Debates') Art 1.

[75] See Consultative Assembly of the Council of Europe, Doc 75 Appendix I ('Introductory Note to the Study of the Problem of the Creation of a European Patents Office').

[76] See generally Consultative Assembly of the Council of Europe, Observations of Governments or of Experts on the Draft Proposal for a Convention on the Creation of a European Patent Office proposed by the Consultative Assembly of the Council of Europe, SG/D (50) 3 (20 July 1950).

[77] See, eg Letter from the Norwegian Ministry of Foreign Affairs, SG/D (50) 3, 49 (23 May 1950); see also Consultative Assembly of the Council of Europe, Report of the Netherlands Experts, SG/D (50) 3, 47 (4 May 1950); Remarks by the Luxembourg Official Expert on the draft on the possible creation of a European Patents Office, SG/D (50) 3, 89 (undated).

[78] See, eg Memorandum from the Swedish Government on the Draft Proposal for a Convention on the Creation of a European patent Office, SG/D (50) 3, 64 (undated).

[79] See United Kingdom, Proposal for Establishment of a Western European Patent Office, SG/D (50) 3, 51 (23 May 1950).

Nonetheless, work on the Longchambon proposal commenced: in 1950 a Committee of Experts was formed to study the proposal,[80] and in 1951 it convened a Working Party to undertake, among other things[81]

> a comparative study of the laws and regulations governing industrial rights in member states and, more particularly of the [illegible word] and juridical decisions on the subject of novelty and patentability and, when circumstances permit, to propose measures of unification.

By June 1951, the Secretariat-General of the Committee, Mr Gajac of France, had made a comparative study of the laws. Noting their divergent conceptions of 'patentability', he suggested that a rule might nonetheless be formed on the requirement for an object having a 'patentable nature',[82]

> that is to say referring:
> - to its character as a real invention (which excludes discovery of an existing phenomenon)
> - to its 'industrial' character (patents for methods and principles)
> - to its conformity with the requirements of public order . . . which excludes the patentability of inventions contrary to law or to morality or again to inventions for 'materials' (chemical substances, foodstuffs, pharmaceutical products) or vegetable novelties.

By contrast, the Rapporteur-General of the Committee, Mr Finniss of France, wrote in a report dated the same month that 'the most essential concepts are far from being unified', with agreement 'even lacking on the notion of the invention'.[83]

Also dated June 1951 is a proposal by Professor Reimer of Germany for a method of unifying national laws, involving the following five central steps.[84] First, the Rapporteur-General would identify the areas for comparative study and draft a questionnaire around them. Second, states would complete the questionnaire, taking account of all aspects of their law, administration, jurisprudence, and

[80] See Consultative Assembly of the Council of Europe, Report on a European Patent Law (Mr A Borels) (15 January 1964) II ('Explanatory Memorandum') [2].

[81] See Committee of Experts, Resolution, CM/WP IV (51) 4 (16 January 1951) 2. The Working Party was to consist of two experts from each of the group Norway, Denmark, and Sweden, the group Belgium, Luxembourg, and Holland, the group Greece, Turkey, Italy, Germany, Great Britain, France, and Ireland, and its first meeting was scheduled for March 1951. See CM/WP IV (51) 4, 2.

[82] See Committee of Experts, Note by the Secretariat General on the Functions and Powers of a European Patents Office, CM/WP IV (51) 8 (18 June 1951) 15–16.

[83] See Committee of Experts, Report on the questions inscribed on the Agenda for the 2nd session of the Committee of Experts presented by MG Finniss, Rapporteur-General, CM/WP IV (51) 11 (25 June 1951) 4.

[84] See Memorandum concerning the establishment of a European Patents Office presented to the Committee of Experts on Patents by the Experts of the Federal Republic of Germany, CM/WP IV (51) 25 (June 1951).

plans for future reform. Third, the Rapporteur-General would present proposals for unification based on responses to the questionnaire. Fourth, delegations would consult with industry and inventors before determining their views with respect to the proposals. And fifth, a plenary session would be convened to agree the basis for a future convention. While this proposal was not immediately taken up, it was clear by July 1951 that the Committee regarded unification as an essential part of its original task.[85] In a meeting that month, it agreed the following substantive points:[86] (a) European patents ought not to be granted for inventions contrary to law, *ordre public* or morality; (b) while some form of (patent or non-patent) protection ought to be available for plants, a unified position with respect to that issue would be difficult to obtain; and (c) further study was required of the patentability of pharmaceuticals, chemical products, and manufactured foodstuffs, on which European States were particularly divided. It also affirmed its commitment to practical solutions, rather than intellectual or juridical ones.[87] Consistent with this, Gajac proposed a liberal rule of patentability in conjunction with a compulsory licensing system (to resolve the 'inconveniences' resulting from that rule), as a means of achieving unification while avoiding the problems of definition that a more precise definition of patentability would entail.[88] As expressed in the minutes of one Committee meeting, unification needed to be attempted, but could 'not become a pretext for inaction'.[89]

Despite this, when discussions on the patenting of products stalled, the Committee of Experts reported to the Council that the Longchambon proposal was not capable of being immediately realized.[90] Affirming its commitment to strengthened European cooperation with a view to eventual unification, the Committee proposed a program of extensive preparatory work, undertaken by a sub-committee (Bureau) of its experts.[91] That proposal was accepted, and the Bureau set to work, following the Reimer method above and beginning with a

[85] See, eg Committee of Experts, Report on the Creation of a European Patents Office presented by MJC de Haan, Vice-Chairman of the Committee of Experts, CM/WP IV (51) 26 (30 November 1951) 2.

[86] See Committee of Experts, Draft Official Report of the meeting of 5th July 1951, CM/WP IV (51) PV7 (6 July 1951) 4–7.

[87] Committee of Experts, Draft Official Report of the meeting of 2nd July 1951, CM/WP IV (51) PV4 (4 July 1951) 3.

[88] Committee of Experts, Criteria of Novelty and Patentability, CM/WP IV (51) 9 (undated) ('Annex containing note by the secretariat-general') 64.

[89] CM/WP IV (51) PV4, 4.

[90] Committee of Ministers, 12th Session, Report to the Committee of Ministers presented by the Committee of Experts on Patents, CM/12 (52) 89 (12 July 1952) 2.

[91] CM/12 (52) 89, 2, 6.

national questionnaire.[92] While no record of the questionnaire remains extant, a note of its headings reads as follows:[93]

A – PATENTABLE INVENTION

- I. Limits of application of the law (industry, agriculture)
- II. General elements of patentability
  1. Novelty

  ...

  2. Technical effect (practicability – accordance with natural laws)
  3. Subject matter
  4. Sufficiency of the description
- III. Influence of prior patent rights (national or foreigner) and prior claiming on the patentability.
- IV. Exceptions to patentability
  1. Substances, especially chemical substances
  2. Foodstuffs, including tobacco and other luxuries
  3. Medicines and medical appliances
  4. Further exceptions, including inventions contrary to law and morality.

The first of these headings—'Limits of application of the law (industry, agriculture),' alluding to the framework of the Paris Convention and its definition of patents as industrial property[94]—is the most important for present purposes, as it is the heading under which inherent patentability was discussed. Summarizing that discussion in 1953,[95] Gajac noted the diverse understandings among European States of the nature of the so-called 'industrial invention'.[96] Given this, he proposed that the Committee not 'attempt to group under general headings the various exclusions' recognized in Europe, but stick instead to identifying their 'common features'.[97] In his study, he summarized those features under five headings—(a) invention and discovery; (b) industry and agriculture; (c) systems, methods, etc; (d) scientific principles and theories; and (e) creations of form—noting the agreement among states that the following could not constitute an 'invention':[98]

- discoveries;
- systems of finance, insurance, accounting, calculating, teaching, publicity, games, and medical treatment (in contrast to the products or apparatus those

[92] See Committee of Ministers, Eighth Meeting of the Ministers' Deputies (Paris, 18th December, 1952): Report to the Committee of Ministers submitted by the Bureau of the Committee of Experts on Patents, CM/12 (52) 149 (17 November 1952) 4–5.

[93] See Committee of Experts, Questionnaire drawn up by the Bureau of the Committee of Experts of the Council of Europe, EXP/Brev (53) 3 (15 January 1953) 2.

[94] See Paris Convention, Art 1.

[95] See Committee of Experts, Comparative Study of Substantive Law in Force in the Countries Represented on the Committee of Experts on Patents presented by the Secretariat-General, EXP/Brev (53) 18 (7 November 1953) 3–7.

[96] ibid, 3.

[97] ibid.

[98] ibid, 4–5 (footnotes omitted).

systems employed, which fell within the boundaries of industry and were (therefore) patentable);
- scientific principles and theories (in contrast to their industrial applications); and
- creations of form (which were protected as designs and industrial models).

Following Gajac's study, Reimer prepared a reform proposal, including the text of a draft convention,[99] with the following patentability provision:[100] 'The object of the application for a European patent must belong to the domain of technology, must be new and must have a quantum of invention.' On the meaning of technology he stated as follows:[101]

It was proposed at Strasbourg to try to define as far as possible the notion of technology, and it was suggested that a definition based on Article 1 Section 3 of the Paris Convention of the Union might be sought. More ample reflection on this matter shows that it is better to maintain the wording of Article 3 Section 1 in its present state. It has already been shown . . . in the remarks relating to Article 3 that the notion of technology is difficult to define completely. Article 1 Section 3 of the Paris Convention of the Union does not define it but only what must be understood by the phrase 'industrial property'. It is therefore not possible to deduce therefrom a definition of technology. Given the difficulty of satisfactorily defining the idea and the diversity of views on the matter in the different countries it would be better to leave each of the examining offices free to assess the matter in its own way. If it should happen that the examining office in one country has granted, in accordance with its own ideas, a European patent which in the view of another country does not belong to the field of technology, Article 6 Section 6 reserves to the second country the right to refuse protection.

Thus, 'the difficulty of satisfactorily defining the idea and . . . diversity of views' on the notion of technology, in combination with his 'minimum' approach to harmonization,[102] led Reimer to oppose the definition or unification of that notion.

[99] See Committee of Experts, Observations on the proposed European Convention on Patents proposed during the meeting of the Bureau of the Committee of Experts held at Strasbourg from the 2nd to 6th February, 1954. Presented by M Reimer, EXP/Brev B (54) 2 (10 March 1954); and for the amended version, Committee of Experts, Proposal for the institution of a European patent with the possible creation of a European Court of Justice. Made by M Reimer, EXP/Brev B (55) 1 (14 February 1955). The Netherlands expert, Mr de Haan, was also asked to prepare a proposal for a report based on Gajac's study, though it seems from the records that he did not. (See Committee of Experts, The Programme for the Committee of Experts on Patents, statement presented by the Scandinavian Experts, Stockholm, September 28, 1955, EXP/Brev (55) 3 (3 October 1955) 3.) He did however prepare one in 1951; (see CM/WP IV (51) 26).

[100] EXP/Brev B (54) 2, 3 (Art 3, s 1). While this particular record of Reimer's provision refers to the requirement for a subject matter having 'the character of an invention', rather than one having 'a quantum of invention', this seems to be an error; other records of the provision refer instead to a requirement for 'a quantum of invention' (see, eg Committee of Experts, Draft of a European Convention relating to Patents of Invention, EXP/Brev (53) 19 (10 November 1953) 4).

[101] EXP/Brev B (54) 2, 4. See also EXP/Brev (53) 19, 4 ('The notion of technology is difficult to define completely. It will not be necessary to define it here: the task of dealing with this matter can be left to the practice of each examining office.')

[102] On this approach, see pages 168–169.

However, the following year he reversed his position. Commenting on an almost identical draft of the provision above, he suggested that '[s]o far as the [domain of technology] is concerned, the interested countries will easily reach agreement in view of the fact that they have always only protected inventions in the field of technology.'[103] While the reason for this reversal is not expressly recorded, there is a record of its context by Reimer himself.[104] That context was the decision of the Committee in 1955 that legal unification could not be achieved, at least in a substantive ('juridical') sense, and that the Committee ought therefore to limit its agenda to creating a European system for granting patents.[105] Consistent with this, it resolved to unify only those aspects of substantive law as would need to be unified to that end.[106] In identifying those aspects, the Committee drew on the headings of Gajac's study to derive the following patentability conditions: industrial character, novelty, technical progress, creative effort, and influence of prior patent rights.[107] It then invited observations from the national experts on the task of unifying each of these conditions. Again, it was under 'industrial character' that inherent patentability was discussed, in replies which may be briefly summarized as follows:

- The Scandinavian experts[108] understood inventions to have a technical character and produce a technical effect, in contrast to the following ('conventionally' unpatentable) objects: 'the treatment of the human body, methods of training [and] such novelties of an intellectual nature (methods of education of a non-technical nature, schemes of financing etc.)'.[109]
- According to the Greek experts,[110] an invention was a subject matter 'aimed at making a material object or industrial product, or at bringing about a new technical result by the application of well known scientific laws', and involving 'a definite creative effort'.[111] By contrast, a discovery comprised a revelation of 'an existing but hitherto unknown product, scientific law, phenomenon

[103] See EXP/Brev B (55) 1, 4.

[104] See ibid, 1–2.

[105] See EXP/Brev B (55) PV1 Final, 1–2.

[106] See ibid; also ibid, ('M Reimer stated that, on reflection, he wished to go back on his opinion that his second proposition did not presuppose any juridical unification').

[107] See Committee of Ministers, Report to the Committee of Experts on Patents to the Committee of Ministers (on the meeting at Strasbourg from the 3rd to the 7th October 1955) CM (55) 137 (7 October 1953) [20]–[21], Appendix I; see also EXP/Brev B (55) PV1 Final, 2.

[108] Committee of Experts, Observations of the Scandinavian Experts: Note on the question of the role and scope of descriptions and claims by the Scandinavian Experts, EXP/Brev (56) 1 (14 April 1956).

[109] ibid, 5–6.

[110] See Committee of Experts, Report by Mr Nicolas Vouropoulos, Greek Expert, on the questions relating to the unification of European patent law selected by the Committee of Experts in October 1955, EXP/Brev (60) 3 (16 August 1960).

[111] ibid, 2.

or set of conditions', including 'scientific theories and laws, financial systems and theoretical or scientific teaching methods', none of which had an industrial character, or was 'capable of yielding, by means of a creative effort, the direct, tangible results required by industry'.[112]

- The Italian experts[113] suggested that after excluding the following categories of subject matter—'inventions of a purely theoretical-scientific nature' (such as 'works of talent' more suited for non-industrial laws); inventions, the results of which were insufficiently certain, or suitable for use on an industrial scale or to leave the experimental laboratory; 'methods of teaching, hygiene, or credit organization, financial systems and the like' (whose 'technical utility [did] not stem from the use of pre-existent natural resources and which [could not] fall within the field of industrial technology'); and discoveries (which lacked industrial character, in the sense of having no applicability in a technological field)[114]—

> it should be easier to define the field of patentability within the framework of the industrial character of an invention, taking 'invention' in the sense of a direct contribution to the domain of technology and hence as a function of the possibility it offers of satisfying specific needs by solving technical problems.

- The Luxembourg experts[115] agreed that scientific discoveries and literary/artistic works were not 'practical inventions', but suggested that it would be 'most difficult to formulate a hard and fast definition' beyond those exclusions, given the imprecision and controversy of national laws.[116]
- Similarly pessimistic were the UK experts,[117] who believed it should be possible to formulate a unified definition of 'invention', but doubted that any uniformity of interpretation could be secured without sacrificing 'sovereignties in jurisprudence'.[118] Further,[119]

> the problem of finding a common definition of 'invention' and establishing uniform practice throughout all the countries concerned to cover any agreed wider field would present even greater difficulty and might prove insurmountable in practice.

[112] ibid.

[113] Committee of Experts, Reply of the Italian Delegation to the Questionnaire Adopted by the Committee of Experts, EXP/Brev (56) 7 (10 August 1956).

[114] ibid, 2.

[115] See Committee of Experts, Comments by the Luxembourg expert on questions dealt with in the Resolution adopted by the Committee of Experts, dated 7th October, 1955, EXP/Brev (56) 9 (26 October 1956).

[116] ibid, 1–2; see also 4.

[117] See Committee of Experts, Observations of the United Kingdom Delegation: Industrial Character, EXP/Brev (56) 3 (3 May 1956).

[118] ibid, 3.

[119] ibid.

With respect to the nature of 'inventions' themselves, the experts noted the widespread exclusion from national laws of 'mere discoveries including natural products and laws, ideas, rules, schemes and knowledge having no industrial application', with the exceptions of Norway, Switzerland, and Germany, which they described as allowing patents for technical discoveries and technical problems.[120]

- The French experts[121] supported a requirement for 'an invention . . . within the field of "industrial techniques",' supplemented by non-exhaustive lists of inventions and non-inventions. As inventions, they described 'methods and processes of industrial manufacture, machines, instruments, tools or mechanical apparatus, industrial products or results, the technical application of a scientific principle that may lead to immediate industrial results', with protection in the latter case confined to the results themselves.[122] As non-inventions, they described 'discoveries (natural laws and phenomena), scientific theories and principles as well as systems and methods', in contrast to 'concrete applications' of the same.[123]
- The Netherlands' experts[124] understood inventions as technical and industrially applicable subject matter, in contrast to[125]

> processes which are not strictly technical, though adapted to industrial application, (e.g. accounting systems, sales methods or architectural projects), and technical processes which are not adapted to industrial application (e.g. medical treatment).

- Finally, the German experts[126] challenged the terms of the Bureau's request by rejecting the independence of the five conditions, and structuring their comments around the following instead: the patentable invention, industrial

[120] EXP/Brev (56) 3, 3, 1–2. This description of German law is consistent with the German experts' reply to the Committee of Experts' 1953 questionnaire (see Reply to the Questionnaire drawn up by the Bureau of the Committee of Experts of the Council of Europe, from the point of view of the German legislation, EXP/Brev (53) 1 (12 January 1953) 2).

[121] See Committee of Experts, Report of the French Delegation on the Unification of European Laws in respect of Patents, EXP/Brev (57) 3 (13 May 1957).

[122] ibid, 6–7.

[123] ibid, 3, 7.

[124] See Committee of Experts, Comments by the Dutch experts on questions raised in the Resolution adopted by the Committee of Experts on 7th October 1955, EXP/Brev (57) 1 (14 June 1957).

[125] ibid, 8–10.

[126] See Committee of Experts, Observations and proposals of the German experts on the unification of general conditions of patentability, EXP/Brev (56) 8 (26 October 1956).

character, novelty, and creative effort. With respect to the requirement for a 'patentable invention',[127] they commented as follows:[128]

The German experts think it unnecessary to specify . . . that a patentable invention must be of a technical character. So far as they know, this criterion is not expressly laid down in any national patent law but only implied; the formulation of the necessary rules in this connection has therefore been left to the jurisprudence of each individual country.

The German experts see no cogent reason for departing from this practice. Should it, however, be thought desirable to specify that the invention must be of a technical character, then they are of the opinion that the term 'technical character' should be defined as precisely as possible. They consider that it might be possible to base such a definition on that given the comments of Friebel and Pulitzer on 'Austrian Patent law' according to which 'technical' refers to all processes applied to living or inert matter in order to shape it, transform it or derive practical results from it.

In addition, the German experts proposed a list of non-inventions—'cases where the question of patentable invention does not arise'—based on the list from Gajac's study. The result was a provision in the following terms:[129]

Paragraph 1

2. No invention shall be patentable when it is simply a question of:
   (a) scientific principles and theories,
   (b) instructions to the human brain, such as accounting systems and rules of games,
   (c) the creation of aesthetic forms,
   (d) the bringing to light of a pre-existing fact (discovery).

Finally, and under the separate heading of 'technical progress', the German experts said the following with respect to the invention:[130]

The German experts consider that it goes without saying – in other words no special stipulation to that effect is required – that the purpose of an invention is to achieve a technical step forward and not backwards. This principle can be seen at work in a decision of the Milan Court of Appeal on 26th May 1953.

The point at issue concerned certain elements of a machine which, though they increased the machine's usefulness, had certain drawbacks. The invention simply consisted in eliminating these drawbacks by doing away with the element in question but at the same time losing the advantages offered by them. The Milan Court rightly regarded this as retrograde from the technical point of view and accordingly refused to recognise it as an invention.

[127] ibid, 2–3.
[128] ibid, 8.
[129] ibid.
[130] ibid, 6–7 (footnotes omitted).

In 1960, these comments were expressly endorsed in a further report.[131]

> Thus, in the German experts' view, an 'invention' for the purpose of European law ought to be conceived as subject matter having technical character and advancing the field of a technical art.

Diagram C Extract from the Committee of Experts, Comparative Study of Substantive Law in Force in the Countries Represented on the Committee of Experts on Patents presented by the Secretariat-General, EXP/Brev (53) 18 (7 November 1953) 3–7 (footnotes omitted), reproduced with permission.

1. Invention and discovery:

As a general rule, a patent can only protect an invention (a creation) and not a discovery, i.e. the simple taking note of a pre-existant [*sic*] reality. The revelation of a natural law cannot be the object of protection any more than a discovered product or natural phenomenon, but a patentable technical indication may be found on a discovery of the scientific expression of an empirically known causality. One can, for example, assure to an industrial activity a character of precision and constancy and hence give rise to a patentable application.

2. Industry and agriculture:

The word 'industry' understood as a defined field of economic activity, may be taken – in France – in a narrow sense (in opposition to agriculture) or in a wide sense – that of human activity oriented towards practical ends. . . . These uncertainties are reflected in the variations found in the national practices.

(a) It is however generally admitted that inventions which are agricultural in their destination although 'industrial' by their method of operation, are amenable to the application of the law . . .

(b) The divergences appear with the purely agricultural activities having no 'industrial' method of operation, i.e. methods or products of breeding or cultivation. . .

3. Systems, methods, etc..:

Though the patentability of inventions arising from the techniques of agriculture give rise to a diversity of replies, the same is not true of those which arise not from 'industrial' techniques but from financial, accounting, commercial, publicising, educational, military, touristic, medical, etc. techniques. All national practices agree in excluding from the field of application of the law monetary systems or systems of insurance, of accounting, of calculating, of teaching, of publicity, etc.. as well as rules of games, or methods of medical treatment.

On the other hand, the manufacture of products or apparatus which these new industrial techniques employ, falls definitely within the boundaries of industry. In this respect, the field of industrial technique is more extensive for 'products' than for methods, the product, even though applicable to other techniques, being necessarily industrial by its manufacture,

[131] See Committee of Experts, Report by Mr Herbert Kuhnemann, German Expert, on the unification of European patent law in respect of the requirements of technical progress and creative effort as conditions of patentability, EXP/Brev (60) 5 (15 November 1960) 2.

while the method can only have this character on condition of putting to work natural resources and not only the faculties of the human spirit. It is for this reason that in contradistinction to 'abstract' curative methods, medical products or instruments are everywhere patentable subject to reservations of public interest or social ethics.

4. Scientific principles and theories:

All practices agree also in excluding from the field of application of the law, purely scientific doctrines, principles or theories, while retaining their industrial applications only . . .

Problem and solution:– One can infer from what has been said the rule that the author of a patentable invention must have been presented with a problem and have indicated a solution to it. It seems however that this rule can be accepted more or less strictly according to the country in question. It is traditional in France to consider as foreign to the French conception and proper to the German (or Swiss) one, the affirmation that the simple fact of stating the problem can constitute a patentable invention, if an expert in the matter is able, by his own method to solve the problem. In fact an affirmation of this kind has no place at all in French jurisprudence and this is evidently closely linked up with that which is expressed in the requirement of a 'creative effort'.

It is convenient to add that the notion of 'technical problem' certainly does not carry a uniform significance everywhere, since the Italian doctrine, for example, makes the absence or existence of a 'problem' of this sort, the criterion for distinguishing a patentable invention from a model of utility.

5. Creations of form:

It is of course a general rule that creations of form escape patent protection to be included in the specific protection of designs and industrial models. Similarly, methods of typographic presentation would be covered only by the laws on artistic property.

6. Finally, apart from its belonging in a general way to the field of 'industrial techniques' the patentable invention must in a more concrete manner present certain qualities of its own, on the nature of which, furthermore, national agreement ceases as soon as their appreciation passes from the plane of 'reality' to that of 'value'.

It is in this way that the rule according to which the invention must be realisable . . . capable of repetition, suitable for carrying out in an industrial undertaking or for putting into practice for the needs of such an undertaking can be considered as of general application, although categories as precisely defined as those in the German law are not found everywhere.

On the other hand a notion such as that of utility . . . is quite foreign to French and Belgian doctrines, which leave on one side all questions of the technical worth of the invention, and a fortiori of its economic worth. Even the decisions which require, in addition to a technical effect (i.e. the immediate efficacity [*sic*] of the invention), a 'result', i.e. the realisation of the ultimate goal sought by the inventor stop short – at least in theory – at a subjective enquiry into intentions to the exclusion of all objective estimation.

It is not possible however, to make of this attitude a characteristic common to the 'law of the Latins' . . .

Of particular interest in the German comments is their suggestion that the requirement for technical progress would be inherent in the need for an invention itself. This suggestion was consistent with an early description by the Committee of its original objective in unifying laws, namely, 'to help inventors [in order] to aid technical progress on that continent'.[132] Despite this, it conflicted with the view of the UK experts,[133] who opposed the recognition of any such requirement as inconsistent with certain states' national laws. In their own comments to the Committee in 1956, they stated that if technical progress were made a requirement of European Patent Law, the UK would not be able to contribute to the project of unification.[134]

Reporting on the experts' comments in a 1960 Memorandum,[135] the Rapporteur-General noted a 'certain uniformity of outlook' in respect of the concept of industrial character. In his opinion, the comments showed European patent systems to be characterized by: (a) 'the extreme terseness' of their legal provisions on industrial character, the usual formula being 'inventions which could be applied for industrial purposes'; (b) the existence of a large body of case law, built up over many years; (c) the remarkable agreement between the structure of case law in different countries over the greater part of the field covered by the concept of industrial character; and (d) the existence, on the other hand, of certain divergences, mainly in the sector of agriculture, and derived largely from considerations of public interest. In light of these comments, his conclusion was that *if* the Committee were to seek to effect a definite harmonization of their systems with a view to resolving differences in the agricultural sphere particularly, and *if* it favoured to that end a definition of 'industrial character', then 'such a definition might in substance . . . follow that proposed by the German Experts'. This included its definition of 'invention' itself, which Finniss described as 'lay[ing] down negative criteria for industrial character'.[136] 'Nevertheless,' he added, 'such an approach 'would certainly not be compatible with the conceptions [of industrial character] currently held in such states as the United Kingdom and the Netherlands.'[137]

Considering the matter in 1961,[138] the Bureau agreed to a threshold requirement for an 'invention', but not to the German definition of that term, on account

[132] See CM/WP IV (51) PV4, 3.

[133] See Committee of Experts, Observations of the United Kingdom Delegation: Technical Progress and Creative Effort, EXP/Brev (56) 4 (11 June 1956) 1–4.

[134] ibid, 4.

[135] Memorandum on the unification of legislation (Item 4 of the Agenda for the meeting of 28th November 1960) by Mr Finniss, Rapporteur-General, EXP/Brev (60) 7 (undated).

[136] ibid, 4–5.

[137] ibid, 5.

[138] See Bureau of the Committee of Experts, Report of the meeting held in Paris on 16th and 17th March 1961, EXP/Brev B (61) 3 (24 March 1961). The Bureau notes its assistance by a Report by Finniss based on his Memorandum above (see Committee of Experts, Report on the unification of certain points of patent law in the countries represented by the Committee of Experts by Mr Finniss, Rapporteur-General, EXP/Brev B (61) 1 (1 March 1961)).

(the Bureau said) of its imprecision. The Committee subsequently accepted this, while affirming its agreement with the national experts that the following subject matter were not themselves 'inventions': 'scientific laws and theories, instructions to the human brain (such as accounting systems or rules of games), [and] creations of form and the mere disclosure of a pre-existing fact (discovery)'.[139] The result was Article 1 of the SPC, requiring an invention for the grant of a patent, but not defining the concept of the 'invention' itself.

In conclusion, it seems clear that when the SPC was signed in 1963, there existed broad agreement among European States with respect to the nature of non-inventions, though uncertainty with respect to the invention itself. For all the experts, an invention related to a practical art, and involved some human action on the physical world. Where they disagreed was with respect to the particular action, and the characterization of the practical art. While the language of 'technical character' was employed widely by the experts, its employment suggested different meanings and implications, and a basis in different patent and legal traditions. In addition, even states for which the concept of 'technical character' was familiar did not necessarily support it as a basis for legal reform. An example is the Scandinavian states, which did not define the 'invention' in their own patent code with reference to the concept of 'technical character', notwithstanding their comments to the Committee above,[140] on account of its ambiguity and increasingly expansive application:[141]

> The expression 'technical' is . . . ambiguous. The word technique covers the method of accomplishing all kinds of performances. It may also be a general term for all manual skills that are used in such performance, e.g. those of the artisan, masseur, pianist etc. The use of the word in the law therefore seems to be of little purpose also with regard to the fact that the word is commonly used in a growing number of applications, cf. the technique of advertising, of selling etc. These fields lie completely outside of the term invention in the patent sense as it has been developed in court and patent office practice.

Finally, for a state such as the UK, 'technical' conceptions of the invention were largely anathema, making them an unattractive basis for a European code.

2. The EEC Patents Working Party, 1961 to 1965

Within a year of its creation, the EEC Patents Working Party had begun its own discussions on a Common Market patent,[142] and produced a Preliminary Draft

[139] See Report by the Committee of Experts to the Committee of Ministers on the meeting held at Strasbourg from 2nd to 5th May 1961, CM (61) 97 (12 June 1961) 6.

[140] See Committee of Experts, The Preliminary Report by the Joint Scandinavian Patent Committees: Draft Patent Law and Extracts from Explanatory Notes, EXP/Brev (62) 2 (2 May 1962).

[141] ibid, 52.

[142] See Patents Working Party, Proceedings of the 1st meeting of the Patents Working Party held at Brussels from 17th to 28th April 1961, IV/2767/61–E (3 May 1961) (1st Working Party Meeting).

1951 — The Committee of Experts on Patents begins to meet. The Secretariat-General, Mr Gajac of France, presents a comparative study of national patent laws in which he identifies the requirement for an 'invention' as common to all states, but an object of 'widespread disagreement' among states.

↓

1953 — Gajac prepares a further comparative study of national laws, based on the responses to a questionnaire inviting states to describe the limits of their patent laws with reference to concepts of industry and agriculture. In the study he:

i) describes industrial character—the requirement that subject matter belong 'in a general way to the field of "industrial techniques"'—as the only patentability criterion common to all states;
ii) notes the variation in ideas by which the limits of the industrial invention are defined;
iii) suggests in consequence that no attempt be made to group under general headings the various exclusions to patentability recognized by states, but rather that the Bureau focus on setting out such common features as can be identified;
iv) summarizes those common features under the following five headings: (1) Invention and discovery; (2) Industry and agriculture; (3) Systems, methods, etc; (4) Scientific principles and theories; (5) Creations of form.

↓

1954 — Professor Reimer of Germany proposes a draft European patent convention limiting patentability to 'the field of technology'. He recommends against attempting to harmonize 'the notion of technology' on account of the diversity of views regarding that concept, and the impossibility of deriving a definition from the Paris Convention.

↓

1955 — Reimer changes his mind regarding the harmonization of the notion of technology, stating that the interested countries should easily reach agreement in view of the fact that they have always only protected inventions in the field of technology.

↓

1956 — The Committee of Experts rejects Reimer's proposal. It agrees that unification would be difficult to achieve in any general sense, and that its focus should be limited to harmonizing such aspects of substantive law as would need to be harmonized to create a centralized granting process. It identifies those aspects as including the industrial limits of the law, and invites observations and comments from states regarding those limits.

The German experts, in response to the Committee's invitation, distinguish between the requirement for industrial character as an issue on which states broadly agree, and those 'cases where the question of patentable invention does not arise'. It proposes that those cases be dealt with by means of the following provision, or an expanded version thereof, based on Gajac's 1953 study:

'2. No invention shall be patentable when it is simply a question of:
(a) scientific principles and theories,
(b) instructions to the human brain, such as accounting systems and rules of games,
(c) the creation of aesthetic forms,
(d) the bringing to light of a pre-existing fact (discovery).'

The experts reject the need to specify the requirement for technical character, suggesting that jurisprudence in respect of that concept is best left to individual states. However, they also propose a definition of technical character should the Committee decide to specify such a requirement. According to that definition, based on Austrian law, 'technical' refers 'to all processes applied to living or inert matter in order to shape it, transform it or derive practical results from it'.

↓

1960

The Bureau of the Committee supports a threshold requirement for an invention, but rejects the proposal for a list of subject matter not constituting inventions. It agrees that the following are not inventions: scientific laws and theories, instructions to the human brain (such as accounting systems or rules of games), creations of form and the mere disclosure of a pre-existing fact (discovery).

The EEC Patents Working Party begins to meet with a view to creating a Common Market patent. It produces a First Preliminary Draft Convention that includes a threshold requirement for an 'invention', with no definition of that term.

↓

1963

The Strasbourg Convention is signed. Article 1 includes a threshold requirement for an 'invention', but that term is not defined.

↓

1964

The EEC Patents Working Party agrees to a joint German–Netherlands proposal to amend its First Preliminary Draft Convention to include a definition of 'invention' based on the German proposal of 1956. The definition reads as follows:

'The following in particular shall not be regarded as inventions within the meaning of paragraph 1:
(a) scientific knowledge and theories as such;
(b) mere discovery of substances occurring in nature;
(c) purely aesthetic creations;
(d) financial or accounting methods, rules for playing games or other systems, insofar as they are of a purely abstract nature;
(e) methods of therapy, including diagnostic methods.'

Diagram D The pre-history of Article 52(2) and (3) of the EPC: the deliberations of the European Committee of Experts on Patents and the EEC Patents Working Party

Convention with the following provision: 'European patents shall be granted for inventions which are susceptible of industrial application and which are new and the realisation of which was not obvious.'[143] An early issue considered by the Party was whether to define 'invention' for the purpose of this provision. Three proposals to that end were considered. The first was to make technical progress an explicit requirement of patentability, with or without a definition of its essential criteria. The second was to replace the word 'invention' with the phrase 'products and processes', in order to restrict the availability of patents to 'practical implementations of [abstract ideas] in certain forms susceptible of industrial application'.[144] And the third was to include the definition proposed by Germany to the Committee of Experts in 1956.[145]

All three proposals were rejected. According to the Working Party, the first and third lacked the clarity and precision appropriate for a definition,[146] and were inherent in the requirements for an invention[147] and patentability.[148] By contrast were the reasons for rejecting the second, which were that limiting patentability to products and processes would be too restrictive, and would require an undesirable departure from the language of 'invention', and the text proposed for the SPC.[149] Hence the conclusion reached by the Working Party, that inherent patentability should remain undefined in the Convention; an approach which, it was noted, had been adopted in many states without 'practical problem'.[150]

However, this conclusion was reversed in 1964, in the course of discussing a power of the future patent office to deny patents for subject matter that were 'obviously not' inventions.[151] In considering the purpose of that power, the Working Party's attention turned to the kinds of subject matter it might exclude from patentability, which led in turn to the kinds of subject matter it ought to exclude. One category of immediate concern was therapeutic methods, which the Working Party agreed ought not to be regarded as inventions, consistent with the situation then

143 See Patents Working Party First Preliminary Draft Convention relating to a European Patent Law, Arts 11 to 29, IV/2071/61, Section 13 (14 March 1961) 2 (Part 1, Section 2, Art 11).

144 1st Working Party Meeting, Section 5, 4.

145 ibid, Section 5, 5; EXP/Brev (56) 8, 8 (s 1(2)).

146 1st Working Party Meeting, Section 5, 5.

147 See Comments on the Patents Working Party First Preliminary Draft Convention relating to a European patent law of 14th March 1961 (Arts 11 to 29) IV/2071/61–E 4 (concerning the first proposal, citing EXP/Brev (56) 8, 7).

148 1st Working Party Meeting, Section 5, 5 (concerning the third proposal). The decision of the Working Party on this point reflected its assumption that the EPC definition of 'invention' would continue to be informed by the laws of the Contracting States.

149 1st Working Party Meeting, Section 5, 4 (citing as an example of the proposal's restrictiveness its possible exclusion of 'inventions in the field of nuclear physics').

150 ibid, Section 5, 5.

151 Patents Working Party, Proceedings of the 12th meeting of the Patents Working Party held at Brussels from 26th February–6th March 1964, 2632/IV/64–E–Final (15 April 1964) (12th Working Party Meeting) Section 10, 22–23.

prevailing in EEC states' national laws.[152] Following this agreement, the German and Netherlands' delegations were asked to draft a list of other 'non-inventions' for consideration.[153] The result was an expanded version of the German list, which was subsequently adopted by the Working Party[154] and incorporated in the following draft provision:[155]

Article 9(2)

The following in particular shall not be regarded as inventions within the meaning of paragraph 1:

(a) scientific knowledge and theories as such;
(b) mere discovery of substances occurring in nature;
(c) purely aesthetic creations;
(d) financial or accounting methods, rules for playing games or other systems, insofar as they are of a purely abstract nature;
(e) methods of therapy, including diagnostic methods.

When the EEC Working Party reconvened in expanded form as the Luxembourg Conference, it was with an amended version of this draft in hand.[156]

3. The Luxembourg and Munich Conferences, 1969 to 1973

From 1969 to 1973, the delegations from the EEC Six were joined by representatives from the countries of the European Free Trade Association and several inter-governmental organizations,[157] for the purpose of drafting a patent convention. With respect to inherent patentability itself, they focused on an amended version of Article 9(2) above[158]—the drafting of which they expressly described as 'based . . . on rule No. 39 of the PCT plan'[159]—and the changes that would be required for it to be adopted. That consideration was undertaken in two separate

[152] 12th Working Party Meeting, Section 10, 22. See also Patents Working Party, Proceedings of the 14th meeting of the Patents Working Party held at Munich from 1st to 12th June 1964, 6498/IV/64–E–Final (1 August 1964) 13–15.

[153] 12th Working Party Meeting, Section 10, 23.

[154] 15th Working Party Meeting, 3.

[155] Patents Working Group, Amendments to the Preliminary Draft Convention relating to a European Patent Law (Arts 1 to 175), 2335/IV/65–E (22 January 1965) Art 9.

[156] Luxembourg Conference, Preliminary Draft Convention for a European System for the Grant of Patents, BR/6/69 (25 July 1969) 17 (table comparing two versions of this text, referred to as the '1965 Draft' and the 'Working Party' text respectively, neither of which is identical to the one extracted above).

[157] 1st Working Party I Meeting, [1].

[158] Luxembourg Conference, Preliminary Draft Convention for a European System for the Grant of Patents, Arts 1 to 41 prepared by working Party I (8 to 11 July 1969) BR/6/69 17 (25 July 1969).

[159] 1st Working Party I Meeting, [22].

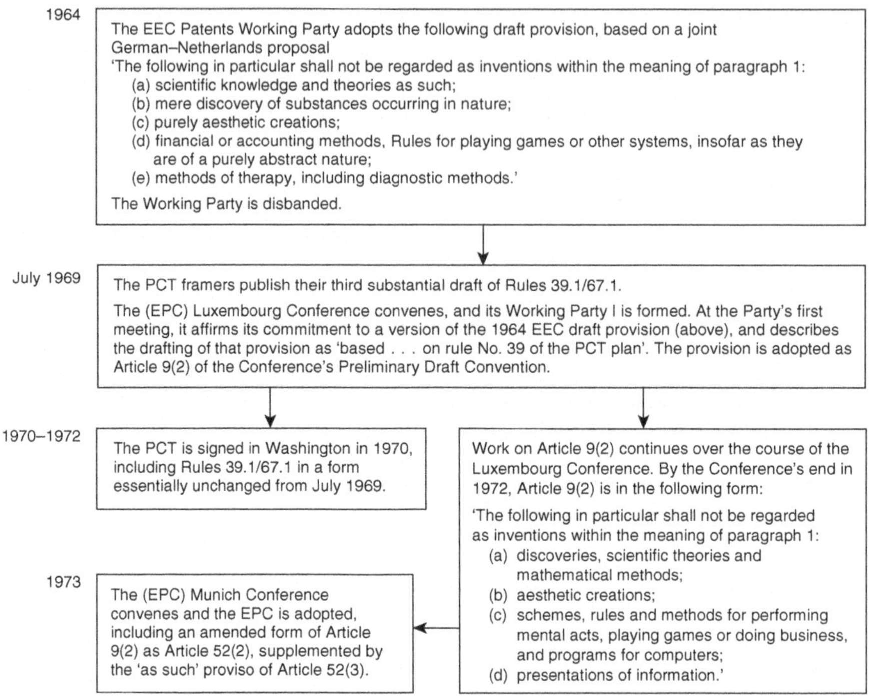

Diagram E The drafting history of Article 52(2) and (3) of the EPC

stages, from 1969 to 1972 (at the Luxembourg Conference[160]) and in 1973 (at the Munich Conference[161]), and focused on the following ten specific issues:

(a) Should Article 9(2) be deleted or moved to the Convention's Implementing Regulations in order to allow for future amendment by (what became) the Administrative Council?[162]

[160] The countries represented at the first meeting of the Luxembourg Conference were the six EEC states, the (then) seven countries of the European Free Trade Association (Austria, Denmark, Norway, Portugal, Sweden, Switzerland, and the UK) and Spain, Greece, Ireland, and Turkey. Those countries were joined in subsequent meetings by Monaco, Yugoslavia, Liechtenstein, and Finland. See K Haertel, 'The Draft Conventions for a European System for the Grant of Patents and for the European Patent for the Common Market' (1970) IIC 289, 291–292.

[161] The countries represented at the Munich Conference were the same countries represented at the Luxembourg Conference, plus four inter governmental organizations and 13 private international organizations, including the International Chamber of Commerce and the International Federation of Inventors' Association.

[162] The Implementing Regulations are amendable by the legislative arm (Administrative Council) of the European Patent Organisation, unlike the text of the EPC, which requires agreement by a Conference of the Contracting States (see EPC, Art 172).

(b) Should the exclusion of plant and animal varieties contained in the Working Party's draft Article 10(b)[163] be incorporated into Article 9(2), or any or all of the Article 9(2) exclusions moved to Article 10?
(c) Should any or all of Article 9(2)(a) to (e) be regrouped?
(d) How should the scientific knowledge and theories exclusion of Article 9(2)(a) be worded?
(e) Should discoveries of naturally occurring substances and presentations of information be excluded?
(f) To which categories of subject matter should the 'as such' proviso of Article 9(2)(a) apply?
(g) What form should the therapeutic methods exclusion of Article 9(2)(e) take?
(h) Should computer programs be excluded?
(i) Should substances of use in therapeutic methods be exempted from the therapeutic methods exclusion?
(j) Should the therapeutic methods exclusion extend also to animals?

In the discussion that follows, the EPC drafters' response to these issues is considered, alongside its implications for their conception of inherent patentability.

(a) Article 52(2) and (3) in Draft: Article 9(2)

The removal of the second paragraph of Article 9, or its relocation to the EPC's Implementing Regulations, was sought repeatedly by the UK delegation throughout the Luxembourg and Munich Conferences.[164] Neither proposal received much support among the other delegations,[165] however, for which the need to ensure flexibility in the face of future scientific and technological developments was less important than 'fixing the conditions for the grant of patents',[166] and ensuring the

[163] The varieties exclusion of the Working Party's draft Art 10(b) reproduced SPC, Art 9(2)(b).

[164] 1st Working Party I Meeting, [22]; 9th Working Party I Meeting, [97]; Luxembourg Conference, Minutes of the 5th Meeting of the Inter-Governmental Conference for the Setting up of a European System for the Grant of Patents Parts 1 and 3 (Luxembourg, 24–25 January and 2–4 February 1972) BR/168/72 (15 March 1972) (5th Luxembourg Plenary Meeting) [25], [28]; Munich Conference, Minutes of the Munich Diplomatic Conference for the Setting Up of a European System for the Grant of Patents (Munich 10 September–5 October 1973), M/PR/1 (Munich Conference Minutes) 27. The UK abandoned its proposal at the Munich Conference (Munich Conference Minutes, [34]).

[165] cf Sweden (1st Working Party I Meeting, [22], but cf Munich Conference Minutes, [33]); the Netherlands (5th Luxembourg Plenary Meeting, [25], [28], but cf Munich Conference Minutes, [28]); Ireland (see Munich Conference Minutes, [27]). Compare also the submissions of the various non-governmental organizations represented at the Luxembourg and Munich Conferences, which unanimously supported the proposal (eg NGO Hearing, [11], [14]; Munich Conference, Comments by CPCCI Standing Conference of the Chambers of Commerce and Industry of the European Economic Community, M/18 (2 April 1973) [3]; Munich Conference, Comments by CNIPA Committee of National Institutes of Patent Agents, M/20 (2 April 1973) [10]; Munich Conference Minutes, [27]).

[166] 1st Working Party I Meeting, [22]. See also 9th Working Party I Meeting, [97]; 5th Luxembourg Plenary Meeting, [25], [28]; Munich Conference Minutes, [29]–[33].

EPC's literal consistency with the draft and final texts of the PCT.[167] In this context, entrusting inherent patentability to the discretion of the Administrative Council was regarded as inappropriate.[168] Indeed, far from removing Article 9(2) or diminishing its status, the majority of delegations supported its expansion, in 1971, to include the additional PCT exclusions of mathematical theories (Article 9(2)(a)), methods of animal treatment (Article 9(2)(e)), presentations of information (Article 9(2)(f)), and computer programs (Article 9(2)(g)).[169] The only concession to flexibility in this context was the rejection at the Luxembourg Conference of a 1972 proposal to make the resulting list of non-inventions an exhaustive one.[170]

(b) The Contents of and Relationship Between Articles 9(2) and 10(b)

Having determined that Article 9(2) should stay in the body of the EPC, the question remained whether it was more appropriate for it to be placed within Article 10 than Article 9. It was recognized from the outset that this depended on the legal basis for each class of exclusion, and in particular, whether the subject matter to which each exclusion related were non-inventions and thus inherently unpatentable, or (potential) inventions the exclusion of which was required on some other ground.[171] A related question that arose early in the Luxembourg Conference was whether the exclusion of plant and animal varieties in the Working Party's proposed Article 10(b) should be moved to Article 9(2) in order to avoid the inference, assumed by the delegations to be incorrect, that they could constitute 'inventions'.[172] The record of subsequent discussions on this issue reveals an apparent reversal of opinion by some delegations, which expressed the view that plant

[167] 1st Working Party I Meeting, [22]; 9th Working Party I Meeting, [97]. The non-governmental organizations objected to retaining Art 9(2) on the basis solely of its consistency with PCT rr 39 and 67 on account of their different purposes (see NGO Hearing, [11]). The Working Party's response to that objection was that the PCT Rules, while concerned only with the requirements of search and examination, 'could nevertheless be regarded as an encouragement towards national harmonization' (see 9th Working Party I Meeting, [95]).

[168] 1st Working Party I Meeting, [22]; 9th Working Party I Meeting, [97].

[169] Luxembourg Conference, Minutes of the 7th Meeting of Working Party I held at Luxembourg from 26th–29th January 1971, BR/94/71 (6 April 1971) (7th Working Party I Meeting) [22]. See also 1st Working Party I Meeting, [22].

[170] 5th Luxembourg Plenary Meeting, [26]. See also Luxembourg Conference, Working Document for the 11th meeting of Working Party I, to be held from 28th February to 3rd March 1972, BR/GT I/145/72 (14 February 1972) (Working Party I Working Document) 2 (recording the view of the Chairman that while Art 9(2) 'should be interpreted strictly, as in any case of exclusion . . . it must not be possible to interpret [its] list . . . as an exhaustive catalogue', and suggesting retention of the words 'in particular' in the introductory clause of the paragraph in order to exclude the 'corollary to such an interpretation, i.e. that everything not listed in Article 9(2) or in Article 10 is to constitute a patentable invention').

[171] 9th Working Party I Meeting, [100].

[172] Luxembourg Conference, Minutes of the 4th Meeting of the Inter-Governmental Conference for the Setting up of a European System for the Grant of Patents (Luxembourg, 20 to 28 April 1971) BR/125/71 (7 July 1971) (4th Luxembourg Plenary Meeting) [16]. See also 5th Luxembourg Plenary Meeting, [37].

and animal varieties were inventions, and that their exclusion should therefore have been by way of Article 10.[173] Equivocation regarding the basis for the varieties exclusion, combined with a desire to follow the form of the Strasbourg Convention, led the Conference to conclude that the exclusion was best placed outside Article 9 after all.[174]

With one exception, all other proposals to recast the basis for the individual exclusions of Article 9(2) by moving them elsewhere were also rejected.[175] The exception concerns therapeutic methods, which were moved, as a result of the Munich Conference in 1973, to their own paragraph within Article 9 on the understanding that they were inherently patentable, albeit still to be denied protection for deemed insusceptibility of industrial application.[176] By contrast, equivalent proposals in respect of computer programs, schemes, and therapeutic applications of known substances were rejected, with the records citing as reasons, 'legal systematisation' and 'the existence of two similar provisions in the SPC'.[177]

(c) The Arrangement of the Exclusions

Following the expansion of Article 9(2) to reflect the wording of the PCT, a proposal was put by the non-governmental organizations (and supported by the Chairman of the Working Party[178]) to conflate certain exclusions, including particularly those dealing with scientific and mathematical theories (Article 9(2)(a)), schemes (Article 9(2)(d)), presentations of information (Article 9(2)(f)) and computer programs (Article 9(2)(g)).[179] That proposal was rejected for all but the last subject matter on the basis that they were distinct.[180] Computer programs, by contrast, were agreed by a majority of delegations to be more appropriately located within the schemes exclusion of Article 9(2)(d),[181] despite strong opposition on this point from the UK.

(d) Scientific and Mathematical Theories

In 1971, a Swiss proposal to change the phrase 'mathematical theories' to 'mathematical methods' was accepted without reported discussion.[182] The following year a further proposal was made by Yugoslavia to redraft the scientific theories

[173] 9th Working Party I Meeting, [100].
[174] ibid. See also 4th Luxembourg Plenary Meeting, [16].
[175] NGO Hearing, [12].
[176] Munich Conference Minutes, [24]–[26].
[177] 5th Luxembourg Plenary Meeting, [27].
[178] Working Party I Working Document, 2.
[179] NGO Hearing, [15].
[180] Luxembourg Conference, Report on the 11th meeting of Working Party I held in Luxembourg from 28th February to 3rd March 1972, BR/177/72 (13 April 1972) (11th Working Party I Meeting) [7].
[181] 11th Working Party I Meeting, [8].
[182] 9th Working Party I Meeting, [90].

exclusion around the concept of scientific discoveries on the basis, primarily, of the jurisprudence connected to that concept in certain Socialist states, and its prior recognition as an autonomous legal category.[183] The view was also expressed that redrafting the exclusion in the way proposed would clarify the distinction between scientific discoveries 'pure and simple'—defined as 'the theoretical description of a newly discovered natural law'—and possible industrial applications of the same.[184] The proposal was rejected on four grounds.[185] First, the unpatentability of discoveries pure and simple was regarded as self-evident and therefore not requiring explicit provision. Second, the amendment was regarded as unnecessary in light of Article 9(2)(a) and (b), which 'covered a large part, if not all, of that category [of scientific discoveries]'. Third, any exact definition of the scope of the scientific discoveries exclusion in Article 9(2)(a) was thought best left to 'jurisprudence'. And fourth, accepting the proposal would have put the EPC at variance with the corresponding wording of the PCT. Despite these considerations, the Conference instructed its Working Party to give the proposal further consideration, which it did in March 1972, expressing its support for a Swiss suggestion that Article 9(2)(a) be amended to refer explicitly to 'scientific discoveries as such'.[186] Considering a further proposal to explicate that wording in light of national legislative practice by addition of the words 'not susceptible of industrial application', the Party observed that 'no qualification would supply complete clarity', making it 'preferable to keep to the original wording proposed by the Swiss'.[187]

(e) Natural Products and Information

In 1971 a proposal by the UK delegation to clarify the meaning of 'mere presentations of information' was rejected at the Luxembourg Conference in order to avoid deviation from the wording of the PCT,[188] and in particular, 'to avoid too precise a definition which might look narrower than the PCT wording'.[189] A similar proposal to restrict the natural products exclusion in Article 9(2)(b) so as not to cover 'materials existing in nature which are isolated and defined for the first time' was also received poorly.[190] After agreeing that patent protection should be available for 'synthetically-produced substances by means of a new process', the Working Party nonetheless decided that precise rules should not be laid down as to the

[183] 5th Luxembourg Plenary Meeting, [30]. That recognition came at the 13th Scandinavian Conference on the Protection of Industrial Property, on which, see TC Johnsen, 'Observations on the Patentability of Computer Software' (1971) 2 IIC 71.

[184] 5th Luxembourg Plenary Meeting, [30].

[185] ibid.

[186] 11th Working Party I Meeting, [5]. See also Working Party Working I Document, 2 (recording the submission of the Chairman that Art 9(2)(b) 'be worded more generally to cover scientific discoveries').

[187] 11th Working Party I Meeting, [5].

[188] 9th Working Party I Meeting, [95].

[189] ibid.

[190] ibid, [91].

protection to be given where the same substance was subsequently discovered to exist naturally.[191] A further proposal to delete the exclusion of natural products altogether on the basis of its self-evidence was also rejected, with reference to the discretion given to decision-makers by inclusion of the word 'mere' in Article 9(2)(b), and the undesirability of making unnecessary changes to the draft.[192] Six months later, however, this decision was reversed, and the natural products exclusion deleted on the basis of the newly worded exclusion in Article 9(2)(a), which, by covering 'scientific theories, discoveries and mathematical methods as such', was thought to have rendered sub-paragraph (b) redundant.[193]

(f) The Restriction of the Exclusions by Use of the Phrase 'As Such'

The use and significance of the phrase 'as such' in Article 9(2)(a) was a topic that emerged several times during the Luxembourg and Munich Conferences with important consequences for future jurisprudence. The first substantive discussion of that issue took place in Luxembourg following the expansion of the paragraph to cover scientific theories, discoveries, and mathematical methods (as such), and was precipitated by a 1972 proposal of the Swiss delegation that the phrase be restricted to scientific theories and discoveries and that it not apply to mathematical methods.[194] The reason for that proposal was the delegation's view that '[a]ny mathematical method as such (pure mathematics) and its application in solving a technical problem (applied mathematics) should . . . be considered to be intellectual activities and therefore excluded from patentability'.[195] The Coordinating Committee agreed, and viewed the Swiss reasoning as equally applicable to scientific theories.[196] As a result, it decided that the phrase 'as such' should apply for scientific discoveries only, which decision was subsequently supported by the Conference.[197]

At the Munich Conference, however, the drafters revised this decision and reinserted the proviso not only for mathematical methods and scientific theories, but for Article 9(2) as a whole.[198] That decision followed the argument of the German delegation that limited use of the restriction 'could lead to the erroneous conclusion that a broad interpretation should be given to items not limited in this way'.[199] Thus, it was argued, the 'as such' proviso should apply uniformly in

191 ibid.

192 ibid.

193 11th Working Party I Meeting, [6].

194 Luxembourg Conference, Minutes of the 3rd meeting of the Co-ordinating Committee (Luxembourg, 23, 24, and 27 June 1972) BR/218/72 (26 September 1972) [4].

195 ibid.

196 ibid.

197 Luxembourg Conference, Minutes of the 6th Meeting of the Inter-Governmental Conference for the Setting up of a European System for the Grant of Patents (Luxembourg, 19 to 30 June 1972) BR/219/72 (26 September 1972) (6th Luxembourg Plenary Meeting) [26].

198 Munich Conference Minutes, [42].

199 Munich Conference, Comments by the Government of the Federal Republic of Germany, M/11 (29 March 1973) [21].

order to prevent its more limited application from producing a distorted reading of the other, unrestricted exclusions. The fact that the means chosen to resolve that distortion created a distortion in the opposite direction by entrenching a more restrictive conception of the exclusions does not appear to have been recognized, except possibly by the non-governmental representatives in their requested clarification of the proposal's impact through a redrafting of Article 9(2) to read '[i]tems which relate exclusively to the following in particular shall not be regarded as inventions'.[200] However, that request was not accepted, with the Conference agreeing instead to a version of the wording contained in the current Article 52(3).[201]

(g) Therapeutic Methods

An additional last-minute change of significance was also made at the Munich Conference on the second substantive proposal put by the German delegation in March 1973.[202] According to that proposal, the therapeutic methods exclusion was to be moved to its own paragraph to reflect its true basis, which, it was said, derived from the requirement for industrial applicability and not inherent patentability after all.[203]

As for the 'as such' proviso of Article 9(2)(a), the delegations' acceptance of that proposal involved a reversal of the drafters' initial position. Indeed, it was precisely a desire to exclude therapeutic methods from patentability that led the EEC Working Party in 1964 to consider inserting a (negative) definition of the 'invention' in reversal of its earlier decision on that issue.[204] In light of this, the subsequent decision by the EPC drafters to exclude therapeutic methods from Article 9(2)(a) is ironic. In addition, it offers an insight into the relationship between the requirements for inherent patentability and industrial applicability respectively.

That those requirements were related was recognized early in the discussions of the European drafters. Hence the justification offered by the EEC Working Party for the 1961 proposal to define inherent patentability in terms of products and processes with reference, among other things, to the effect that doing so would have in emphasizing 'that a patent does not protect an inventor's abstract idea but rather its practical implementation in forms susceptible of industrial application'.[205] Hence also the justification offered for the proposal nine years later to redraft the

[200] Munich Conference, Conference Document, M/66/I (11 September 1973) (proposal by the International Association for the Protection of Industrial Property (IAPIP)).

[201] Munich Conference Minutes, [42].

[202] M/11, [21].

[203] ibid.

[204] The Party subsequently confirmed that the therapeutic methods exclusion should cover diagnostic methods as well, despite the availability of patents for such methods under German law, which it explained with reference to 'the non-availability in Germany of patents for chemical products'. See 15th Working Party Meeting, 4.

[205] 15th Working Party Meeting, 4.

discoveries and natural materials exclusions around the concept of scientific discoveries, which was to emphasize the distinction between discoveries on one hand (not inventions) and possible industrial applications of discoveries on the other (potentially inventions). Finally, and in the specific context of the therapeutic methods exclusion itself, there was the 1972 proposal to restrict that exclusion to humans in order to enable the patenting of 'more typically industrial' methods such as insect sterilization and stock rearing.[206] What each of these examples shows is a concern to limit the definition of the 'invention' so as not to obstruct the patenting of subject matter that had a specific industrial use or application. Further, that concern was consistent with the context of the drafters' work under the Paris Convention, including its conception of patents as industrial property; a conception which provided the explicit analytical framework for the Committee of Experts' own earlier work.

By contrast is the view reflected in the 1971 decision to recast the mathematical theories exclusion in terms of mathematical methods, and in the subsequent removal of the 'as such' proviso from that and the scientific theories exclusion in order to prevent patents being granted for applied mathematics and science. By accepting that not all subject matter supporting an industrial application could and/or should be regarded as inventions, this view severed inherent patentability from the requirement for industrial applicability and, in so doing, confirmed the early view of the EEC Working Party that the objects of patent protection were not products and processes or other 'forms susceptible of industrial application', but rather their underlying inventions.[207] Hence the significance of the German proposal of 1973 concerning therapeutic methods, which was to distinguish between the need for an invention and industrial applicability respectively, by making it clear that while therapeutic methods might possibly be regarded as inventions, they were not susceptible of industrial application, and were accordingly incapable of supporting a patent.[208]

(h) Computer Programs

The status of computer programs also changed in the course of the drafters' deliberations. Initially the view taken was that the patentability or otherwise of computer programs could not be prejudged because of the uncertainty of their position given the (then) 'present state of developments'.[209] The subsequent decision to adopt the wording of the PCT, however, led to the introduction for

206 5th Luxembourg Plenary Meeting, [32].

207 1st Working Party Meeting, 4.

208 See also ibid.

209 Luxembourg Conference, Minutes of the 2nd Meeting held at Luxembourg on 13th to 16th January 1970, BR/26/70 (30 January 1970) 14. See also 1st Working Party I Meeting, [22]; 9th Working Party I Meeting, [96].

discussion of a new paragraph excluding 'computer programmes'.[210] That discussion took place in October 1971, when the UK argued strongly in favour of the proposed exclusion, supporting its argument with a definition of 'computer programs' as non-inventive mathematical methods.[211] Despite counter-arguments put forward by other delegations, the exclusion was eventually approved, albeit without the UK's definitional flourish which, it was felt, would 'tie the hands of the [EPO] and of the national courts'.[212]

Opposition to the computer programs exclusion was strong, however, leading to a 1972 proposal to regroup it with the schemes/plans exclusion of Article 9(2)(d) for the explicit purpose of giving decision-makers greater discretion when determining its scope.[213] With the exception of the UK, the proposal received broad support among the various delegations on the ground that it 'would, as a matter of fact, make for the exclusion of computer programs as such, while allowing precedents to be used to assess the patentability of any related inventions'.[214] The reference here to the exclusion of computer programs 'as such' reflects the submission of Haertel in response to the 1972 proposal that the exclusion be limited to 'computer-programmes as such,'[215] and confirms both the equivalence of the original phrases 'purely', 'mere', and 'as such' in Article 9(2), and their intended effect in maximizing the discretion of future decision-makers by confining the scope of the exclusions in Article 9(2). For the UK delegation, however, the proposed relocation threatened to have the opposite effect. Specifically, by creating an ambivalent text it threatened to support a misreading of the exclusion as allowing the protection of programs 'by indirect means', and in so doing to hinder rather than promote 'the free development of precedents of paramount importance in this still very uncertain field'.[216]

The final discussions on the computer programs exclusion took place at the Munich Conference in 1973, and focused on whether the phrase 'computer program' should be defined to clarify the restrictive scope of the exclusion and, in particular, to prevent its application in respect of a program's 'structures or

210 7th Working Party I Meeting, [22].

211 9th Working Party I Meeting, [96].

212 ibid. The decision not to define 'computer programmes' was one reason for the omission of a definition from the PA. See Hansard HL (series 5) vol 383, col 110 (9 May 1977).

213 NGO Hearing, [15], [18].

214 11th Working Party I Meeting, [8]. Use of the phrase 'as such' in this context was not a reference to the literal terms of the computer programs exclusion being proposed, which, following the agreement of the delegations on this issue, read 'schemes, rules and methods for performing purely mental acts, playing games or doing business, and programs for computers'. See Luxembourg Conference, Draft Convention Establishing a European System for the Grant of Patents (Stage reached on 20 May 1972) BR/199/72 (25 May 1972) Art 50(2)(c).

215 Working Party I Working Document, 2.

216 11th Working Party I Meeting, [8]. cf the UK delegation's position here with the proposal during the parliamentary debates on the PA to define 'computer program' in a restrictive manner so as not to exclude from patentability computer hardware, namely, as 'one or more recorded symbols or signals representing a sequence of instructions which is capable of causing a computer to perform a useful function' (Hansard HL (series 5) vol 379, col 1409 (15 February 1977); see also Hansard HL (series 5) vol 381, cols 841–844 (29 March 1977)).

algorithms' and 'unforeseen technologies'.[217] The conclusion reached was that any such attempts at further defining the exception would be futile, and that the EPO would 'simply have to be relied upon ... to interpret this expression unequivocally'.[218] The concerns of delegations persisted, however, shifting from the formulation of an explicit definition of 'computer program' to use of that phrase without further elaboration.[219] The proposal on this second issue was to replace 'computer programs' with 'data handling systems' in recognition of the fact that 'computer program' denotes a more complex system than *dataenverarbeitungsanlage* (in the German text) or *ordinateur* (in the French text), and might thus support an unduly broad construction of the exclusion as a whole.[220] While agreeing in substance with this argument, the UK delegation was opposed to the change in nomenclature, insisting that the definition of 'computer program' 'could safely be left to the practical work of the [EPO]';[221] a view with which the Conference finally agreed.[222]

(i) Substances for Use in Therapeutic Methods

The question whether patents should be denied for substances used in therapeutic methods was first raised in October 1971, when it was suggested that 'therapeutic uses' should be removed from the scope of the therapeutic methods exclusion in order to ensure continued investment in medical research.[223] That argument was rejected by the Conference on the ground that the interests of encouraging investment in medical research did not justify the deviation from national practice that permitting patents for therapeutic uses would entail.[224] Six months later, however, the decision was reversed, again in the face of opposition by the UK delegation, and the therapeutic methods exclusion amended expressly to permit the patenting of any 'substance intended for a specific indicated purpose'.[225] An additional proposal by the French delegation to move the resultant (permissive) provision to Article 10 was by contrast rejected, because of the departure it involved from the SPC.[226]

(j) The Application of the Therapeutic Methods Exclusion to Animals

As has been seen, in October 1971 the drafters made an in-principle decision to adopt the wording of the PCT and thereby expand the list of subject matter

[217] Munich Conference Minutes, [17].
[218] ibid, [18].
[219] ibid, [19]–[22].
[220] ibid, [19].
[221] ibid, [20]. cf the account given during the parliamentary debates on the PA (Hansard HL (series 5) vol 383, col 110 (9 May 1977).)
[222] ibid, [22].
[223] 9th Working Party I Meeting, [92].
[224] ibid.
[225] 11th Working Party I Meeting, [9].
[226] ibid.

enumerated in Article 9(2) to include (among other things) computer programs, mere presentations of information, and methods of treating animals. As has also been seen, the UK supported the first and second of those exclusions in principle, but cast itself in opposition to the majority of other delegations with its additional proposal that the presentations of information exclusion be clarified, and that the computer programs exclusion be retained in its own paragraph and not diluted by conflation with the other exclusions of Article 9(2)(c). To similar effect was its position in respect of the veterinary methods exclusion, which the UK opposed with the argument that the therapeutic methods exclusion should be confined to humans in recognition of the practical difficulty in distinguishing between 'methods of breeding in animals and methods protecting animals from disease'.[227] That argument was rejected by a majority of other delegations, which viewed the exclusion of animal methods as important because of the link between human cures and experimentation on animals, and because of the number of countries in which an equivalent exclusion was recognized.[228] The same proposal was raised and rejected for a second time the following year.[229]

Despite this, concerns regarding the animal methods exclusion persisted, prompting a third proposal by the Netherlands to clarify it in a way that would confirm its inapplicability in respect of non-therapeutic and non-diagnostic treatments such as methods of sterilizing insects and stock rearing.[230] That proposal was also rejected, this time on the ground that the proposal in its then existing form would not in any case have covered such treatments, making the clarification unnecessary.[231] The UK, however, remained unconvinced, and again sought clarification of the exclusion in March 1973, this time by agreement among the drafters that the word 'therapy' was 'concerned with the treatment of illness or disease and would not extend, in the case of animals, to treatments effected with a view, for example, to increasing the quantity or quality of the ultimate product'.[232] That agreement was provided, and the drafters' intentions in relation to the animal methods exclusion confirmed and recorded in the minutes of the meeting.[233] In emphasizing the limits of the animal treatment exclusion, the drafters appear to have been concerned with ensuring, rather than permitting, future recognition of the suitability for protection of non-excluded subject matter, consistent with an overall view of Article 9(2) as a positive statement of what could constitute an invention rather than a negative statement of what could not.

[227] 9th Working Party I Meeting, [94].
[228] ibid.
[229] 11th Working Party I Meeting, [9].
[230] 6th Luxembourg Plenary Meeting, [27].
[231] ibid.
[232] Munich Conference, Comments by the United Kingdom Government, M/10 (29 March 1973) [6].
[233] Munich Conference Minutes, [23].

Diagram F Resolution of issues concerning Article 52(2) and (3) of the EPC at the Luxembourg and Munich Conferences

Issues	Luxembourg Conference	Munich Conference
I. Deletion or relocation of Article 9(2)	Proposal to delete or relocate paragraph 2 rejected to ensure certainty and consistency with the PCT.	As at the Luxembourg Conference.
II. Location of Article 9(2) and Article 10 exclusions	Proposal to move the varieties exclusion of Article 10 to Article 9(2) rejected to ensure consistency with the SPC and with reference to its proper legal basis.	Proposal to remove various exclusions from Article 9(2) rejected for all but the medical treatments exclusion (see below) on the grounds of their proper legal bases.
III. Grouping of Article 9(2) exclusions	Proposal to conflate the exclusions covering scientific and mathematical theories, schemes, and presentations of information rejected with reference to their distinct nature. Proposal to conflate the computer programs and schemes exclusions accepted.	
IV. Scientific deductions and theories	Proposal to redraft mathematical 'theories' as 'methods' accepted. Proposal to limit the 'scientific discoveries' exclusion to 'scientific discoveries as such' accepted, with a further proposal expressly to exclude discoveries 'not susceptible of industrial application' rejected.	
V. Discoveries of natural materials, and presentations of information	Proposal to clarify the presentations of information exclusion rejected to ensure consistency with the PCT. Proposal to limit the natural products exclusion to permit the patenting of materials isolated and defined for the first time rejected to ensure flexibility. Proposal to delete the natural products exclusion as self-evident rejected and then accepted following revision of the scientific and mathematical theories exclusion, above.	

(Cont'd)

Diagram F Resolution of issues concerning Article 52(2) and (3) of the EPC at the Luxembourg and Munich Conferences (*Cont'd*)

Issues	Luxembourg Conference	Munich Conference
VI. 'As such' restriction	Proposal to limit the 'as such' restriction of Article 9(2)(a) to preclude the patenting of mathematical methods and scientific theories applied to solve a technical problem accepted to preclude the patenting of 'intellectual activities'.	Proposal to extend the 'as such' proviso to cover paragraph 2 as a whole accepted to avoid misinterpretation of its more limited application.
VII. Medical treatments		Proposal to remove the medical treatments exclusion from paragraph 2 accepted with reference to its proper legal basis.
VIII. Computer programs	Proposal to create a computer programs exclusion accepted to ensure consistency with the PCT, despite strong objections. Proposal to conflate the computer programs and schemes exclusions accepted to ensure flexibility.	Proposal to clarify the computer programs exclusion as not covering the structures or algorithms of programs or unforeseen technologies rejected. Proposal to clarify the term 'computer program' to protect the exclusion from wide interpretation rejected.
IX. Substances for use in medical treatments	Proposal to restrict the medical treatments exclusion expressly to permit the patenting of substances for use in medical treatments initially rejected to ensure consistency with national laws but subsequently accepted. Proposal to move the resultant provision to Article 10 rejected.	Proposal to move the (permissive) provision to Article 10 accepted, consistent with the relocation of the medical treatments exclusion, above.
X. Veterinary treatments	Proposal to create a veterinary treatments exclusion accepted to ensure consistency with the PCT and national laws and to facilitate medical research.	Proposal to record the intended scope of the veterinary methods exclusion (as limited to the treatment of illness or disease) accepted.

4. Conclusion

The deliberations at the Luxembourg and Munich Conferences in relation to the ten issues set out above record the drafters' intentions with respect to Article 9(2) generally as well as their specific understanding of the meaning and scope of the EPC model of inherent patentability at the time of its adoption in 1973.

(a) The Intent Behind Article 52(2) and (3) of the EPC

The primary concern of the drafters in reviewing Article 9(2) was to ensure its consistency with the SPC and, with greater ramifications for inherent patentability, Rule 39.1 of the PCT. That concern, however, was not driven by any particular understanding of inherent patentability itself, but rather by a desire to facilitate inter-jurisdictional harmonization. Indeed, there is a surprising lack of recorded discussion among the delegations regarding the purpose of Article 9(2) and Rule 39.1 and their relationship inter se. Such was the drafters' apparent lack of concern with that issue that variations designed to clarify the meaning of the latter in the context of the former were repeatedly and explicitly eschewed by the drafters.

The concern to ensure consistency between the EPC and existing international agreements (including in particular the SPC) was apparent before the PCT was concluded in 1970, but was until then of secondary importance to the separate and competing objectives of flexibility and legal certainty. When the EEC Working Party considered the issue in 1961, flexibility was regarded as paramount, and provided the primary justification for omitting an express definition of the 'invention'. By contrast, in 1964 its focus shifted to ensuring legal certainty and defining in precise terms the purpose and scope of the proposed power of examiners to exclude non-inventions.

It was in the context of this tension between flexibility and certainty that Rule 39.1 of the PCT was adopted, with its own list of threshold exclusions. Understood in that context, its significance for the larger and more diverse group of EPC drafters becomes apparent: it drove a line through the certainty/flexibility dilemma by asserting as a third and overriding objective, conformity with the text of the PCT under the renewed[234] guise of facilitating harmonization. Hence the view expressed repeatedly throughout the Luxembourg and Munich Conferences that the interests of harmonization demanded literal entrenchment of the PCT provisions, even at the cost of future flexibility and certainty of meaning, and notwithstanding its different purpose and basis. Further, the significance, if any, of the fact of Rule 39.1's derivation from a text which had been rejected for inclusion

[234] Harmonization (with the SPC) had earlier been raised by the EEC Patents Working Party as a reason for rejecting certain proposed definitions of inherent patentability (see at page 144).

in the SPC before being supported for inclusion in a Common Market patent convention, does not seem to have been considered by the drafters.

The Member State most strongly and consistently opposed to the literal transposition of the PCT provisions into the EPC was the UK, albeit not it seems out of objection to the PCT Rule itself, but rather out of concern to ensure a more flexible solution than the wholesale entrenchment of that Rule would permit. Indeed, according to the UK, inherent patentability should be left entirely to the discretion of future decision-makers, or at least be capable of variation by means other than formal amendment of the EPC text. In the event that the EPC did entrench the proposed (PCT) provision, the UK position was that its meaning should at least be clear, and proposed to that end certain definitional and explicatory provisions.[235] The general response to such proposals was that attempts to define inherent patentability precisely would be futile and self-defeating, and would render the EPC unacceptably inflexible. By this second response, the drafters adopted the very argument invoked at an earlier stage by the UK delegation to support the omission of Article 9(2) altogether. By the first response, they supported the initial view of the Working Party that explicit clarification of inherent patentability was impossible.[236] They also went further, however, by implying that attempts to clarify the PCT-adapted list of exclusions would be self-defeating, because of the risk that varying their terms would cause them to be misconstrued by future decision-makers as reflecting an intention to vary their meaning, thus undermining the drafters' professed goal of harmonization.

This general position of the drafters was departed from to support amendment of Article 9(2) in three specific contexts. The first involved the computer programs exclusion, which was placed with the schemes and plans exclusion in Article 9(2)(d) to ensure that it be read restrictively and the discretion of future decision-makers preserved. The second involved the 'as such' proviso of the original Article 9(2)(a), which was extended across Article 9(2) as a whole in order to avoid any future misunderstanding as to the intention behind its more limited application. And the third involved the therapeutic methods exclusion, which was moved out of Article 9(2) and recast as a non-inherent patentability-related exclusion to preserve that Article's doctrinal integrity.

Two conclusions can be drawn from these exceptions. The first is that consistency with the PCT was an objective grounded in pragmatism as much as in principle. This is supported by the impact of all three of the exceptions referred to, which was to entrench in different ways a more restrictive reading of Article 9(2) than that suggested by Rule 39.1 itself. It is also supported by the justifications offered for the first and second of the exceptions, including particularly the final extension of the 'as such' proviso. As noted, the reason for that extension, similar to

[235] This was exactly the view taken by the German national experts in their 1956 opinion regarding a requirement for technical character (see page 137).

[236] 1st Working Party Meeting, 5.

the reason for relocating the computer programs exclusion,[237] was to prevent selective use of the proviso from supporting an (erroneous) inference that the paragraphs to which it did not apply were intended for broad construction. Rather than clarifying Article 9(2), however, recasting the provision in this way made it more ambiguous, obscuring the drafters' intentions by overcompensating for the perceived distortion and turning a text they regarded as vulnerable to unduly broad construction into a text vulnerable to unduly narrow construction. Indeed, it seems reasonable to say without lapsing into presentism that by emphasizing the restrictive nature of Article 9(2) overall, the proviso helped transform that paragraph from an inclusive and narrowly cast list of what could *not* be patented to an exhaustive and widely cast list of what *could* be patented. In this way, the drafters' attempt to negotiate a middle ground between the competing objectives of certainty and flexibility through entrenchment of a restrictive list of exclusions that would minimize intrusion on the EPO's future decision-making power was undermined.

The second conclusion supported by the drafters' variation of Article 9(2) in defiance of their primary objective of harmonization is that they had a genuine and competing concern to distinguish clearly between inventions and non-inventions and, more specifically, between subject matter excluded for its failure to constitute an invention and subject matter excluded on other, non-inherent patentability grounds. Initially that concern was reflected in their discussions regarding the basis for the animal and plant varieties exclusion of the Working Party's draft Article 10(b), and whether any of the Article 9(2) exclusions related to non-inventions and were thus better placed in Article 10. In relation to the former, the view was taken that varieties were not non-inventions and thus did not belong in Article 9(2). In relation to the latter, the view was taken that therapeutic methods were also not non-inventions and therefore also did not belong in Article 9(2).[238] Whether this latter view reflected a desire for doctrinal coherency over and above consistency with the PCT, or a desire to imply a more expansive conception of the invention itself, is not clear, though some support for the latter view can be seen in the general tenor of the proposal from which it arose.[239]

[237] The reason in question being that placement of the exclusion in Art 9(2)(d) was required to prevent its location in the original Art 9(2)(e) from suggesting incorrectly that it was intended for broad construction. The debate concerning the replacement of the phrase 'computer program' with 'data handling systems' similarly proceeded from the argument that the breadth of the former relative to the German and French equivalent expressions might support an inference that the exclusion as a whole was intended for broad construction.

[238] A similar argument made by one of the non-governmental organizations in relation to schemes failed (see NGO Hearing, [12]).

[239] See M/11, [21] in which the proposal concerning therapeutic methods was presented with the proposal concerning the 'as such' proviso.

(b) *The Meaning of Inherent Patentability at the Time of the EPC's Conclusion*

Also revealed in the preparatory documents is a disagreement within the EEC Working Party of 1961 as to the proper conception of inherent patentability and, in particular, as to whether inventions were distinct from the tangible forms in which they were susceptible of industrial application. Reflected in that disagreement was a wider uncertainty, apparent at each stage of the European discussions, regarding the relationship between the requirements of inherent patentability and industrial applicability. Specifically, were all industrially applicable subject matter inventions, or was the nature of a subject matter *qua* 'invention' not prejudiced by its susceptibility of industrial application?[240] As has been seen, the statutory context of the SPC as a special agreement within the meaning of the Paris Convention, and the definition of patents in Article 1 of that Convention as forms of industrial property, encouraged an assumption throughout the 1950s that the threshold limits of patentability were appropriately conceived in industrial terms. As has also been seen, that assumption was not reflected in the SPC's final text, which clearly distinguished the two requirements.[241] Similarly for the EPC drafters, the fate of the therapeutic methods exclusion suggested an understanding of the inherent patentability and industrial applicability requirements as distinct, if not separate.

A further uncertainty of a similar nature involved the relationship between conceptions of the 'invention' and technical progress. In 1961 the members of the EEC Working Party rejected a proposal for the recognition of a requirement for technical progress, citing in doing so the opinion of the German delegation to the Committee of Experts in 1956—affirmed in 1960—that such a requirement was implicit in the need for an invention itself. However, while this may have been true for the laws of the EEC Six, it was not regarded as true for other future EPC states. Hence the significance of the Working Party's failure to refer to the opinion expressed by the UK in 1956, that the notion of 'technical progress' was not an accepted (nor uniformly understood) requirement of European law, and would therefore be an insupportable inclusion in any European system.

This discussion raises, too, the relationship between inherent patentability and the separate concept of 'technical character'. It seems clear from the *travaux* that in discussing a Common Market patent, the EEC Working Party assumed that inventions would be technical in nature.[242] In addition, that assumption was consistent

[240] Compare the views expressed by the Working Party in its initial deliberations on inherent patentability (1st Working Party Meeting, 4–5) with the subsequent acceptance at the Luxembourg Conference that the concept of 'an invention' did not encompass all subject matter susceptible of industrial application. That acceptance accorded with the view of the UK's Patents Liaison Group and Banks Committee (see the discussion at pages 182–189).

[241] See SPC, Arts 1, 3.

[242] This is clear from their discussion of issues other than inherent patentability, such as the scope of the plant and animal varieties exclusion (see, eg 1st Working Party Meeting, 46).

with the observations and comments of many of the national experts submitted to the Committee of Experts during the 1950s,[243] and with the initial proposal of Reimer in 1954 that the availability of patents be restricted to the 'domain of technology'. However, it is also clear that those experts' use of similar language to explicate the concept of inventions belays differences in understanding and jurisprudence, which was precisely the reason given by Reimer in 1954 for recommending against definition or unification of the concept of 'technology'. Further, some states which supported a 'technical character' theory of inventions before the Committee of Experts in the late 1950s subsequently changed their position on that issue. Hence the Scandinavian states' decision in 1962 that the ambiguity and potential breadth of that concept made it an appropriate basis for a legal definition of the 'invention'. Finally, and importantly, neither the EEC Six nor the European experts of the 1950s were per se responsible for the EPC, and there is no evidence in the recorded discussions of the two EPC preparatory Conferences themselves to support the conclusion that those who were so responsible intended that Article 9(2) would resolve to a requirement for 'technical character'. On the contrary, at the Luxembourg Conference it was agreed that not all solutions of technical problems were, or should be, inventions; the example given there being applications of mathematical methods and scientific theories in the solution of a technical problem, which the Conference accordingly excluded from patentability by removing their 'as such' limitation. While the subsequent decision to reinstate that limitation for each of the subject matter of Article 9(2) reversed that removal, it did so for the reasons outlined above, and not to extend patentability to cover such applications themselves.

While the 11th hour proposal and decision on the 'as such' proviso did not purport to reflect any explicit variation in the drafters' understanding of what might constitute an invention, variations in that understanding were apparent throughout the drafting proceedings. Thus, plant and animal varieties and therapeutic methods both went from being regarded as non-inventions to (potentially)[244] inventions, albeit without impacting their (un)patentability, while computer programs went from being regarded as possibly inventions to non-inventions, and from possibly patentable to definitely unpatentable on that basis.

Finally, the *travaux préparatoires* suggest an increase in the importance ascribed to policy considerations, or at least a change in emphasis regarding the policies to be observed, in setting the future boundaries of the European patent system. For example, the view was initially taken that the policy interest in encouraging investment in medical research did not justify the deviation from national practice that permitting patents for therapeutic methods would have involved.

[243] cf, eg EXP/Brev (56) 2–3.

[244] While no final agreement regarding the basis for the varieties exclusion was reached by the drafters, the initial assumption that varieties were not inventions was challenged by some delegations during the Luxembourg Conference.

Subsequently, however, the relative importance of those policies was revised by the drafters' decision that the need to ensure the availability of patents for substances when used in therapeutic methods justified an explicit provision to the effect that such treatments were actually, as distinct from potentially, patentable.[245] The exception to this was, of course, harmonization, which was consistently regarded by the drafters as a paramount objective in formulating Article 9(2).

In conclusion, it seems clear that in 1973 when the EPC was signed, the conception of the invention supported by the drafters was very different from that of UK law. This can be seen from their treatment of the relationship between conceptions of the invention on one hand and industrial character and technical progress on the other. By distinguishing clearly between these concepts, and rejecting a view of the invention as advancing the technical (or other) arts, the drafters distanced themselves from the requirement of earlier UK law for an ingenious subject matter of practical utility directed to advancing the industrial arts, notwithstanding the UK experts' apparent view to the contrary. Further, just as the drafters conceived the invention as lacking any industrial or progressive element, so too they conceived it as lacking any (other) policy element. The implication was that in the era of the EPC, the threshold requirement for an invention would play an important but limited role. Against this background, it is interesting to consider the drafters' discussions regarding chemical product patents and equiparating patent claims.

D. The European Drafters' Intent Regarding Chemical Product Patents and the Requirement for Equiparating Claims

1. The Inherent Patentability of Chemical Products

(a) The European Committee of Experts on Patents, 1951 to 1963

The patentability of chemical products arose early in discussions of harmonization, after being identified by the European Committee of Experts on Patents as an issue on which state laws diverged.[246] Indeed, of the group of states surveyed in 1951, eight (Germany, Luxembourg, Belgium, Denmark, Ireland, Norway, Sweden, and Switzerland) prohibited patents for chemical products, five (France, the UK, Greece, Turkey, and Switzerland) allowed them, and one (the Netherlands) allowed limited protection through claims for chemical products-by-process. Of the eight that prohibited chemical product patents, a majority did so by express provision, suggesting policy as the basis for the prohibition;[247] with a notable exception in the case of Denmark, whose delegate to the Committee of Experts

[245] 11th Working Party I Meeting, [9].
[246] See CM/WP IV (51) 9, 19–20.
[247] See EXP/Brev (57) 3, 5.

on Patents reported in 1951 that 'by the laws of his country a product was not considered to be an invention'.[248]

Nonetheless, the question considered by the Committee in 1951 was not whether chemical products possessed the constitutive properties of inventions, which it seems to have been accepted that they did, but whether they ought nonetheless to support a per se exclusion. Following the results of the survey above, the members of the Committee were asked to 'set aside preoccupations of a national character' in considering that issue, and focus on the desirable position at law.[249] Despite this, they could not agree:[250]

> The Committee of Experts considered the question of the exclusion from the scope of patent protection of categories such as chemical and pharmaceutical products and manufactured foodstuffs. The majority of the Committee agreed that such exclusion should no longer obtain, being of opinion that any adverse consequences arising from the removal of exclusions could be avoided by a system of compulsory licenses. Certain members of the Committee however felt themselves unable at present to form a definite opinion and in consequence, the Committee considered that the matter should be made the subject of further study by the members with a view to discussion and examination at a later meeting.

The further study was completed by the end of the year with the aim not of proposing a solution per se, but of exploring in depth the relevant issues so that the Committee might reach an informed position.[251] In it, several issues were raised against patentability. One was that chemical products might not, in principle, be appropriate for a patent, on the ground that the protection conferred by a product patent—valid on the basis of a single industrial application of the product[252]—might exceed the patent's contribution to the art.[253] The reason was similar to that given by Buckley LJ in *Kopp v Rosenwald Bros*[254]—that chemical products do not involve the process element of a mechanical product—except that the focus of the study was on methods of applying a chemical product rather than on methods of manufacture per se.[255] Specifically, the study suggested that chemical products were less likely to be of use without further acts of invention, which acts might be precluded if the products were patented. Hence its further suggestion, that patents

248 CM/WP IV (51) PV7, 6.

249 See ibid.

250 See Committee of Experts, Resolutions adopted by the session of the Committee of Experts on Patents held at The Hague on 2nd to 9th July 1951, CM/WP IV (51) 17 final (7 July 1951) 8.

251 See Committee of Experts, The Patentability of Products, CM/WP IV (51) 27 (30 November 1951) 2, 6.

252 See ibid, 5. The study also referred to the possibility of speculative product claims resulting in obstructive patents the revocation of which might, for reasons of cost, be a less attractive choice than payment of a royalty.

253 See ibid, 4–6.

254 (1900) 19 RPC 205, Ch, discussed at page 83.

255 CM/WP IV (51) 27, 4.

ought not to be available for chemical products as such, or inventive activity would be at risk of being stifled.[256]

Against this, the study recognized that 'unless the patent for the product is to be in fact only a patent of 'application' and not 'of product' it should cover not only the application or applications indicated by the inventor but also all possible applications'.[257] It also noted, consistent with this,[258]

> that the invention of the new application or even of the new process was only rendered possible by the discovery of the product, and that in consequence, the inventor of the process or application should in equity pay tribute to the inventor of the product who, even though he may not have realised all the implications of his discovery, has made a permanent contribution to technical progress and should not find himself dispossessed by the effect of a late subsidiary discovery.

Added to this were further points of a strictly pragmatic nature, namely, the desirability 'in the interests of administrative simplicity' of treating chemical substances in the manner of other products, and allowing any resulting social inconvenience to be resolved by a compulsory licensing system.[259] In support of treating chemical products as patentable per se, three further points were made in the study, reminiscent of the views of the 1931 Sargant Committee:[260]

> [T]he injustice which [denying protection to chemical substances appears] to involve for the discoverers of substances, on the whole more important than certain mechanical novelties which are fully protected; the uncertainty of the physico-chemical tests which decide whether or not the substance is a true chemical product or species; as well as the differentiation, often not easy, between the product and the process, have given rise to a literature too abundant and too well-known to be reviewed here.

In support of allowing compulsory licensing in respect of such patents, the study referred to the harm that those patents might cause on account of their importance to later inventors, and their irreplaceability to researchers within the chemical field.[261]

> Thus, a new mechanical part, for example a special kind of screw or valve, may well have many different applications, but in general they are immediately foreseeable once the nature and function of the device have been explained, whereas, although the process of 'tailoring molecules' i.e. designing a new chemical product with a view to achieving certain definite properties for a particular application, may be said to have begun, it is still in general the case that a new chemical compound is first prepared (perhaps with some definite object in view, perhaps by accident) and is then applied to a variety of different uses.

[256] CM/WP IV (51) 27, 4.
[257] ibid, 5.
[258] ibid.
[259] ibid, 3.
[260] ibid, 2–3.
[261] ibid, 4.

By 1952, the patentability of products remained a central issue for the drafters; one of two on which their efforts at unification were focused.[262] Reflecting again their commitment to succeeding in those efforts, the drafters were asked to shelve their personal views and 'do everything possible' to reach an agreement.[263] One response to that request was a joint proposal by the UK and German delegations covering three contentious categories of products: chemical substances, foodstuffs, and pharmaceuticals.[264] The argument made in the proposal was that patents should be available for each type of product, with any harm to the public interest resulting from such patents addressed by a compulsory licensing system such as that which existed in the UK already. The Committee, however, was not convinced, and all but repeated its earlier resolution.[265]

In the event, that was essentially the end of the matter, at least for the purpose of the Strasbourg Convention.[266] The reason is that in 1955, the drafters of that Convention decided to exclude chemical products from their agenda, on account of the public interest issues which their patenting raised.[267] Following that agreement, the drafters focused on a more limited form of harmonization that would preserve national sovereignty on all but fundamental patentability grounds,[268] while giving each country by express provision 'the power of refusing protection'[269] in respect of certain categories of claim.

In 1961, the drafters considered the terms of that reservation, and restricted it to pharmaceutical products and horticultural processes, and by a later draft, to

262 ibid.

263 See EXP/Brev (52) 1, 2.

264 See Committee of Experts, Draft by delegations of Germany and United Kingdom, EXP/Brev (52) 3 (16 January 1952).

265 Committee of Experts, Report of the Committee of Experts on Patents to the Committee of Ministers, EXP/Brev (52) 8 rev (22 January 1952) 2.

266 It is not clear how much further consideration took place. The only record of such consideration is in the form of the experts' responses to a second questionnaire, revealing the same degree of divergence as in 1951, and a passage in a Note on harmonization, supporting limited protection for products on evidentiary grounds, as a means of ensuring protection for methods (see Committee of Experts, Note on the Standardisation of Laws, included in the Agenda for the Fourth Meeting of the Committee of Experts on Patents, EXP/Brev (52) 23 (26 June 1952)). According to the Note, while the disagreement over chemical product patenting was one of principle, even those drafters opposed to such patenting supported a form of limited protection, which was consequently regarded as a feasible compromise position. In 1953, states' individual responses to a third questionnaire indicated that while many countries excluded chemical substances from patentability, the exclusion was limited to chemical substances 'as such', with indirect protection for the same via patents for processes for creating a substance and presumptions that products were created by application of the patented process (see the replies to EXP/Brev (53) 3, contained in EXP/Brev (53) 1ff, including especially EXP/Brev (53) 18, 17–18.)

267 See, eg EXP/Brev (57) 3, 5. See also J Pila, 'Article 53(b) EPC: A Challenge to the *Novartis* Theory of European Patent History' (2009) 72 MLR 436, nn 50–52.

268 See Pila, ibid, nn 59–60.

269 EXP/Brev (53) 19, 6. Grounds of revocation were also provided for, which did not extend to insufficiency or fair basis.

'food or pharmaceutical products, or other chemical substances and . . . alloys'.[270] After discussion, the reference to chemical substances and alloys was deleted from the text, leaving the version that was signed in 1963, explained in the following contemporaneous note:[271]

The exclusions provided for in Article 2 are the only ones that Contracting States are entitled to maintain permanently in their legislation. With regard to the most controversial question of all, for which the proposed solutions are many and varied, namely the patentability of chemicals, foodstuffs and pharmaceuticals, the draft aims at ensuring uniform national legislation in the direction of the widest patentability. However, on two particularly difficult points (patentability of food and pharmaceuticals and of agricultural and horticultural processes) it was found necessary to allow Contracting States to postpone bringing their legislation into line with the Convention by means of a reservation.

The implication to be drawn from this is that by 1963, the Strasbourg drafters had changed their view of the appropriateness of harmonizing European laws with respect to the patenting of chemical substances. Having previously excluded substances from their harmonization agenda, the drafters now included them, on the basis (as stated in the memorandum above) that they did not raise sufficient difficulties to justify compromising their commitment to 'ensuring uniform national legislation in the direction of the widest patentability' in the context of 'chemicals, foodstuffs and pharmaceuticals'. It followed that to the extent that chemical substances satisfied the Convention's patentability requirements, they would need to be accorded patent protection.[272]

(b) The Luxembourg and Munich Conferences, 1969 to 1973

The EPC history followed a similar trajectory, with the question of patents for chemical products occupying a similarly central place in the drafters' discussions. This is apparent from the following explanation by Dr Kurt Haertel—Chairman

[270] See Committee of Experts, Preliminary draft Convention on the unification of certain points of substantive law on patents for inventions, Article 6(1) in Committee of Experts, Report of the Committee of Experts to the Committee of Ministers on the meeting held in Strasbourg from 2nd to 5th May 1961, CM (61) 97 (12 June 1961) 22 (Appendix V).

[271] Committee of Experts, Report of the Committee of Experts to the Committee of Ministers on the meeting held in Strasbourg from 10th to 13th July 1962, CM (62) 160 (8 August 1962) 4.

[272] An Austrian proposal of 1963 to extend the permanent exclusions to cover chemical substances, foodstuffs, beverages, drugs, disinfectants, and therapeutical processes, and to define more restrictively the role of the claims in determining the scope of protection conferred by a patent (see Committee of Experts, Unification of Laws Convention, Amendment suggested by the Austrian Delegation, EXP/Brev (63) 7 (10 May 1963)) was rejected. Following that rejection, the Austrian Government made a declaration criticizing the draft Convention as going beyond the unification of technical definitions in its imposition of an obligation to protect in particular chemical products and foodstuffs, 'since such a protection would be novel for most European countries': 'It concerns the creation of new kinds of patents rather than a true unification' (Committee of Experts, Report of the Committee of Experts to the Committee of Ministers on the meeting held in Strasbourg from 14th to 17th May 1963: Annex II (Declaration by the Austrian delegation) CM (63) 101 (27 May 1963) 16).

of the EPC Luxembourg Conference—of their original vision of harmonization, and its implications for chemical product patents:[273]

What effect does the European patent have in those countries for which it has been granted? It will have the same effect as a national patent granted by the national patent office. It is a logical consequence of this approach that a European patent can be declared invalid in one country even if it was validly granted by the European Patent Office. An example can best illustrate this. The European patent law provides for the patentability of chemical and pharmaceutical products. If such a patent has been granted for a country which does not provide protection for chemical and pharmaceutical products, then the European patent could be declared invalid in that country by a national nullity proceeding. This has been called the 'minimum approach'.

As Dr Haertel went on to explain, at the January 1970 session of the Conference the EPC drafters abandoned this approach in favour of a 'maximum' approach instead.[274] In the mind of Dr Haertel at least, this raised a doubt over the drafters' earlier agreement to support protection for chemical products:[275]

This concept implies that the European patent can only be declared invalid in the States for which it was granted, by reasons provided for in the European patent law, and not by those provided for in the domestic patent laws. . . . Whether the Convention Draft will eventually contain the minimum approach or the maximum approach, will depend upon the results of discussions with various interested international organizations. But even if the interested parties opt for the maximum approach, as this author believes they will, that solution as such will not be incorporated in the Convention. In this event, the Signatory States will have the right, at least during a period of transition, to be totally exempt from the maximum approach, or to exclude certain areas of technology, for example, that of chemical and pharmaceutical products, from the maximum approach.

In the result, a temporary reservation was proposed for the Convention that exempted EPC Signatory States from according protection for chemical products.[276] Originally modelled on the Strasbourg reservation,[277] that provision

273 Haertel, 'The Draft Convention for a European System for the Grant of Patents and for the European Patent for the Common Market', 297.

274 See Luxembourg Conference, Minutes of the meeting of Working Party I, Luxembourg, 7th–11th September 1970, BR/49/70 (26 October 1970) [90]. See further Pila, 'Article 53(b) EPC: A Challenge to the *Novartis* Theory of European Patent History', nn 45–47 and accompanying text.

275 Haertel, 'The Draft Convention for a European System for the Grant of Patents and for the European Patent for the Common Market', 297–298.

276 This account confirms that the EPC drafters understood and intended that chemical substances would be inherently patentable under *European* (cf national) law. Further evidence of that understanding and intention can be derived from their limited discussion of such substances (in the context of pharmaceuticals and diagnostic methods), and their omission of the same from permanent exclusion. See also BR/218/72, [5] (envisaging 'a system which allowed the patentability of new substances').

277 See Committee of Experts, Second Preliminary Draft of a Convention Establishing a European System for the Grant of Patents, 1971. See also BR/125/71, [83] (recording the drafters' agreement not to extend the possibility of making a reservation regarding patentability to chemical products on the grounds that the list as it was conformed with the text of the SPC, and was 'already a compromise solution which took into account the special situation in certain countries').

was extended at the 1973 Munich Conference to apply in respect of chemical substances. The reasons are apparent from the records of the debates.[278]

First, the drafters' commitment to a maximum solution created a tension between unifying standards of patentability on one hand, and ensuring the greatest number of signatories to the Convention on the other.[279] While states with developed chemical industries were unfazed by the prospect of chemical product patents, states with developing chemical industries were not, and insisted on a right to reserve their positions for a minimum ten-year (renewable) period. For some developed states and non-governmental organizations, the situation was complicated by a belief that the premise of that objection was misconceived; ie that patents would stimulate not impede industrial growth. Consistent with this, they regarded the position of developing states as 'represent[ing] a failure to understand the function of a patent system in a modern economy'.[280] Supporting that argument was the suggestion by some that patentable inventions in chemicals 'accounted for between 35 and 40% of all inventions' due to the 'extremely high . . . cost of chemical research and development'.[281] Further, while it was accepted that chemical product patents had only recently begun to be granted by industrialized nations, the reason for their historical denial was said to be not 'the level of industrialisation . . . but the wishes of the chemical industry itself'.[282] According to the European Industrial Research Management Association (EIRMA):[283]

> The reasons for such limitations no longer existed and it was difficult to argue that patents issued to foreign companies for chemical products could slow down the economic progress of less advanced countries. This was further underlined if account were taken of the ever increasing degree of industrial cooperation within Europe and the possibility of preventing abuses by means of compulsory licensing.

This was consistent with an additional point, that the political and social difficulties created for governments by allowing patents for food and pharmaceutical products—to which the draft reservation also related—did not apply with respect to chemical products:[284]

> The United Kingdom delegation also felt that it would not be in a country's interests to exclude as large and important an industrial sector as chemical products as such from protection while the processes for manufacturing such products could be patented.

[278] See Munich Conference Minutes. For an account of the discussion of these issues at the Luxembourg Conference, see BR/49/70, [100]; BR/125/71, [83]–[85]; NGO Hearing, [161]–[162]; 5th Luxembourg Plenary Meeting, [159]–[162]; BR/218/72, [13]–[14]; 6th Luxembourg Plenary Meeting, [54]–[55].

[279] Munich Conference, Minutes of the Conference, Munich 10 September to 5 October 1973: Minutes of the Proceedings of the Main Committee I, M/PR/II, [1008].

[280] ibid, [1007].

[281] ibid, [1014].

[282] ibid, [1009].

[283] ibid.

[284] ibid, [1008]. This represented a decoupling of issues which had historically been treated together, along with the patenting of plant and animal varieties. For a discussion, see generally Pila, 'Article 53(b) EPC: A Challenge to the *Novartis* Theory of European Patent History'.

Countries which, for social reasons, wished to protect the public both as regards the price of such products and manufacture of an adequate proportion of them by local undertakings could solve the problem by compulsory licensing. Finally, the United Kingdom felt that Article 166, as it stood, took sufficient account of the legitimate interests of those countries which were not yet able to fully accept the system established by the Convention.

Finally, the EIRMA 'pointed to the distortions of the discriminatory effect which [a reservation] would have on research-based industries, or at any rate, those located outside the territory of the Member States of the Communities.'[285]

In addition to rejecting each of these arguments, and the suggestion of discrimination on which they were based, developing states made two central claims. The first was from the nature of harmonization, as a long-term project that could only be accomplished over time, in stages.[286] And the second was from its wider political context, including the need to have regard to developing states' efforts to accommodate maximum harmonization.[287]

In the event, the last of these factors proved to be decisive, and after a long and detailed discussion on the issue, the decision was taken by the Munich Conference to extend the reservation to chemical products.[288] That decision was described at the time as representing a compromise 'between the fullest possible harmonisation of laws governing patentability and the accession of the greatest possible number of states';[289] or in the words of the German delegation, 'a point of connection between those states which supported the maximum solution and those which could not fully do so'.[290] The result was Article 167(1)(a) of the EPC, which read as follows:[291]

Each Contracting State . . . , at the time of signature or when depositing its instrument of ratification or accession, may reserve the right to provide that:

(a) European patents, in so far as they confer protection on chemical, pharmaceutical or food products, as such, shall, in accordance with the provisions applicable to national patents, be ineffective or revocable; this reservation shall not affect protection conferred by the patent in so far as it involves a process of manufacture or use of a chemical product or a process of manufacture of a pharmaceutical or food product.

(c) Conclusion

From the discussion above, it is clear that products in general were assumed to be inherently patentable, but that there were concerns among many of the European drafters with respect to the patenting of chemical products. While those concerns were described as being public policy concerns, the policies were specific to the

[285] M/PR/II, [1019].

[286] ibid, [1003].

[287] ibid, [1004].

[288] It was agreed by the drafters that the reservation would apply only in respect of the products *as such* (see ibid, [1157]).

[289] ibid, [1008].

[290] ibid, [1107].

[291] Consistent with its terms, this provision was deleted as part of the 2000 EPC Revision.

patent system, namely, whether the justifications for granting patents were valid in respect of chemical products, and whether the differential treatment of such products would discriminate unfairly against chemical inventors. Importantly, the justification for the grant of a European patent was conceived not in terms of 'technical progress', as for the Committee of Experts in 1951,[292] but rather in terms of industrial development. Thus, the question considered by the EPC drafters was whether allowing patents for chemical products would impede or facilitate industrial growth. Because of the position taken by the delegations, the resolution of that question was all but inevitable: the fact that doubts regarding the impact of chemical product patents were confined to developing states, made it natural to resolve them by a temporary reservation that would last until such time as those states' chemical industries had developed.

Interestingly, the question of chemical product patents was assumed by the EPC drafters to be an all or nothing question: as the 'invention' would be the chemical product, the patent would protect the product as such.[293] Not considered was the possibility that a product *qua* invention might be something other than a discrete physical artefact, nor (relatedly) that the restriction of patentability to patentable *inventions* might require that a subject matter be conceived in a certain way, having regard to the policy objectives of the system. This can be seen also from the drafters' separate treatment of patent scope, in which the focus was on the technical construction of claims, and not on its underlying policy basis, namely, the consideration required for a patent grant, and its implications for the conception of individual inventions. This was despite the drafters' recognition that the monopoly conferred by a European grant would have to be proportionate to the benefit secured, or the patent would be vulnerable to national revocation.

2. The Requirement for Proportionate Claims

In 1952, the Committee of Experts noted the divergence that existed in national regulations relating to the requirement for patent claims.[294] Further, according to the Secretariat-General's 1953 Comparative Study of national laws, the UK was the only country which expressly allowed revocation for lack of fair basis.[295] Given this, and the drafters' support for a conception of the function of claims that sat 'somewhere between' the UK and continental conceptions[296]—namely, determining '[t]he extent of the protection conferred by the patent' with interpretive

[292] See CM/WP IV (51) PV4, 3.

[293] With the possible exception of the Austrian delegation (see n 272).

[294] EXP/Brev (52) 23, 3–4.

[295] EXP/Brev (53) 18, 38–43.

[296] See C Robinson, 'Patent Protection for Chemical Products in Canada, Great Britain and the United States' (1972) 2 IIC 139 (describing the claims in pre-1977 UK law as defining 'the fences . . . of the monopoly, rather than as, [in Germany] being directed to the central idea of the invention—the *Erfindungsgedanke*').

assistance from the specification[297]—it is unsurprising that the lists of revocation grounds contained in the various draft Conventions did not extend to lack of fair basis itself.[298] Nonetheless, the UK delegation was committed to preserving the rule of *Mullard Radio Valve Co Ltd v Philco Radio & Television Corp*,[299] and insisted that that rule be recognized by the drafters as a basis for revocation of a grant. In so doing, it remained true to its earlier commitment, expressed to the Committee of Experts in 1956, to ensuring that the monopoly conferred by a European patent would not 'exceed the consideration . . . [given] in return':[300]

> [The UK] Courts have always refused to countenance wide and speculative claims, particularly in relatively unexplored fields, based on a disclosure in a small area of the field. Such claims would give an inventor a monopoly out of proportion to the actual inventive contribution which he has made, would be an unwarrantable interference with further research and invention in the field, and would contravene the principle underlying the UK law that the monopoly granted to an inventor should not exceed the consideration which he gives in return to the public. This requirement of the UK law may lead in practice to difficult decisions since the question is basically one of degree, but the UK experts consider that a comparable requirement must, in the general interest, be included in any unified law.

In 1970 the EPC drafters agreed:[301]

> When drafting sub-paragraph (b) [allowing revocation of a patent for insufficiency], the Working Party took into account the balance to be sought between two kinds of problem: the public interest in knowing the exact scope of a patent, on one hand, and the continuity of the decisions taken by the European Patent Office on the other.
>
> . . . It was noted that sub-paragraph (b) could apply, in conjunction with Article 71, to the case raised by the British delegation concerning 'speculative claims', the scope of which would only be evident at a later date, when an inventor is able to make a product covered by the patent, by a process not referred to therein.

[297] See CM (62) 160, [10].

[298] See, eg EXP/Brev (53) 19; CM (61) 97, 20–22 (Appendix V); BR/40/70, 2–3 (discussed in BR/49/70, 30–31); Second Preliminary Draft, Article 133; Munich Conference, Preparatory Documents Drawn up by the Inter-Governmental Conference for the setting up of a European System for the Grant of Patents (Government of the Federal Republic of Germany, 1972) 144–146 (Art 138).

[299] [1936] 2 All ER 920, HL, discussed at pages 86–87, 97–98.

[300] EXP/Brev (56) 2, 9. See further EXP/Brev (56) 2, 7 (describing the UK law of claims as 'differ[ing] radically from [that] of some European countries, according to which the Courts may seek outside the claims for the real invention, i.e. for the monopoly or "forbidden territory"', and continuing: 'In the opinion of the United Kingdom experts the United Kingdom practice is the correct one. . . . Having regard to the views repeatedly expressed by the United Kingdom Courts, it is considered unlikely that any contrary practice would be entertained.')

[301] BR/40/70.

3. Conclusion

It can be seen from this discussion that the drafters of the European patent regime supported both the inherent patentability of chemical products and a prohibition against non-equiparating claims. The question remains as to what this meant, and how the principles would operate inter se. While there is no express discussion of this recorded in the *travaux*, the drafters clearly understood that a patent for a product would protect the product as such, independent of the method by which it was made. Given this understanding, it seems reasonable to presume that they intended that any requirement for equiparating claims would be subject to a conception of the invention as including chemical and other products. Still, it is remarkable that the issue was not expressly discussed by the European drafters after 1951. This is particularly so given that it was being discussed elsewhere in Europe, at the time of the Luxembourg and Munich Conferences, by other European patent specialists, including European patent judges. Further, an indication of their views is given by the report of a 1971 conference.

E. Other European Perspectives on Chemical Product Patents: A Question of Non-Discrimination

In 1971, the German Association of Comparative Law held its annual Conference for Comparative Law in Mannheim, with representatives attending from both civil and common law countries.[302] The working session for the industrial property and copyright law section of the Association took place on 24 September, and was conducted under the theme 'Fundamental Questions of Protection for Chemical Inventions'.[303] Three options for protection were discussed at the Conference: (a) absolute protection for chemical products; (b) utility-linked protection for chemical products; and (c) an intermediary solution allowing (a) or (b), depending on the product's contribution to the art. The speakers were from Canada, Sweden, and Germany, and reported the availability of protection for chemical product patents in their countries as follows:[304]

> Mr Robinson illustrated in his talk that absolute product protection of chemical inventions is recognised in Canada, Great Britain and the United States.[305] In contrast to this, it was shown from Mr Lewin's presentation that the Scandinavian countries, after exhaustive

[302] The proceedings of the Conference are reported at 'Information: Mannheim Conference on Fundamental Questions of Patent Protection for Chemical Inventions' (1973) 3 IIC 357 (Mannheim Conference Proceedings).

[303] ibid, 357.

[304] ibid, 358.

[305] See Robinson, 'Patent Protection for Chemical Products in Canada, Great Britain and the United States'.

discussion of the pros and cons, decided in favour of utility-linked product protection in their uniform patent statutes which became effective on January 1, 1968. In Germany, the question is still not finally settled.[306]

Discussion of the options at the Conference was thorough. Those who supported utility-linked protection did so largely on the basis of patent theory, arguing that utility-linked protection would encourage more vigorous inventive activity and induce patentees to disclose more completely their inventions.[307] By contrast, those who supported absolute protection did so largely on the basis of industrial practice, doubting the relevance of theoretical arguments for the issue.[308] While discussants were split over the ideal result, 'strong notice' was taken of an 'intermediary proposal' made by members of the German judiciary:[309]

> [A]bsolute product protection is certainly appropriate for a compound which is inventive with regard to its structure. Such a product which has previously not existed and which has not been suggested also absolutely enriches technology because it exhibits, as such, novelty, inventive step and technical advance. On the other hand, if inventive quality is due to a new product only in view of its unexpected effects, the product as such having been suggested, then only utility-linked product protection is appropriate for this product.

By tying the scope of monopoly rights to the technical contribution made by the product, this proposal was consistent with the 1956 UK comments regarding claim scope,[310] with the 1956 and 1960 German comments regarding inventions and technical progress,[311] and with the Committee of Experts early characterization of the purpose of the European patent system as being to aid technological progress on the continent.[312] Nevertheless, it was rejected. Indeed, according to the published report of the Conference, the only real point of agreement among participants was the importance of treating all chemical inventions on an equal footing:[313]

> [Supporters of absolute protection] greeted the repeal of the prohibition against product protection effected by the Interim Law as the removal of a discrimination against chemical inventions. However, if only a utility-linked product protection were now to be granted, that would mean a new discrimination because product patents in all other areas of technology would grant protection without limitation to particular uses of the protected article. In response, proponents of utility-linked product protection emphasized that product inventions in other areas of technology also are not protected absolutely, but instead are protected only within the framework of their intended use. An inventive idea is always

[306] See K Nastelski, 'Product Protection for Chemical Inventions in Germany' (1972) 3 IIC 267.

[307] Mannheim Conference Proceedings, 359.

[308] ibid. Some supporters of absolute protection also believed theory to be on their side.

[309] ibid, 363. This proposal was criticized by some participants for being based on an inordinately difficult distinction.

[310] See EXP/Brev (56) 2, 9.

[311] See ibid, 8, 6–7; EXP/Brev (60) 5, 2.

[312] See CM/WP IV (51) PV4, 3.

[313] Mannheim Conference Proceedings, 360.

characterized by a problem and its solution. Both circumscribe the scope of protection. Thus, an article is not infringing or anticipatory if another problem is solved with it. The problem to be solved by an invention, in a patent law sense, implies the utility link of its protection.

Commenting on this aspect of the Mannheim discussion, the authors of its report concluded as follows:[314]

The fact that proponents of both absolute and utility-linked product protection referred to the argument of placing chemical inventions on the same footing as other inventions showed that there exists here an important field for further investigation. It should be clarified to what extent product inventions are generally protected and in what respect product inventions in the chemical arts differ from article inventions in the other arts, possibly even to such an extent that new, independent patent categories must be sought for chemical inventions.

A year after the report of the Mannheim Conference was published, the 'field of further study' recommended in this passage became even more important as a result of the introduction of Article 27.1 of TRIPS. As seen above, that Article requires that patents 'be available for any inventions, whether products or processes, in all fields of technology, provided that they are new, involve an inventive step, and are capable of industrial application'. Further, 'patents shall be available and patent rights enjoyable without discrimination as to the place of invention, the field of technology and whether products are imported or locally produced'.

As has also been seen, in 2000 the EPC was amended to reflect these principles.[315] As a result of that amendment, Article 52(1) of the EPC states as follows: 'European patents shall be granted for any inventions, in all fields of technology, provided that they are new, involve an inventive step and are susceptible of industrial application.' The same change has not been made to section 1(1) of the Act, with the result that it differs from its EPC counterpart. In the face of this difference, however, the UK courts are likely to do as they have done with section 1(2) itself, and defer to the text of Article 52 of the EPC.[316] In combination with section 60 of the Act, which defines infringement with respect to method *and* product patents, this confirms the centrality of the TRIPS non-discrimination principles in contemporary European law and jurisprudence. Given that centrality, and the recurrence of discrimination as a theme in the history above,[317] it seems

[314] Mannheim Conference Proceedings, 360.

[315] See 'Basic Proposal for the Revision of the European Patent Convention' Doc CA/100/00 e 37 ('Article 52(1) EPC has been brought into line with Article 27(1), first sentence, of the TRIPs Agreement with a view to . . . making it plain that patent protection is available to technical inventions of all kinds').

[316] See, eg *Symbian v Comptroller General of Patents* [2008] EWCA Civ 1066, [6].

[317] With respect to s 38A(1), as long as the reason for introducing that section was as recorded in Hansard—to reduce German use of the UK patent system—it seems likely that it would be regarded today as offending Art 27.1 of TRIPS; see *Canada—Patent Protection of Pharmaceutical Products, Request for the Establishment of a Panel by the European Communities* (12 November 1998) WT/DS114/5, [7.92] ('[i]t is quite plausible . . . that the TRIPS Agreement would want to require

appropriate to consider the meaning of that concept, and its implications for the patentability of chemical products. Specifically, it seems appropriate to consider whether the courts could legitimately deny or restrict protection for a chemical product as such on the basis of either their conception of products *qua* invention or their application of the requirement for equiparating claims.

A preliminary issue is what principles ought to be applied in considering the issue of discrimination. The TRIPS Agreement has been approved by a decision of the EC Council,[318] and held to be a European Community Treaty within the meaning of section 1(2) of the European Communities Act 1972, with the result (according to Jacob LJ in *Pozzoli SPA v BDMO SA*[319]) that it ought to be construed in a purposive and teleological manner, similar to the manner of the EPC itself.[320] However, the applicable jurisprudence remains unclear. In particular, it is unclear whether a UK court considering the matter ought to look beyond European and international legal principles. The answer, it is submitted, is that it ought to, for the issue is less compliance with the TRIPS Agreement itself, or even the terms of the EPC, than the proper application of the 1977 Act.

If the courts were to look to national law, one source of potential value would be the Disability Discrimination Act 1995. According to section 24 of that Act, a person discriminates against a disabled person if:[321]

> (a) for a reason which relates to the disabled person's disability, he treats him less favourably than he treats or would treat others to whom that reason does not or would not apply; and (b) he cannot show that the treatment in question is justified.

In its focus on reasons that 'relate to' the prohibited ground, this test would seem to require a similar causal connection to that required by Article 52(1).[322] Further, a comparison of the House of Lords and WTO decisions in *London Borough of Lewisham v Malcolm*[323] and *Canada–Patent Protection of Pharmaceutical Products*[324] respectively suggests that similar principles govern the application of section 24 as govern the application of Article 27.1 itself. Extrapolating from *Malcolm*, whether a legal principle is discriminatory depends on the underlying reason for the principle and its relationship to the relevant prohibited ground, and potentially also on its comparative impact. According to the WTO Panel hearing the *Canada* case, the

governments to apply exceptions in a non-discriminatory manner, in order to ensure that governments do not succumb to domestic pressures to limit exceptions to areas where right holders tend to be foreign producers.')

318 See [1994] OJ L336/1.

319 [2007] EWCA Civ 588, [4].

320 See page 110, including n 10.

321 This is notwithstanding the UK courts' treatment of the EPC as the relevant source of law in this area; ibid.

322 Art 27.1 prohibits 'discrimination *as to* . . . the field of technology'.

323 [2008] UKHL 43.

324 *Canada—Patent Protection of Pharmaceutical Products: Complaint by the European Communities and their member States, Report of the panel* (17 March 2000) WT/DS114/R.

word 'discriminate' must be given its ordinary meaning, thus covering direct and indirect discrimination but not extending to all forms of differential treatment.[325] In particular, 'Article 27.1 does not prohibit bona fide exceptions to deal with problems that may exist only in certain product areas',[326] noting that 'preoccupation with the effects of a statute in one area does not necessarily mean that the provisions applicable to other areas are a sham, or of no actual or potential importance', and that the statute is consequently a (discriminatory) sham.[327] On the other hand, 'to the extent the prohibition of discrimination does limit the ability to target certain products in dealing with certain of the important national policies referred to in Articles 7 and 8.1, that fact may well constitute a deliberate limitation rather than a frustration of purpose.'[328]

Applying these principles, it seems that neither a threshold exclusion of product per se claims resulting from a conception of products *qua* invention, nor a test of claim validity such as that in *Olin Mathieson Chemical Corp v Biorex Laboratories Ltd*,[329] would discriminate as to a subject matter's field of technology, nor as between products and methods themselves, regardless of whether it impacted disproportionately on chemical products and claims. This is because the reason for the test would be unrelated to such matters, even if the fact which triggered its application—for example, the disproportionality of rights conferred by a patent—were due to the chemical (or structural) nature of the claim.[330] This is important, for it suggests that neither pre-1977 conceptions of the 'invention' as a method rather than an artefact, nor Graham J's test in *Olin Mathieson*, ought to offend the principles of non-discrimination that run through the Act and European patent jurisprudence. Indeed, one could argue that those principles *require* such a test. That argument is consistent with the understanding of the requirement for an invention implicit in that test, as anchoring the system to its original purpose, and with the understanding of that purpose itself as the reference point for ensuring the technological neutrality of the patent system.

325 *Canada—Patent Protection of Pharmaceutical Products*, [7.94].

326 ibid, [7.92].

327 ibid, [7.104].

328 ibid [7.92]. Art 7 in particular provides as follows: 'The protection and enforcement of intellectual property rights should contribute to the promotion of technological innovation and to the transfer and dissemination of technology, to the mutual advantage of producers and users of technological knowledge and in a manner conducive to social and economic welfare, and to a balance of rights and obligations.'

329 [1970] RPC 157, Ch, discussed at pages 97–98.

330 The reason is the nature of the relevant comparator, which, applying the principles from *Malcolm* [2008] UKHL 43 at least, would be a non-chemical product that, because of the scope of its associated claims, and the impact of a patent on the field of research, raises issues of disproportionate monopoly protection. This analysis is consistent with the interpretation of Art 52(2) of the EPC supported in J Pila, 'Dispute over the Meaning of 'Invention' in Art. 52(2) EPC—The Patentability of Computer-Implemented Inventions in Europe' (2005) 36 IIC 173, 185–187. See also G Dinwoodie and R Dreyfuss, 'Diversifying Without Discriminating: Complying with the Mandates of the TRIPS Agreement' (2007) 13 Michegan Telecommunications and Technology L Rev 445, 451–453.

Before concluding, it remains to consider the intention behind section 1(2) of the Patents Act, and the legislative context in which that section was itself introduced.

F. The Intention Behind Section 1(2) of the UK Patents Act

As has been seen, the UK delegation had a strong presence in the EPC's preparation, and frequently found itself in opposition to the other delegations on the issue of inherent patentability. The central sources of tension between the UK and other EPC states were the relative importance of flexibility and certainty in formulating a legal definition of the invention, and the differing requirements of those objectives in the context of individual provisions and subject matter. Important in this regard were two specific positions of the UK delegation that were repeatedly expressed throughout the debates. The first was that computer programs, narrowly defined should be excluded from the scope of inherent patentability, and the second was that the interests of ensuring the EPC's literal harmony with the PCT should not override matters of substance. Similarly in discussions of patent invalidity, the UK's agreement to a compromise position whereby a claim would determine the extent of protection but be interpreted in light of the specification, did not dissuade it from asserting the fundamental point that 'the monopoly granted to an inventor should not exceed the consideration which he gives . . . to the public'.[331] Implicit was a recognition that however important harmonization might be, it could not be the single determinant of reform. Also implicit was a recognition of the fundamental question at the heart of the system: at what point does the monopoly conferred by a patent exceed the consideration it secures for the public?

Against this background, it is interesting to consider the discussions that were being had among UK reformers with respect to patents during the period of the EPC debates. In addition to contextualizing the UK's position in those debates, those discussions provide the background to the Patents Act, including its model of inherent patentability. Two immediate questions that arise in that context are why the UK changed the wording of Article 52(2) and (3) when transposing them into section 1(2) of the Patents Act, and what effect it intended harmonization to have on pre-1977 patent jurisprudence?

1. The Patents Act in its Legislative Context

The Patents Act was the culmination of a national process of reform that began in 1961 with the establishment of the Patents Liaison Group (PLG), and that ended in 1977 with the enactment of the Patents Bill 1977[332] and its accompanying

[331] EXP/Brev (56) 2, 9.

[332] cf Hansard HL (series 5) vol 379, cols 234–237 (24 January 1977).

Explanatory and Financial Memorandum.[333] During the sixteen-year period between those two events, three reports into the UK patent system were published by the Government that reflect, collectively, the majority of the ground work undertaken at the national level in preparation for the Act. They are the Report of the Patents Liaison Group published in 1965,[334] the Report of the Banks Committee published in 1970,[335] and a White Paper of 1975 entitled *Patent Law Reform*.[336]

2. The PLG Report, 1965

The establishment of the PLG in 1961 was precipitated by the UK's desire to enter the EEC in combination with the then recent proposals for the creation of a Common Market patent and EEC patent system. Hence its initial terms of reference, which were to formulate views on those proposals and, following the conclusion and signature of the SPC in 1963, to report on the changes that would be required to the Patents Act 1949 in order for that Convention to be ratified.[337]

From the outset, the PLG recognized the 1949 Act's definition of 'invention' as a central matter for review under its terms of reference,[338] on account particularly of its archaic language, and the restrictive and uncertain manner in which it had theretofore been applied.[339] Those factors, combined with the definition's exclusion of inventions 'contrary to law' in contradiction of Article 2(a) of the SPC, led it to conclude that a fresh start to inherent patentability was required, involving a new and unambiguous definition that would resolve contemporary uncertainties in a manner consistent with the SPC and reflective of contemporary legal thought.[340] The PLG also emphasized the need to make clear, in formulating that definition, which aspects of the historical case law would continue to apply.[341]

Despite this, when it came to consider the precise form that the proposed 'new and unambiguous' definition should take, the PLG demurred, eventually deciding

[333] Explanatory and Financial Memorandum to the Patents HL Bill 1976–1977. On the legislative background to PA, s1(2), see especially Hansard HL (series 5) vol 379, cols 1402–1411 (15 February 1977); vol 381, cols 841–847 (29 March 1977); vol 383, cols 108–110 (9 May 1977).

[334] United Kingdom, *United Kingdom Patent Law: The Effects of the Strasbourg Convention of 1963* (Cmnd 2835, 1965) (PLG Report).

[335] See n 37. For a discussion, see GW Tookey, 'The British Patent Systems: A Survey of the Principal Recommendations of the Banks Committee' (1971) 2 IIC 117.

[336] United Kingdom, *Patent Law Reform* (Cmnd 6000, 1975) (White Paper).

[337] See PLG Report, 3.

[338] ibid, [6]–[7].

[339] ibid, [6]–[7], [32]. The PLG was particularly concerned that the courts not interpret the existing definition as excluding agricultural subject matter from patentability in contradiction of the SPC (see ibid, [32]).

[340] See PLG Report, [32]. The PLG's criticisms of the historical definition of the 'invention' were similar to those made by the Swan Committee in 1947. See United Kingdom, *Patents and Designs Acts: Final Report of the Departmental Committee* (Cmd 7206, 1947) [122]–[128].

[341] PLG Report, [6].

that an exhaustive statement of inherent patentability should not be attempted. The reasons given were as follows:[342]

> The field covered by science and technology is even now extremely wide and, with the progress of time, the boundary of patentable subject matter will inevitably increase. It is considered wiser therefore not to seek to put a limit on what should be regarded as an 'invention'. The group recommends therefore that the new Act should not define 'invention' with any greater particularity than is necessary to conform to the [SPC]. Although it would be considered useful if somewhere in the Act it were stipulated that claims may relate to a substance, product, apparatus or process, the group does not believe these terms should be employed as part of any definition of 'invention'. The group recognises that the deletion of the existing definition and its replacement by wording to the effect that patents may be granted for inventions which can be made or used in any kind of industry, including agriculture, will give rise to some uncertainty until the courts have established some case law. But it is convinced that it is better to follow this course, which has the merit of being flexible instead of restrictive.

Thus, the continued use of the term 'invention' to denote inherently patentable subject matter was never questioned by the PLG, and was left undefined in order not to exclude unwittingly new types of subject matter that might arise from future and unforeseen scientific or technological advancements, and thereby fetter unintentionally the scope of the patent system. As a result, and consistent with the UK's position during the EPC debates, concerns of legal certainty were allowed to outweigh the demands of flexibility.

However, having eschewed the idea of an express definition, the PLG nevertheless recommended the insertion of a list of subject matter 'such as are recognisable as legitimate limitations of the meaning of "invention" as used in the [SPC]'[343] for the explicit purpose of restricting the uncertainty that removing the existing definition would otherwise entail.[344] That list, which the PLG emphasized as being non-exhaustive and intended 'not to prejudice unduly, by implication, the generality of the area of patentability', was as follows:[345]

> (i) Methods of calculation; theoretical ideas and scientific principles.
> (ii) Business schemes (e.g. methods of office management, product promotion etc.).
> (iii) Commercial, financial and propaganda schemes.
> . . .
> (iv) Treatment of human beings.
> . . .
> (v) Designs or arrangements in which the novelty resides solely in appeal to the eye.
> (vi) Designs or arrangements which serve only to convey information and in which the novelty resides solely in the information conveyed.

[342] ibid, [35].
[343] ibid, [36].
[344] ibid.
[345] ibid, [42].

Little substantive justification was offered by the PLG for these exceptions. Regarding (iii), the Group referred to the historical unpatentability of schemes nationally and internationally, and the importance of confirming that position expressly in order to 'avoid uncertainty and to prevent any attempt to obtain patents for [schemes] on the grounds that they are of use in industry'.[346] Regarding (iv), the PLG distinguished 'non-therapeutic methods, e.g. cosmetic treatment', which they regarded as inherently patentable.[347] Regarding (v), it referred to the importance of avoiding overlap between patent and industrial design protection.[348] Finally, regarding (vi), the PLG stated simply that this exception 'excludes printed sheets and analogous cases'.[349]

The only other aspect of the PLG's report of immediate relevance is procedural, and concerns its recommendation that the express powers of the Comptroller-General in section 10(1)(c) of the 1949 Act to refuse a patent application[350] be retained and extended to include 'any apparatus or process consisting of a collocation of known integers or steps each of which effects its known function and contributes to a merely aggregate result'.[351] Having made that recommendation, however, the PLG expressed the view that the resulting power of exclusion would relate to 'obviousness rather than "invention"'[352]—adding that the Comptroller 'should have the power to deal with both during examination'.[353]

3. The Banks Report, 1970

The PLG Report of 1965 formed the basis for a further and considerably more extensive review of the UK patent system that was commissioned by the UK Board of Trade in 1967 and undertaken by a new Committee headed by Mr Banks, a former Deputy Chairman of the British Petroleum Company. The Committee's initial terms of reference were to[354]

> examine and report with recommendations upon the British patent system and patent law, in the light of the increasing need for international collaboration in patent matters, and, in particular, of the UK Government's intention to ratify the [SPC].

The result was the publication in 1970 of the second official report on the UK patent system within five years, this time dealing exhaustively with all of its aspects, including inherent patentability and fair basis of claims.

[346] PLG Report, [42].
[347] ibid.
[348] ibid.
[349] ibid.
[350] See Patents Act 1949, s 10(1)(c).
[351] PLG Report, [43].
[352] ibid.
[353] ibid.
[354] Banks Report, xvii. The Banks Committee's terms of reference were expanded two months after its formation.

With respect to inherent patentability, the Committee proceeded in two parts.[355] First, it set out the history and judicial interpretation of the definition of 'invention' in the 1949 Act, and considered the amendments that would be required to that interpretation for ratification of the SPC. And second, it considered whether UK law would be improved by adopting the PLG's recommendation that 'invention' remain undefined in future legislation, and what, if any, statutory inclusions and exclusions in that context ought to be permitted. On the first of these points, the Committee identified five categories of subject matter as having 'never been considered to be an "invention"' within the meaning of the Act:[356]

(i) Methods of calculation; theoretical ideas and scientific principles.
(ii) Business schemes (e.g. methods of office management and product promotion).
(iii) Commercial, financial and propaganda schemes.
(iv) Treatment of human beings.
(v) Designs or arrangements which serve only to convey information and in which the novelty resides solely in the information conveyed, e.g. normally printed sheets and mere presentations of information.

The report also noted the considerable uncertainty that existed regarding the inherent patentability of certain other subject matter, including specifically computer programs and agricultural, horticultural and biological processes,[357] and suggested that such uncertainty was unlikely to be resolved under the 1949 Act on account of the very few cases concerned with inherent patentability that ever reached the High Court.[358] Indeed, it was precisely this absence of judicial authority that was suggested by the Committee to have enabled a Patent Office practice of granting patents for computer programs in deference to the high threshold which had then recently been established for rejecting applications.[359] Implicit in that suggestion was a belief by the Committee that computer programs were not inherently patentable, which, while seemingly inconsistent with the High Court of Australia's liberalizing decision in *National Research Development Corp v Commissioner of Patents*,[360] was consistent with the line of UK cases that immediately followed that decision involving information-related subject matter. The effect of those cases, in combination with the Committee's understanding of computer programs as mathematical methods, supported the view that such programs were not patentable, despite their potential industrial use, which view was also expressed by the PLG, and the UK delegation to the Luxembourg Conference.[361]

355 Banks Report, [210]–[243].

356 ibid, [213].

357 ibid, [218].

358 ibid.

359 ibid, [482]. On the undesirable influence of Patent Office practice on patentable subject matter, see also Hansard HL (series 5) vol 379, cols 290–291 (24 January 1977) (regarding the PO practice of refusing to protect contraception, discussed at pages 88–90).

360 (*NRDC*) (1959) 102 CLR 252, discussed at pages 90–95.

361 See also MW Molyneaux, 'A History of Patenting Software-Related Inventions in the United Kingdom: Part II' [1996] Comp and Telecommunications LR 51, 51; L van Raden, 'Technology Dematerialised: Another Approach to Information-Related Inventions' [1996] EIPR 384, 385.

After considering the state of existing law, the Banks Committee turned to consider the PLG's recommendation that no express definition of 'invention' ought to be included in future legislation. Noting the longstanding criticisms of the definition in the Statute of Monopolies 1623,[362] and the possibility of using ratification of the SPC as an opportunity to make 'a new start',[363] the Committee proceeded to reject the PLG's recommendation, concluding that the existing definition ought to be retained.[364] In its view, the need for flexibility in the face of future scientific and technical developments had been, and promised to continue to be, well served by that definition,[365] with the result that abolishing it 'might throw doubt on the entire body of case law' developed since 1623, resulting in 'even greater uncertainty than [existed] at present'.[366] That view, combined with the Committee's recommendation that a new, specialist Patents Court be created to hear appeals from the Comptroller,[367] determined its recommendation that the then existing 'manner of new manufacture' definition ought to be retained, albeit subject to a new proviso, alterable by statutory instrument or alternative means other than primary legislation,[368] stipulating 'for the avoidance of doubt, that certain specified matter is either within or outside the definition'.[369] In relation to the content of that proviso the Committee recommended as follows. First, agricultural and horticultural processes ought to be expressly included within the definition of the 'invention',[370] consistent with the Australian decision in *NRDC* and the PAT's decision in *General Electric Co Ltd's Application (Patent)*.[371] Second, 'methods of veterinary treatment, both curative and diagnostic', ought to be excluded,[372] and 'processes for treating human beings with known substances' not made the subject of an express inclusion.[373] Third, computer programs—'in whatever form . . . presented e.g. a method of programming computers, a computer when programmed in a certain way and where the novelty or alleged novelty lies

[362] 21 Jac I c 3; see Banks Report, [224].

[363] ibid.

[364] ibid, [226].

[365] In support the Committee referred to *NRDC* (1959) 102 CLR 252, HCA and *Swift and Co v Commissioner of Patents* [1960] NZLR 775, NZSC, stating: 'We have no reason to suppose that the United Kingdom Courts will not in future take the liberal view of patentable inventions which was taken by the Australian and New Zealand Courts [in those cases]' (Banks Report, [226]). That statement contradicted the PLG's conclusion on this point.

[366] Banks Report, [226].

[367] ibid, [227].

[368] ibid, [229].

[369] ibid, [227].

[370] ibid, [231].

[371] [1961] RPC 21, PAT.

[372] Banks Report, [235].

[373] ibid, [240]. The reason for the Committee's recommendation in relation to uses of known substances in therapeutic methods was that permitting protection for such uses would extend patentability beyond its existing scope to permit patents for therapeutic methods, and thereby put the UK in conflict with the majority of other countries and in possible breach of the SPC's requirement for industrial applicability (see ibid, [239]–[240]).

only in the program'[374]—ought to be excluded in order to resolve the uncertainty created by the Patent Office's liberal granting practices, and to affirm generally the historical exclusion from UK law of all 'methods of mathematical calculation or sets of instructions for carrying out such calculations'.[375] In reaching this last conclusion, the Committee referred also to the practical difficulties associated with searching the prior art of computer programs,[376] emphasizing the need for the Government to revisit the proposed exclusion in the future in order to take account of international developments.[377] This was consistent with the tenor of the Committee's conclusion overall, which expressed confidence that its recommendations would satisfy the three objectives emphasized by the PLG in its own report five years earlier, namely, to resolve existing uncertainties, to meet the requirements of the SPC, and to ensure the UK's future capacity to respond to new developments and international trends.[378] It also supported the PLG view of inventions as not including all objects of industrial use.

Reflected in this account is the concern of the Committee with the institutional wellbeing of the UK system. That concern was also apparent from the Committee's initial statement of the essential problem it faced, which was 'how to ensure an efficient national system which is fitted to take the maximum advantage of developments in international relationships on patenting, as and when these may occur'.[379] However, the Committee also considered more substantive patent law matters, including (in chapter 2 of its Report) 'the value of patents'.[380] In its discussion of that issue, it noted the absence of available information regarding the economic impact of patents,[381] and resigned itself in consequence to basing its assessment of their value 'on a study of the history of the development of the patent system in this country and elsewhere and on the representations based on a wide range of experience which were made to us both in writing and orally'.[382] Before doing so, however, the Committee affirmed as 'the essential test of the value of patents . . . whether they contribute anything, and if anything, how much, to the wellbeing of the country'.[383] This concern with the contribution of patents to the public interest was further reflected in its discussion of the statutory requirement for fairly based claims, as well as its concluding statement in chapter 2. Regarding the former, the Committee supported amending the section 4(4) requirement of the 1949 Act to strengthen the necessary correspondence between claims and their

[374] Banks Report, [487].

[375] ibid, [483]. cf the definition of 'computer program' proposed for inclusion in the PA, reproduced in n 216.

[376] ibid.

[377] ibid, [487].

[378] ibid, [229].

[379] ibid, xviii.

[380] ibid, [40]–[68].

[381] ibid, [44].

[382] ibid, [46].

[383] ibid, [43].

description in order 'that a claim should not necessarily be justified merely by an equivalent statement in the description of the invention if both are evidently wider than what has actually been invented'.[384] Once amended, the provision would require 'that the claim should not be unduly wide having regard to the matter disclosed in the complete specification'.[385] Regarding its conclusion, the Committee stated as follows:[386]

> There remains the qualification that the complexity of the administration of the system, which arises partly from the surge of technological discovery and partly from the concept that a patent is a piece of property, with its legal connotations, ought not to be allowed to detract materially from the value of patents to the national interest.

4. The White Paper, 1975

As seen above, in the five-year period between publication of the Banks Report in 1970 and the Government's White Paper in 1975, the CPC and EPC were each written and concluded. In the same year as signing the CPC and thereby committing itself to developing a single standard for European patentability, the UK Government published the White Paper with the express intention of 'set[ting] out the broad scope of the Government's proposals [for a new patent law] as a basis for early public discussion'.[387] Given this, the Paper would seem to offer the best insight into the intention behind the Act, including its principles of inherent patentability,[388] and that it does is apparent from its introduction, as follows:[389]

> The Government accepts most of the main recommendations of the Banks Committee and intends to introduce legislation during the next Parliamentary Session revising the patent system:
>
> (i) so that it is in tune with modern trends and sufficiently flexible to accommodate future changes in technology;
> (ii) to streamline procedures for obtaining patents;
> (iii) to establish a Court specially equipped to hear patent cases;
> (iv) to take account of international developments in which successive British Governments have played a leading part in the interest of our industry and commerce.

[384] Banks Report, [533].
[385] ibid.
[386] ibid, [68].
[387] White Paper, [6].
[388] See also United Kingdom, *Patent Law Reform: A Consultative Document* (1975) [38]–[41] (described in its Introduction as a supplement to the White Paper).
[389] White Paper, [5].

Diagram G UK reports on inherent patentability, 1965 to 1975

	PLG Report 1965	Banks Report 1970	White Paper (and Patents Act 1977)
Terms of reference	Formulate views on international developments and (after 1963) report on the changes required for compliance with the Strasbourg Convention.	Report on the changes required to comply with the Strasbourg Convention.	Formulate proposals for public discussion.
Explicit concerns	Ensure legal certainty and consistency with the Strasbourg Convention.	Ensure the flexibility required to accommodate future scientific and technical developments whilst limiting the risk of future legal uncertainty.	Reflect modern trends, whilst ensuring sufficient flexibility to accommodate future technological and international developments.
Proposal	Omit an express definition of 'invention' but insert the following non-exhaustive list of non-inventions: • Methods of calculation; theoretical ideas and scientific principles • Business schemes • Commercial, financial, and propaganda schemes • Treatments of human beings • Designs or arrangements in which the novelty resides solely in appeal to the eye • Designs or arrangements which serve only to convey information and in which the novelty resides solely in the information conveyed.	Retain existing (manner of new manufacture) definition of 'invention' and insert the following list of non-inventions (to be amendable by something less than full-scale legislation such as a statutory instrument): • Agricultural and horticultural processes (inventions) • Methods of veterinary treatment, both curative and diagnostic (not inventions) • Computer programs, in whatever form presented (not inventions).	Omit an express definition of 'invention' but insert the following non-exhaustive list of (non-)inventions: • Discoveries, scientific theories and mathematical methods as such (non-inventions) • Literary, dramatic, musical, and artistic works and other aesthetic creations as such (non-inventions) • Schemes, rules, or methods for performing a mental act, methods of playing a game or doing business, and computer programs as such (non-inventions) • Presentations of information as such (non-inventions) • Microbiological processes and the products thereof (inventions).

(Cont'd)

Diagram G UK reports on inherent patentability, 1965 to 1975 (*Cont'd*)

	PLG Report 1965	**Banks Report 1970**	**White Paper (and Patents Act 1977)**
Existing law	*Not inventions*: • Subject matter proposed for exclusion, above • Agricultural subject matter and inventions contrary to law • Possibly also computer programs (status uncertain).	*Not inventions*: • Subject matter proposed for exclusion (status uncertain) • Methods of calculation; theoretical ideas and scientific principles • Business schemes • Commercial, financial, and propaganda schemes • Treatments of human beings • Designs or arrangements which serve only to convey information and in which the novelty resides solely in the information conveyed • Therapeutic methods.	[Not considered, but general acceptance of the Banks Report.]

Of these three objectives—relevance, flexibility, and consistency with international trends—it was the second that was emphasized in the Paper's discussion of inherent patentability:[390]

Throughout the history of our patent system the emphasis has been on enabling patents to be granted for contributions to knowledge which have industrial application. For many years invention has been defined as 'manner of new manufacture' and although this definition has been given a flexible interpretation, it remains uncertain whether it is sufficient to embrace all future changes in technology which are deserving of patent protection. For this reason, the Government proposes to abolish the existing definition and to provide that patents may be granted for any invention which is new, involves an inventive step and is susceptible of industrial application. There will also be a non-exhaustive list of exclusions from patentability, subject to amendment by Order in Council. Following Banks, this list will exclude, amongst other things, computer programs and methods for treating human or animal bodies by surgery or therapy.

How the exclusion of such subject matter could be reconciled with a view of the patent system as existing to protect 'contributions to knowledge which have industrial application' was not explained, and nor were the implications of that view for the PLG and Banks Committee's conceptions of inventions as not including all objects of industrial use.

Publication of the White Paper resulted in the introduction of the Patents Bill 1977 during the subsequent Parliamentary Session of the House of Lords. That Bill included what would become section 1 of the PA, and was described in its accompanying Explanatory and Financial Memorandum as giving effect to the White Paper and 'certain international patent treaties', confirming, as the section's principal objectives, flexibility, relevance, and consistency with international conventions.[391] In addition, while the realization or otherwise of the first and second of those objectives was left entirely in the hands of decision-makers, the third was given explicit recognition in section 130(7), reproduced above at page 123.

5. Conclusion

The history of the national reform process that led to the enactment of the Patents Act reveals the shared concern of the PLG and Banks Committee to formulate a flexible model of inherent patentability that would not only be consistent with modern trends and responsive to future scientific and technological advances, but that would also clarify the position of certain classes of subject matter and thereby remove the uncertainties of existing law. It also reveals, however, the different views as to what form the model should take, despite each group's agreement on the scope of inherent patentability at the time of their reports. Of note in that regard is the similarity of the 1965 model of the PLG with the model subsequently enshrined in the EPC, and the similarity of the 1970 model of the Banks Committee with the model repeatedly advanced by the UK delegation at the Luxembourg and

[390] White Paper, [12].

[391] The same objectives are reflected in the parliamentary debates on PA, s 1(2); for the references, see n 333.

Munich Conferences. For example, while the PLG recommended the omission of an explicit definition of the 'invention' in preference for a non-exhaustive list of schemes and other inherently unpatentable subject matter similar in form to that contained in Article 52(2) of the EPC, the Banks Committee recommended retaining the Statute of Monopolies-based definition in a readily amendable instrument with an additional provision confirming the exclusion from that definition of computer programs and methods of veterinary treatment.

In contrast to the PLG and Banks Committee, the overriding concern of the UK Government in formulating its White Paper and subsequent legislation five years (and two European Conventions) later was to ensure consistency between the UK and European principles of inherent patentability, thereby fulfilling its commitment to creating a harmonized patent system with uniform standards of protection. Overall, this connection between the Patents Act and the EPC must be taken to have been the central aim of the UK Legislature in removing the Statute of Monopolies-based definition, despite its professed support for the Banks Report and recognition of objectives beyond harmonization, thus of greater importance than the understandings of the reformers regarding the impact of that change on substantive law. Indeed, even putting the Legislature's intent aside, there remains the fact of the Executive's ratification of the EPC, and the political significance it creates for the development of divergent standards of patentability by the UK courts, which significance was recognized by the UK parliamentary drafters.

Nevertheless, it would be wrong to read the Patents Act as having been intended to displace the old regime. Among other things, such a reading sits uneasily with the parliamentary record, as well as with the commissioning of the Banks Report, and the subsequent extension of its terms of reference, so soon before the EPC's conclusion. Still, it seems clear from the legislative history that, at the time of its enactment in 1977, section 1(2) of the Patents Act was viewed as reflecting the status quo of pre-1977 law; a view consistent with the assertion of one of the UK experts involved in framing the SPC, PCT, and EPC soon after the Patents Act's commencement, that the changes introduced by that section were 'more apparent than real'.[392] That belief, reflected also in the reform initiatives discussed above, must be taken at least in part to have influenced the UK Legislature's commitment to the EPC, thus leaving obvious potential for future problems in the event that the EPO diverged from pre-1977 UK law in its construction of Article 52. Were such a divergence to occur, the UK courts would be in the difficult position of needing either to accept section 1 as having initiated a fresh start to inherent patentability, or to sacrifice the objective of legislative harmony in contradiction of the Act's tenor, if not terms, in order to preserve the historical continuity and doctrinal integrity of its own Statute of Monopolies-based jurisprudence. Compounding these issues was the virtual absence of principled policy consideration throughout

[392] See E Armitage [Comptroller-General of Patents], 'The New British Patent Legislation' (1978) 9 IIC 207, 210.

the European debates, and the political and doctrinal vacuum into which Article 52(2) and (3) were consequently introduced. Specifically, if the UK courts were to abandon pre-1977 jurisprudence in determining issues of inherent patentability, it was not clear what they might reach for in its place.

G. Conclusion

The EPC model of inherent patentability is a political construct that owes its existence and form to the Contracting States' pursuit of a harmonized system for the granting of patents. To this extent, it stands a long way from the pre-1977 UK model, which was a judicial construct derived from four centuries of common law jurisprudence that, at the time of the EPC's commencement in 1977, had attracted relatively little political attention since 1623.[393] In addition, and unlike its EPC counterpart, under this latter model the need for an invention was a threshold requirement in a comprehensive patent system that regulated all aspects of a patent's life, and not merely the pre-grant phase regulated by the EPC.

The potential significance of these differences for substantive understandings of inherent patentability is great, and is a theme that returns continually in contemporary European patent jurisprudence. Underpinning that significance, however, is a further difference that derives from the substance of the UK and EPC models of inherent patentability themselves, as they existed at the time of the EPC's commencement in 1977. On their face, Article 52(2) and (3) of the EPC reflect a less flexible and otherwise more restrictive understanding of inherent patentability than the pre-1977 UK conception. Further, in the 1960s and early 1970s when the most important discussions on those provisions were taking place, the implications of the Australian High Court's decision in *NRDC* for traditional and emergent technologies in UK law were still being unravelled, leaving the status of many of the subject matter referred to in Article 52 indeterminate. Principal among those subject matter were biological materials, computer programs and other information-related subject matter, and methods of medical treatment. As has been seen, it was likely the indeterminate position of those subject matter that fuelled the UK drafters' desire for legislative certainty amid flexibility and harmonization during the decades in question.

The result of that desire, however, was to encourage the adoption of a model of inherent patentability which is unlike the Statute of Monopolies-based model. Its difference is the product of two specific factors. The first is the negative terms in

[393] Aside from the reviews discussed here see, among others, United Kingdom, *Patents and Designs Acts: Final Report of the Departmental Committee* (Cmd 7206, 1947) [122]–[128]; United Kingdom, *Report of the Commissioner Appointed to Inquire into the Working of the Law Relating to Letters Patent for Inventions* (5974, 1865) ix–x C (1st series). For a discussion, see Banks Report, [17]–[27]; Tookey, 'The British Patent Systems: A Survey of the Principal Recommendations of the Banks Committee', 117–118.

which Article 52(2) and (3) are couched, and the second is their focus on specific categories of subject matter, which, in combination, support a formalistic definition in contrast to the philosophical and policy-based definition of pre-1977 legislation. Further, the fact that Article 52(2) and (3) of the EPC are a definition rather than, for example, a list of patentability exclusions, creates the impression that the paragraphs are connected inter se, and grounded in a coherent, positive conception of the invention. Hence the question presented for determination by the EPO and decision-makers of each of the EPC's Contracting States in 1977, namely, what is the conception of inherent patentability that Article 52(2) and (3) support, and what factors other than the literal terms of those provisions may and ought to be considered in answering that question? Further, what are the implications of the answer to that question for the conception of individual subject matter *qua* (patentable) invention? Adding to the UK courts difficulty in answering these questions were two further (doctrinal and policy) factors. The first was the provision in section 125(1) of the Act of a second definition of the 'invention', as follows:

> [A]n invention for a patent for which an application has been made or for which a patent has been granted shall, unless the context otherwise requires, be taken to be that specified in a claim of the specification of the application or patent, as the case may be, as interpreted by the description and any drawings contained in that specification, and the extent of the protection conferred by a patent or application for a patent shall be determined accordingly.

And the second was the challenge confronted by the European and UK drafters themselves: how to 'fix the conditions for the grant of patents' while ensuring the flexibility required to respond to extraneous (political, technological, and jurisprudential) developments.[394]

In concluding, it is appropriate to return to the questions raised at page 179 regarding the reasons for the use of different language in transposing Article 52(2) and (3) into the Patents Act, and the intended impact of that transposition on pre-1977 jurisprudence. Notwithstanding some ambivalence on these issues, consistent with the drafters' dual aims of harmonizing law with international conventions and legislating law for the UK itself,[395] Hansard records the drafters' intent that any conflict between these aims be resolved in favour of harmonization, at least in the context of section 1(2) itself. That intent is apparent from the reasons given for rejecting proposals to clarify the restrictive operation of that section by defining the term 'computer program' in paragraph (c)[396] and deleting the phrase 'among other things' in the section's introductory phrase.[397] It is also apparent from the

[394] See pages 147–148 (with respect to the European drafters; n 333 (with respect to the UK drafters).

[395] See Hansard HL (series 5) vol 379, cols 1407–1408 (15 February 1977) per Lord Oram, vol 379, cols 236–237 (24 January 1977) per Lord Elwyn-Jones (LC); Explanatory and Financial Memorandum to the Patents HL Bill 1976–1977.

[396] See Hansard HL (series 5) vol 381, cols 841–844 (29 March 1977), vol 383, col 110 (9 May 1977). The definition is reproduced in n 375.

[397] See ibid, vol 379, cols 1403–1408 (15 February 1977).

understanding of section 130(7) of the Act, as committing future UK courts to the harmonious development of patentability and patent validity principles, not 'more or less' or 'roughly', but in so far as would be possible having regard to the different drafting styles and texts of national EPC transpositions.[398] And finally, it is apparent from the rejection of a proposal to amend section 1(5) to recognize the right of the Secretary of State to prescribe different criteria for the grant of a patent under UK and EPC law, 'to take account of developments in technology in this country which may not have been reached in [other EPC] Member States'.[399] As expressed in the House of Lords, such an amendment was 'undesirable because it invites the Secretary of State to break the basic obligation of the European Patent Convention, that is to treat European patents in the same way as British patents'.[400]

398 See ibid, vol 381, cols 854–855 (29 March 1977) per Lord Elwyn-Jones (LC).
399 ibid, vol 379, col 1410 (15 February 1977) per Lord Lyell.
400 ibid, vol 379, cols 1410–1411 (15 February 1977) per Lord Oram.

5

Article 52(2) and (3) of the EPC According to the European Patent Office

A. Introduction

The drafting history of Article 52 of the European Patent Convention[1] (EPC) reveals the central objective in framing that provision to have been the harmonization of its literal terms with the terms of other international agreements, including particularly Articles 1 and 2(b) of the Strasbourg Convention[2] (SPC) (for Article 52(1)), and Rule 39.1 of the Patent Cooperation Treaty[3] (PCT) (for Article 52(2)). To the extent that deviations from that objective were permitted, they came at the eleventh hour with the aims of ensuring two things: first, that the drafters not be misconstrued as having intended Article 52(2) to be interpreted broadly; and second, that the exclusions of Article 52(2) be confined to non-inventions. As seen in chapter 4, the first of those aims accounted for the extension of the 'as such' proviso of Article 52(3) beyond scientific discoveries to cover all Article 52(2) subject matter, and for the relocation of the computer programs exclusion from its own provision to Article 52(2)(c). The second accounted for the removal of the medical treatments exclusion from Article 52(2) to Article 52(4), and its recasting from an exclusion of inherent patentability to an exclusion of industrial applicability.

Also revealed in the drafting history of Article 52 of the EPC is the drafters' equivocation regarding the conceptual basis for inherent patentability and the constitutive properties of inventions themselves. Indeed, a central effect of the drafters' emphasis on literal harmony as the governing objective in formulating Article 52 was to obviate the need to consider either of those issues in depth. Their discussions were, however, nevertheless instructive, in revealing the following two sources of principal uncertainty. The first was the suitability for patent protection of medical treatments, plant and animal varieties, essentially biological processes,

[1] Convention on the Grant of European Patents (European Patent Convention, as amended) (Munich, 5 October 1973, 13 ILM 268).

[2] Convention on the Unification of Certain Points of Substantive Law on Patents for Inventions (Strasbourg Convention) (Strasbourg, 27 November 1963, ETS 47).

[3] Patent Cooperation Treaty (Washington, 19 June 1970, 9 ILM 978).

and computer programs. And the second was the relationship between inherent patentability on one hand, and each of industrial applicability, technical character, technical progress, and public policy on the other. As has been seen, the first of those questions was resolved with the drafters' agreement, in deference largely to the SPC and PCT, that medical treatments, plant and animal varieties, and biological processes were inventions,[4] in contrast to computer programs, which they agreed were not. By contrast is the second question, for which no resolution is discernible from the preparatory documents. On the contrary, the drafters' deliberations in respect of specific subject matter reflect an understanding of inherent patentability as a requirement with no unifying conceptual or policy foundation.

It was against this background of incomplete deliberation and limited understanding that the European Patent Office (EPO) was charged with applying Article 52. In this chapter the results of that application are considered, in three parts, with reference to the body of law generated by the EPO's Boards of Appeal. The first part focuses on the Technical Board of Appeal's (TBA's) treatment of inherent patentability to 1988, and studies the earliest six cases involving Article 52(2) and the various substantive and interpretive issues they raise. The second focuses on the eleven years of decision-making subsequent to those cases, and considers the TBA's attempt during that period to explicate the meaning of Article 52(2) for computer-related subject matter. And the third focuses on developments within the law of inherent patentability since the end of that period, and the shift they reflect from the legalistic approach of early TBA decisions to an approach informed by policies of harmonization. Hence the conclusion of the chapter, which is that in the thirty-plus years of their existence Article 52(2) and (3) of the EPC have come full circle, with the perceived requirements of harmonization providing today the same escape from principled doctrinal and normative discussion with respect to inherent patentability as it did during the EPC debates themselves. Whether this position can be maintained, however, is unclear, and may depend among other things on the influence of national states' jurisprudence, and interpretations of the requirements of other international instruments, including Article 27.1 of the TRIPS Agreement.[5]

B. Different Approaches for Different Categories of Subject Matter: 1977 to 1987

The meaning of inherent patentability under the EPC was not considered in detail until December 1985 with the TBA's decision in *CHRISTIAN FRANCERIES/*

[4] Albeit inventions that are excluded from patentability on other, non-inherent patentability grounds. See EPC, Arts 52(4), 53(b).

[5] Agreement on Trade-Related Aspects of Intellectual Property Rights (TRIPS Agreement), Marrakesh Agreement Establishing the World Trade Organization, Annex 1C, Legal Instruments—Results of the Uruguay Round, vol 31 (15 April 1994, 33 ILM 81).

Traffic Regulation.[6] Between publication of that decision and September 1988, five further cases involving appeals from decisions of the Examining Division under Article 52(2) were decided, each reflecting a different conception of inherent patentability from the last.[7] As a result, by the late 1980s the meaning of Article 52(2) was a matter of considerable uncertainty.

1. Inherent Patentability in Light of the Business Methods Exclusion

(a) *The Exclusion of Abstract Methods Producing an Immaterial Good:* CHRISTIAN FRANCERIES/Traffic Regulation

Christian Franceries involved an appeal from a decision of the EPO's Examining Division to refuse a patent for certain procedures and devices for regulating traffic flow.[8] The TBA rejected the appeal, agreeing with the Division that the subject matter for which a patent was claimed was a method for doing business as such within the meaning of Article 52(2)(c). In so doing it emphasized two points:[9] first, that the precise terms of Article 52(2)(c) are of secondary significance to their general effect in excluding, at the very least, all abstract and speculative plans, principles, and methods; and second, that the exclusion thus interpreted embraces any method or device the sole purpose of which is to implement an Article 52(2)(c) activity. Applied to the system of traffic regulation at issue, those points were held to support a denial of patentability on the ground that the aim of the system was to control the movement of goods and persons, which was a purpose 'characteristic of an activity whose object is economic transactions and exchanges' so as to make it a method for doing business as such excluded by Article 52(2)(c).[10] The fact that certain characterizing parts of the system contained material or technical elements did not alter this conclusion. Similarly, the fact that several of the claims were directed to physical devices was also not dissuasive, for, properly construed, those claims were concerned not with devices at all, but rather with 'methods for realisation whose characteristics are the same as those mentioned in the claims for the processes'.[11] As a consequence, the applicants' product claims were held to be a different formulation of its process claims and to be inherently unpatentable on the same analysis.[12]

[6] T_16/83 (*CHRISTIAN FRANCERIES/Traffic Regulation*) [1988] EPOR 65.

[7] They are: T_51/84 (*STOCKBURGER/Coded Distinctive Mark*) [1986] EPOR 229; T_208/84 (*VICOM/Computer-Related Invention*) [1987] EPOR 74; T_26/86 (*KOCH & STERZEL/X-Ray Apparatus*) [1988] EPOR 72; T_366/87 (*STERNHEIMER/Harmonic Vibrations*) [1989] EPOR 131; T_115/85 (*IBM/Computer-Related Invention*) [1990] EPOR 107.

[8] See [1988] EPOR 65, 67.

[9] See ibid, 70.

[10] See ibid, 68.

[11] ibid.

[12] See ibid, 71.

Considered in its entirety, the decision in *Christian Franceries* supports three basic principles. The first is that Article 52(2)(c) should be interpreted purposively as requiring, at the least, that an invention, when properly construed as a whole and with reference to its underlying objective, not be abstract or speculative. The second is that any method or system concerned solely with conferring economic advantage or facilitating economic transactions is inherently unpatentable within the meaning of Article 52(2) thus interpreted, regardless of the specific material or technical means by which it is implemented. And the third is that claims directed to a device characterized by its function and not its material form must be assessed for patentability purposes as directed to that function itself. Reflected in these principles is a restrictive reading of inherent patentability, which the TBA supported in *Christian Franceries* with reference to the requirement for industrial applicability, and its preclusion of patents for methods of transport and other 'service activities' not producing a material good.[13]

(b) The Exclusion of Subject Matter not Described as Requiring Technical Means: STOCKBURGER/Coded Distinctive Mark

The TBA's purposive interpretation of Article 52(2) in *Christian Franceries* returned a negative conception of inherent patentability as excluding, at the least, transactional and other service activities, regardless of the material or technical means required for their implementation. In the second case in which the meaning of that Article was considered, however, a shift was apparent towards a more liberal reading of inherent patentability premised on a positive understanding of Article 52(2) as denoting a requirement for technical means. That case is *STOCKBURGER/Coded Distinctive Mark*,[14] which involved an application directed to an anti-counterfeiting process for devising and recording coded marks on physical objects, and to the objects bearing the marks thus recorded.[15] The TBA rejected the appeal and upheld the Examining Division's finding of inherent unpatentability on the basis that the claim was for an activity excluded by Article 52(2)(c), being a process capable of implementation by a mix of cerebral and other non-technical means.[16] While still focusing on the claimed subject matter as a whole, the TBA in *Stockburger* was concerned less with the purpose of the process to be implemented than with the technical means required for its implementation. Hence its conclusion, that the applicant's intended use of certain technical apparatus (such as a printer) in marking the goods was insufficiently reflected in the claims to be regarded as essential to the subject matter overall. The result, it implied, was a need to determine the question of inherent patentability with reference to the subject matter conceived

[13] See [1988] EPOR 65, 70–71.
[14] T_51/84 (*STOCKBURGER/Coded Distinctive Mark*) [1986] EPOR 229.
[15] See ibid, 230.
[16] See ibid, 231.

independent of that apparatus. Thus conceived, the process and products represented a 'cerebral activity' capable of implementation by non-technical means and were therefore excluded from patentability by Article 52(2)(c).

2. Inherent Patentability in Light of the Computer Programs Exclusion

(a) *The Patentability of Computer Programs as Technical Methods for Producing Physical Entities:* VICOM/Computer-Related Invention

The shift from a negative to a positive definition of inherent patentability was completed in the third of the TBA's decisions involving Article 52(2), which was also the first in which the EPC model of inherent patentability was applied in respect of a computer program, ie the decision in *VICOM/Computer-Related Invention*,[17] which remains one of the most important and frequently cited decisions involving Article 52(2) to date.[18]

The patent claims in *Vicom* were directed to methods for digitally processing images and related apparatus that were refused a patent for constituting a mathematical method and/or computer program as such within the meaning of Article 52(2)(a) and (c).[19] While the TBA considered each ground of refusal separately and on its own merits, its conclusion turned largely on its view of the mathematical methods exclusion of Article 52(2)(a).

In considering that exclusion the TBA distinguished between two types of subject matter. The first was mathematical methods or algorithms carried out on numbers and producing a numerical result, which it held to be inherently unpatentable as an abstract concept producing no direct technical effect.[20] And the second was the same methods or algorithms when[21]

> used in a technical process . . . carried out on a physical entity (which may be a material object but equally an image stored as an electric signal) by some technical means [which 'might include a computer comprising suitable hardware or an appropriately programmed general purpose computer'] implementing the method and provid[ing] as its result a certain change in that entity.

According to the TBA, methods fitting this description were beyond the scope of Article 52(2)(a) and were thus inventions within the meaning of Article 52(1).

Applying that distinction, the TBA concluded that the methods for image processing before it were inherently patentable. In particular, while such methods were described in mathematical terms and took as their underlying idea a

[17] T_208/84 (*VICOM/Computer-Related Invention*) [1987] EPOR 74.

[18] See n 29.

[19] See [1987] EPOR 74, [VI].

[20] ibid, [5].

[21] ibid.

mathematical method, they used technical means (in the form of a computer) to produce a technical effect (in the form of physical images) in a manner susceptible of industrial application.[22] The significance of the last of these considerations for the methods' inherent patentability was two fold. First, it confirmed the technical nature of the problem to which the methods were directed,[23] and thus, on the TBA's reasoning, of the process in which the methods were used.[24] And second, it denoted the permissible object of protection itself, which was the subject matter represented by the manner of application, meaning the object susceptible of industrial application and thus the 'technical process' embodying the claimed invention.[25] By emphasizing the technical means of the methods' implementation this approach was consistent with *Stockburger*. By emphasizing the methods' industrial applicability it was consistent with *Christian Franceries*.

According to the TBA, this resolution of the mathematical methods objection was determinative also of the second ground of refusal under Article 52(2)(c). Specifically, the technical nature of the methods earned the applicants the right to protection of the computer means by which those methods were implemented.[26] In relation to the programs themselves the TBA held as follows:[27]

> The Board is of the opinion that a claim directed to a technical process which process is carried out under the control of a program (be this implemented in hardware or in software), cannot be regarded as relating to a computer program *as such* within the meaning of Article 52(3) EPC, as it is the application of the program for determining the sequence of steps in the process for which in effect protection is sought. Consequently, such a claim is allowable under Article 52(2)(c) and (3) EPC.

In relation to the hardware its reasoning was analogous:[28]

> Generally claims which can be considered as being directed to a computer set up to operate in accordance with a specified program (whether by means of hardware or software) for controlling or carrying out a technical process cannot be regarded as relating to a computer program *as such* and thus are not objectionable under Article 52(2)(c) and (3) EPC.
>
> In arriving at this conclusion, the Board has additionally considered that making a distinction between embodiments of the same invention carried out in hardware or in software

22 [1987] EPOR 74, [5]–[7].

23 The technical nature of the problem to which the *Vicom* methods were directed was confirmed by their use in investigating the properties of the imaged objects or designing an industrial article (see ibid, [3]).

24 The permissible objects of protection in *Vicom* were the methods for obtaining and reproducing images of a physical or simulated object. According to the TBA, however, the same reasoning and conclusion held also for methods aimed at enhancing or restoring images (see ibid, [3]).

25 See ibid, [16].

26 See ibid, [12], [15].

27 ibid, [12].

28 ibid, [15]–[16]. Note the TBA's further finding (ibid, [17]) that 'claims directed to apparatus for carrying out a certain function should [not] be limited to the apparatus when indeed carrying out this function', on the ground that such a limitation 'would result in an undue limitation of the possibilities of the patent owner to assert his rights'.

is inappropriate as it can fairly be said that the choice between these two possibilities is not of an essential nature but is based on technical and economical considerations which bear no relationship to the inventive concept as such.

Generally speaking, an invention which would be patentable in accordance with conventional patentability criteria should not be excluded from protection by the mere fact that for its implementation modern technical means in the form of a computer program are used. Decisive is what technical contribution the invention as defined in the claim when considered as a whole makes to the known art.

The central principle of these passages is that a carrier of a recorded program and its implementing hardware represent the material forms of embodiment of the program's underlying mathematical method. As a result, and assuming the embodiments have no impact on the method's inventive concept, a claim directed to such a carrier or hardware needs (on the TBA's reasoning) to be assessed for inherent patentability purposes as though it were directed to the method itself. This result was consistent with the view of inventions from *Christian Franceries* as any tangible manifestation of an abstract (inventive) concept presented in an industrially applicable form, and was important for its expression of that view to support the opposite conclusion in the realm of computer-related technology. Of additional interest is the TBA's emphasis on the physical nature of the visual images, which brought the methods within the *Christian Franceries* understanding of industrial subject matter as non-speculative methods producing a material good.

The meaning of *Vicom* and its significance for computer-related subject matter have been construed by subsequent Boards and courts in divergent terms with important consequences for European law.[29] Immediately after its publication some commentators argued that its effect was to render the computer programs exclusion of Article 52(2)(c) redundant by accepting as an invention any program that produces or changes a physical entity, where 'physical entity' includes the

[29] In the UK, cf, eg *Merrill Lynch Inc's Application* [1989] RPC 561, CA, 569 (describing *Vicom* as having turned on the effect of the methods in causing 'a substantial increase in processing speed'); *Hitachi Ltd's Application* [1991] RPC 415, SE, 420 (describing *Vicom* as having turned on the methods' use of technical means to manipulate and produce physical entities in the form of a digital image); *Fujitsu Ltd's Application* [1996] RPC 511, SE, 517 (describing *Vicom* as having turned on the methods' 'more efficient restoration or enhancement of the technical quality of an image', citing T_65/86 (*IBM/Homophone Checker*) [1990] EPOR 181, [21]); *Fujitsu Ltd's Application* [1997] RPC 609, CA, 616 (describing *Vicom* as having turned on the way in which the images were produced). The TBA has taken a more uniform view of the case, as authority for the proposition that technical processes carried out on a physical entity such as an image stored as an electric signal are inherently patentable (see, eg T_935/97 (*IBM/Computer Programs*) [1999] EPOR 301, [7.1], consistent with the early rationalization of that case in *IBM/Homophone Checker*, above). Despite this, *Vicom* has been relied on to support different and at times conflicting results (cf, eg T_931/95 (*PBS PARTNERSHIP/Controlling Pension Benefits System*) [2002] EPOR 52, [4]; T_164/92 (*BOSCH/Electronic Computer Components*) [1995] EPOR 585, [5.1]; T_110/90 (*IBM/Editable Document Form*) [1995] EPOR 185, [5]; T_22/85 (*IBM/Document Abstracting and Retrieving*) [1990] EPOR 98, [13]; T_38/86 (*IBM/Text Clarity Processing*) [1990] EPOR 606, [17]; T_52/85 (*IBM/Semantically-Related Expressions*) [1989] EPOR 454, [5.5]; T_790/92 (*IBM/Editing Business Charts*) (TBA 29 October 1993) [4.6]).

electric signals generated in the course of a program's physical interaction with its implementing apparatus.[30] However, this reading of the decision is not correct for two reasons. First, the TBA required more than a program's production of, or change to, a physical entity (ie a technical effect): it required also the existence of a technical problem and process as evidenced by the subject matter's industrial applicability. And second, the TBA did not decide that all electric signals are physical entities, but rather that all objects represented by electric signals are physical entities. While the practical import of these qualifications might appear minimal, at the time they represented real restrictions to the scope of inherent patentability by precluding, for example, the protection of computer games and methods for producing or treating data for their failure, respectively, to support an industrial application and produce or change a physical entity.[31]

Despite this, there is one aspect of the TBA's reasoning in *Vicom* that fuels the impression of that case as having deprived both the computer programs exclusion and the other exclusions of Article 52(2) of substance, namely, its construction of the relationship between the mathematical methods and computer programs exclusions of Article 52(2)(a) and (c) respectively. In its decision, the TBA treated those paragraphs as discrete in the sense of requiring that subject matter opposed under both be assessed first as a mathematical method and only then as a computer program. When combined with the TBA's view of operable programs as satisfying the 'technical means' limb of technical character, however, this approach produces the following result: the use of a computer program in the implementation of a (mathematical) method strengthens the method's claim to inherent patentability, which claim, once established, precludes the method's exclusion from patentability as a computer program. Thus, and consistent with the TBA's treatment of the 'cerebral methods' exclusion in *Stockburger*, reducing an otherwise inherently unpatentable method to electronic form increases significantly the likelihood of the method, and the computer program comprising such form, being regarded as inventions within the meaning of Article 52(1). It is the counter-intuitiveness of this result in light of the explicit classification of computer programs as non-inventions under Article 52(2) that explains the impression of *Vicom* as having abolished the computer programs exclusion in all but form. This impression was the basis for the appeal in the next case involving Article 52(2).

[30] See, eg S Davies, 'Computer Programs Claims: The Final Frontier for Software' [1998] EIPR 429, 431; T_26/86 (*KOCH & STERZEL/X-Ray Apparatus*) [1988] EPOR 72, [3.3] (appellants' submissions).

[31] See [1987] EPOR 74, [3], [7].

(b) *The Patentability of Computer Programs as Part of a Technical Apparatus:* KOCH & STERZEL/X-Ray Apparatus

KOCH & STERZEL/X-Ray Apparatus[32] involved a patent application for a computer-controlled X-ray apparatus that was characterized in the application by the computer's ability, grace of its controlling program, to determine the technical specifications required to ensure the machine's optimum exposure.[33] Two arguments were put in opposition to the application, each by a different appellant but both premised on the same approach to inherent patentability. According to the first, the application should have been rejected by the Examining Division on the ground that the only difference between the apparatus claimed and the prior art consisted in the use of new but inherently unpatentable computer software to control a known apparatus.[34] The second went further, claiming that the TBA should reject the decision in *Vicom* on the basis that it undermined the meaning of Article 52(2) by conceding as inherently patentable all computer programs.[35] Like the first argument, this claim sought an alternative interpretation of Article 52(2) in reliance on German federal case law, by which subject matter would be denied patentability if the 'essence' of its teaching was 'a pure computer program'.[36]

Those submissions, and the view of inherent patentability on which they were based, were explicitly rejected. According to the TBA, any test of inherent patentability premised on a need for the essence of the subject matter's teaching to be technical would be misguided in law and practice.[37] In law, it held, all that is required by the EPC is that the teaching of the subject matter be technical overall, regardless of the field in which the subject matter lies, or the parts of the subject matter from which its essential contribution to such teaching derives. In practice, any test focused on the subject matter's (technical) essence would require a 'weighting' of its technical and non-technical components, which would not only be difficult to undertake, but which would also result in a situation in which patents might be denied for a method or product the technical components of which are novel and inventive but ultimately subordinate to its non-technical features.[38]

Explicated in the context of computer programs, this reasoning supported the TBA's distinction between a programmed computer involving no more than the '[transformation of] mathematical values into electric signals with the aid of natural forces' on one hand, and a program controlling an apparatus 'so as technically

[32] [1988] EPOR 72.
[33] [1987] EPOR 74, [IX].
[34] ibid, [X], [3.2].
[35] ibid, [XI], [3.3].
[36] ibid. For a comparative analysis of the treatment of software patents under German and EPC law, see W Tauchert, 'Patent Protection for Computer Programs—Current Status and New Developments' (2000) 31 IIC 812–824; also M Lehmann, 'The Legal Protection of Computer Programs in Germany: A Summary of the Present Situation' (1988) 19 IIC 473–487.
[37] [1987] EPOR 74, [3.4].
[38] ibid.

to alter its functioning' on the other.[39] According to the TBA, while the first is inherently unpatentable on the basis that 'the electric signals . . . amount to no more than a reproduction of information' and cannot for that reason be regarded as involving any technical effect per se, the second is inherently patentable, provided only that they are claimed in combination with the apparatus they control.[40] The subject matter before the TBA was held to satisfy this distinction. In the TBA's own words:[41]

> Claim 1 relates neither to a computer program on its own and divorced from any technical application, nor a computer program in the form of a recording on a data carrier, nor a known, general-purpose computer in combination with a computer program. It is in fact an X-ray apparatus incorporating a data processing unit operating in accordance with a routine which produces a technical effect in the X-ray apparatus.

Thus, it was held, a computer program is inherently patentable if it produces an alteration in the technical functionality of a physical apparatus, over and above the alteration of electric signals inherent in the program's implementation by a known, general-purpose computer.

Like *Vicom*, *Koch* is a difficult case that has attracted little critical attention. The source of its difficulty lies in the TBA's failure properly to have explicated the implications of its reasoning for *Vicom* and the meaning of Article 52(2) itself. In relation to the former, the emphasis in *Koch* on technical teaching as the characterizing feature of inventions is similar to the emphasis in *Vicom* on inventive concept, and implies a similar distinction between the substance of an invention and its material forms of embodiment. Contradicting that view, however, is the TBA's further distinction between computer programs claimed in isolation—which it regarded as inherently unpatentable presentations of information as such within the meaning of Article 52(2)(d)—and the same programs when applied to alter the technical functionality of a physical apparatus. The suggestion implicit in this latter distinction, that the capacity of a computer program to support a patent depends entirely on its impact on the technical functionality of a physical apparatus, is inconsistent with *Vicom*, where the technical effect of the program derived from the program's instructional content itself. This raises the uncertainty of *Koch* for Article 52(2) more generally, which derives from the tension that exists between the basic premise of that case, that the need for an invention is a discrete requirement of patentability determinable with reference to the technical character of the subject matter as a whole, and the TBA's specific finding in contradiction of *Vicom*, that a general-purpose apparatus that derives its novel technical features from the conventional functionality or content of a computer program will never support a patent.

39 [1987] EPOR 74, [3.3].
40 ibid.
41 ibid, [3.1].

(c) *The Patentability of Computer Programs as Solutions to a Technical Problem:* IBM/Computer-Related Invention

The contradictory indications of *Koch* likely account for the practice of the Examining Division, following that case, of refusing patents for subject matter the technical character of which was regarded as deriving from a computer program. That was the situation in the next appeal to the TBA involving Article 52(2)(c), which was *IBM/Computer-Related Invention*,[42] where the TBA upheld as inherently patentable a computerized method for automatically displaying messages relating to the internal conditions of a computer.[43] While conceding that the 'basic idea' of the method might have resided in the program itself, the TBA was of the view that the claim was directed not to such idea per se, but rather to its 'use in the solution of a technical problem'[44]—namely, the 'giving [of] visual indications automatically about conditions prevailing in an apparatus or system'[45]—which use could not be considered a computer program as such within the meaning of Article 52(2)(c).[46] In support of its finding the TBA relied exclusively on *Vicom*, including specifically its principle that[47]

> an invention which would be patentable in accordance with conventional patentability criteria should not be excluded from protection by the mere fact that for its implementation modern technical means in the form of a computer program are used.

Despite this attempt to relate the two cases, the TBA's decision in *IBM* is difficult to reconcile with *Vicom* in both reasoning and outcome. In reasoning, it suggests that any computer-implemented method is inherently patentable if it solves a technical problem, regardless of whether it also produces a 'technical effect'. And in outcome, it seems clear that the methods of *IBM* do not produce such an effect within the meaning of *Vicom*; their effect being the automatic generation of text messages, which lack the physicality of the images of *Vicom*. Hence the significance of *IBM*, which lies in its reliance on *Vicom* to support a more liberal reading of inherent patentability. Pursuant to that reading, any programmed computer can support a patent if it involves a solution to a technical problem, where 'technical problem' includes any problem relating to the internal functioning of a computer, regardless of whether it produces the material or physical object required in *Christian Franceries* and *Vicom*, or possesses the technical non-conventionality required (for general-purpose computers) in *Koch*. By this reasoning the TBA confirmed the role attributed to the computer programs exclusion in *Vicom* as delimiting the non-computer programs exclusions of Article 52(2), by, for

[42] T_115/85 (*IBM/Computer-Related Invention*) [1990] EPOR 107.
[43] See ibid, [VII]. See also T_42/87 (*KEARNEY/Computer-Related Invention*) [1997] EPOR 236.
[44] [1990] EPOR 107, [9].
[45] ibid, [7].
[46] ibid, [9].
[47] ibid, [10].

example, enabling the protection of abstract and mathematical methods (in the *Christian Franceries* and *Vicom* sense) when presented in automated form.

IBM reflects the most restrictive reading of the computer programs exclusion of Article 52(2)(c) to its date. It does not, however, reflect the widest reading of inherent patentability overall, which is contained in an appeal involving the aesthetic works exclusion of Article 52(2)(b) that was delivered immediately before *IBM* but after *Koch* in July 1988.

3. Inherent Patentability in Light of the Aesthetic Works Exclusion

The subject matter at issue in *STERNHEIMER/Harmonic Vibrations*[48] was a process for selecting acoustic or electronic vibrations that had been refused a patent by the Examining Division in reliance on Article 52(2)(a) and (b).[49] The TBA allowed the appeal and overturned the Division's decision on the basis of the following three principles of inherent patentability, each of which it derived explicitly from its reading of Article 52(2). The first is that the need for an invention is a discrete requirement of patentability imposed by Article 52(1) independent of the 'usual conditions of patentability' required by that Article.[50] The second is that an examination of the non-restrictive list of ineligible subject matter contained in Article 52(2) reveals as their common characteristic,[51]

> the characteristic of not constituting or implying the use of technical methods to produce a concrete technical effect, for example to make it possible to modify the structure of behavior, or to determine the properties of an object, environment or physical phenomenon.

And the third is that the effect of the 'as such' proviso of Article 52(3) is to suggest that the subject matter's inclusion of non-technical elements will only preclude patentability where the subject matter does 'not also include other elements which are themselves of a technical nature'.[52] On the basis of these principles, and with explicit and exclusive reference to *Koch*, the TBA formulated the test of inherent patentability under Article 52 as follows:[53]

> [I]t is necessary in determining whether the subject-matter of a claim is an invention for [*sic*] purposes of Article 52(1) of the EPC to assess this subject-matter as a whole and to examine whether it uses technical means or produces technical effects. If it does, then this subject-matter constitutes an invention for the purposes of Article 52(1) of the EPC, and any simultaneous use of non-technical means does not remove the technical nature of the teaching considered as a whole.

[48] T366/87 (*STERNHEIMER/Harmonic Vibrations*) [1989] EPOR 131.
[49] See ibid, [V].
[50] See ibid, [4.1].
[51] ibid.
[52] ibid.
[53] ibid.

Applied to the subject matter at issue this requirement for the use of technical means or the production of technical effects led to a finding of inherent patentability. Specifically, while the claims at issue did not disclose or imply any technical means for selecting or using the vibrations, they did disclose technical means in the form of an electronic instrument for modulating, amplifying, and harmonizing the vibrations once produced. As a result of such disclosure the TBA concluded that[54]

> [t]he operation for regulating the means of producing the vibrations thus defined is therefore of a technical nature, so that the process in Claim 1, of which this technical operation is one stage, does indeed constitute an invention for the purposes of Article 52(1) of the EPC.

This decision supports the most expansive understanding of inherent patentability under the EPC considered to this point, pursuant to which any method can support a patent if described in the patent claims as involving the use of a specific mechanical or functional device at any stage of its implementation, regardless of the novelty of its result, the conventionality of its means, or its susceptibility of industrial application overall. In addition, while some parts of the TBA's decision reflect a further requirement for 'a concrete technical effect', the very broad definition of that requirement as encompassing any modification of the structure of behaviour or determination of the properties of an object, environment, or physical phenomenon[55] goes beyond both the *Vicom* requirement for a new or altered physical entity, and the *Koch* requirement for a machine with new or non-conventional technical features, and appears to concede as 'technical' all human acts upon the physical world along with most human discoveries. Indeed, the most significant restriction to inherent patentability recognized by the TBA in *Sternheimer* derives not from its view of inventions per se, but rather from its finding that inherent patentability is a discrete and threshold requirement of patentability. The effect of that finding was to require that subject matter not constituting an invention be refused a patent independent of its capacity to satisfy the other conditions of Article 52(1).[56] On the other hand, once the existence of an invention was established, concepts of inherent patentability would have no role to play.

4. Conclusion

By September 1988 when the decision in *IBM* was published, the TBA had given detailed consideration to the meaning of inherent patentability on six occasions in relation to four of the specific exclusions of Article 52(2). The preceding analysis supports the following conclusions in respect of that consideration.

[54] ibid, [4.2].
[55] ibid, [4.1].
[56] See also *IBM/Computer-Related Invention* [1990] EPOR 107, [13].

As initially formulated in the context of the business and mental methods exclusions, an invention was understood by the TBA to be a subject matter that did not implement or comprise an abstract activity or other item excluded by Article 52(2). As subsequently formulated in the context of computer programs and aesthetic works, an invention was a subject matter having technical character.

The initial understanding of inventions as non-abstract subject matter was expressed in terms of two specific tests, in *Christian Franceries* and *Stockburger* respectively. Pursuant to the first, a subject matter the sole purpose of which was to implement an activity or item excluded by Article 52(2), being an abstract activity or other item insusceptible of industrial application, was inherently unpatentable regardless of its use of technical or material means. Pursuant to the second, a subject matter that was capable of being performed cerebrally using non-technical means was inherently unpatentable under Article 52(2)(c). The application of these tests returned in each case a finding of inherent unpatentability. However, their formulation and the manner in which they were applied reflected different views of the importance, when considering Article 52(2), of a subject matter's use of technical means and an applicant's use of technical terminology when describing its subject matter in the patent claims. In *Christian Franceries*, for example, the TBA concluded that a subject matter aimed at implementing an inherently unpatentable activity could not be saved from exclusion by its use of technical means, and emphasized, in reaching that conclusion, the need to look beneath an applicant's claims in order properly to identify the alleged invention in respect of which inherent patentability was required to be shown. In *Stockburger*, by contrast, it suggested that a subject matter described in the claims as dependent for its realization on technical means was not excluded from patentability merely because it otherwise involved an inherently unpatentable activity.

A similar tension between the substance and form of a claim is reflected in the remaining four decisions involving Article 52(2) from this period, albeit in the context of a positively cast conception of that Article expressed in terms of the following specific tests of inherent patentability:

1 Does the subject matter involve the use of technical (eg computer) means to achieve a technical (ie physical) effect in a manner susceptible of industrial application (*Vicom*)?

2 Does the subject matter use a mix of technical and non-technical means to achieve a technical result such as—in the case of a computer program—a non-conventional alteration in the technical functionality of a physical apparatus (*Koch*)?

3 Does the subject matter involve the use of technical (eg computer) means to address a technical problem (*IBM*)?

4 Does the subject matter involve the use of technical (eg mechanical) means to achieve a technical (ie human) effect (*Sternheimer*)?

In answering the first of these questions the TBA in *Vicom* adopted an approach similar to that in *Christian Franceries*, by distinguishing the substance of an alleged invention (comprised of its industrially applicable inventive concept) on one hand from its material forms of embodiment (not affecting that concept) on the other. In *IBM* and *Sternheimer*, by contrast, the TBA showed a greater preparedness to accept the claims' reference to technical means as sufficient evidence of the subject matter's technical character overall. Different again was the approach in *Koch*, where the TBA's view of computer programs as patentable only when claimed in conjunction with a physical apparatus the technical functionality of which was affected by the program in a new or non-conventional manner reflects aspects of both tendencies, as well as a general ambivalence regarding the relevance, when determining inherent patentability, of the source and extent of the subject matter's 'technical teaching'. At the heart of that ambivalence lay a deeper uncertainty concerning the relationship between inherent patentability on one hand and, on the other, the non-threshold conditions of patentability imposed by Article 52(1), and the production by the subject matter of a physical object. The same uncertainty is apparent in the other cases decided during this period. On the relationship between inventions and the requirements for novelty, inventiveness, and industrial applicability, for example, there are the different constructions of Article 52(2) by differently constituted Boards of Appeal as denoting in whole or in part a requirement for industrial applicability (in *Christian Franceries* and *Vicom*), and reflecting a discrete and threshold requirement independent of the other conditions imposed by Article 52(1) (in *Sternheimer* and *IBM*). So, too, the relationship between inherent patentability and the physical world was differently construed by the TBA in its early decisions. While *Christian Franceries* and *Vicom* suggest a requirement for the production by a method of a material or physical object, in *Stockburger*, *IBM*, and *Sternheimer*, the TBA took a more liberal view of inherent patentability focused on solving technical problems (in *IBM*) and using technical means (in *Stockburger* and *Sternheimer*), without any additional need for a new product per se.

Three further conclusions are warranted. First, that the difficulties in applying Article 52(2) and (3) were exacerbated by the technical complexity of the subject matter in issue. Second, that the courts' approach to that application changed as they moved from considering the business methods exclusion[57] to the computer programs exclusion, consistent with a restriction in the Article's application. And third, that the apparent unity in their approach following that change was belied by substantial doctrinal inconsistency.

[57] While the headnote summary of *Stockburger* describes the TBA's decision as having turned on the business methods exclusion of Art 52(2)(c), the TBA's description of the subject matter as a 'cerebral activity' suggests the mental methods exclusion.

C. Article 52(2) and (3) of the EPC as a Requirement for Technical Character: 1988 to 1999

From its initial consideration of inherent patentability in *Christian Franceries*, the TBA accepted Article 52(2) as a definitional rather than a deeming provision and sought to extrapolate from its negatively-cast terms a uniform conception of the constitutive elements of an invention. From *Stockburger* its conclusion was that an invention within the meaning of Article 52(1) is any subject matter that is technical. Two questions immediately arise: what was the basis for that requirement, and what did 'technical character' require?

Considering the first of these questions, commentators in the late 1980s and early 1990s referred to the EPO's 1985 Guidelines for Examiners.[58] According to paragraph 2.2 of those Guidelines, an 'invention' within the meaning of Article 52(1) was a subject matter the 'contribution' of which was 'of a technical character'.[59] As seen in chapter 4, however, the only source of law binding on the EPO is the text of the EPC itself, with the result that the Guidelines are not a valid basis for the requirement itself. It is possibly for this reason that the TBA, when considering this issue immediately after *IBM*, referred not to the EPO Guidelines but rather to the nature of the Article 52(2) exclusions, the 'long-standing legal practice' of the EPC's Contracting States, and the emphasis on technicality in Rules 27–29 of the EPC's Implementing Regulations.[60] As justifications, however, these factors are insufficient.[61] That the paragraphs of Article 52(2) do not obviously resolve to a requirement for a technical contribution is demonstrated by the contradictions in the TBA's own interpretation of them, and their reference to computer programs, the technical character of which is difficult to deny. Similarly, Rules 27–29 of the Implementing Regulations require that inventions be described and claimed by an applicant in terms of their technical features, which is different

[58] [1985] OJ EPO 173, 177–179. See, eg J Worthy, 'Software Patents in the UK After Merrill Lynch' [1989] EIPR 380, 381; B Sherman, 'The Patentability of Computer-Related Inventions in the United Kingdom and the EPO' [1991] EIPR 85, 92. See also *Merrill Lynch Inc's Application* [1988] RPC 1, SE, PAT; aff'd [1989] RPC 561, CA; JAH van Voorthuizen, 'The Patentability of Computer Programs and Computer-Related Inventions under the European Patent Convention' (1987) 18 IIC 627, 628.

[59] [1985] OJ EPO 173 (Part C, Chapter IV, 1.2(ii)).

[60] *IBM/Document Abstracting and Retrieving* [1990] EPOR 98, [3]–[4]. See also Sherman, 'The Patentability of Computer-Related Inventions in the United Kingdom and the EPO', 92–93; T_790/92 (*ESSWEIN/Automatic Programmer*) [1991] EPOR 120, 125 (describing the subject matter in Art 52(2) as 'inventions [that] constitute the solution to [abstract, commercial, or recreational] problems, whose formulation may nonetheless require the exercise of an inventive step', and the Article itself as making it 'clear . . . that the European legislator intended that the problem to be solved [for the purpose of inventive step] be of a technical nature', consistent with a view of the patent system as aiming to 'accelerate technological development').

[61] cf Sherman, 'The Patentability of Computer-Related Inventions in the United Kingdom and the EPO', n 67.

from saying that a subject matter is constituted entirely by such features. This leaves the 'long-standing legal practice' of UK law in chapters 2 and 3, and of European States, and the further implication of the TBA's comments that the technical character theory of Article 52(2) was supported by the Contracting States themselves. However, and as the study of the EPC *travaux* in chapter 4 has shown, this was not the case. Specifically, neither the 'long-standing practice' of EPC Contracting States, nor the views of the EPC drafters themselves, provides clear support for a definition of Article 52(2) and(3) as denoting a requirement for technical character.

There remains the second of the questions identified above: what was required for a subject matter to be technical in nature? In the second era of the TBA's decision-making from 1988 to 1999, the uncertainty surrounding this question came to the fore, as the TBA sought to explicate the meaning of its early decisions, and conceptions of technical character generally, in the context primarily of computer-related subject matter.

1. The Whole Contents and Contribution Theories of Article 52(2) and (3)

The majority of inherent patentability cases brought in the eleven years following *IBM* involved an application directed to a programmed general-purpose computer that was refused a patent by the EPO's Examining Division in reliance on the computer programs and/or mental methods exclusions of Article 52(2)(c). The TBA's decisions in those cases support in reasoning or conclusion two general theories of technical character that have been denoted in their varying formulations as the 'whole contents' and 'contribution' theories respectively.[62] The first requires that a subject matter be technical in the sense of reflecting any or all of a technical effect, a solution to a technical problem, and technical means. By contrast is the second, which sets a more onerous standard by requiring that a subject matter's technical character reside in its contribution to the prior art in the form of a non-conventional result in a field of activity not excluded by Article 52(2). The result of applications of the latter approach during the period in question was the denial of patents for computer systems that: (a) depended for their novelty or inventiveness on a non-technical or conventional source, such as the instructional content or conventional operation of a program,[63] the meaning of data,[64] or the

[62] cf Sherman, 'The Patentability of Computer-Related Inventions in the United Kingdom and the EPO', 85–86; J Newman, 'The Patentability of Computer-Related Inventions in Europe' [1997] EIPR 701, 702; *Aerotel Ltd v Telco Holdings* [2006] EWCA Civ 1371, discussed at pages 314–318.

[63] See, eg *IBM/Document Abstracting and Retrieving* [1990] EPOR 98; *IBM/Text Clarity Processing* [1990] EPOR 606; T_158/88 (*SIEMENS/Character Form*) [1992] EPOR 69; T_833/91 (*IBM/External Interface Simulation*) [1998] EPOR 431; T_204/93 (*AT&T/Computer System*) [2001] EPOR 300.

[64] See, eg *IBM/Semantically-Related Expressions* [1989] EPOR 454; *IBM/Homophone Checker* [1990] EPOR 181.

intellectual input of a computer programmer[65] and/or (b) produced a result in a field covered by Article 52(2), such as linguistics, mental processing, or computer programming.[66] The result of applications of the former, by contrast, was the allowance of patents during the same period for individual systems in spite of, or without regard to, their fulfilment of the same criteria.[67] Hence the existence from 1988 to 1999 of a series of cases that are difficult to reconcile.[68] In addition, far from conceding the divergence of authority thus created, the TBA routinely obscured it by a mix of generic terminology and passing reference to prior decisions in a manner that suggested a single line of practice consistent in principle and correct in fact. The effect was to exacerbate the uncertainties of early TBA case law by masking them and perpetuating them in a new line of cases.

The same uncertainties are reflected in the treatment during this period of informational and aesthetic subject matter other than computer programs, the inherent patentability of which was considered by the TBA in four main cases: *BEATTIE/Marker*,[69] *FUJI/Coloured Disk Jacket*,[70] *BROSELOW/Measuring Tape*,[71] and *BBC/Colour Television Signal*.[72] In all but the last of these cases the TBA rejected the approach to Article 52(2) supported for similar subject matter in *Sternheimer* in preference for the principle, consistent with the contribution theory above, that a subject matter distinguished from the prior art by its informational or aesthetic content alone was excluded from patentability by Article 52(2)(d) and (b) unless that content had a secondary technical effect that advanced the prior art

[65] See, eg *IBM/Document Abstracting and Retrieving* [1990] EPOR 98; *IBM/Text Clarity Processing* [1990] EPOR 606.

[66] See, eg *IBM/Semantically-Related Expressions* [1989] EPOR 454; *SIEMENS/Character Form* [1992] EPOR 69; *AT&T/Computer System* [2001] EPOR 300; *IBM/Homophone Checker* [1990] EPOR 181.

[67] See, eg *KEARNEY/Computer-Related Invention* [1997] EPOR 236; T_59/93 (*IBM/Rotating Displayed Objects*) (TBA, 20 April 1994); T_6/83 (*IBM/Data Processor Network*) [1990] EPOR 91; *BOSCH/Electronic Computer Components* [1995] EPOR 585; *IBM/Editable Document Form* [1995] EPOR 185; T_236/91 (*TEXAS INSTRUMENTS/Language Understanding System*) [2000] EPOR 156; T_769/92 (*SOHEI/General-Purpose Management System*) [1996] EPO 253.

[68] cf, eg *IBM/Editable Document Form* [1995] EPOR 185 with *IBM/Document Abstracting and Retrieving* [1990] EPOR 98, *IBM/Text Clarity Processing* [1990] EPOR 606, and *AT&T/Computer System* [2001] EPOR 300; *TEXAS INSTRUMENTS/Language Understanding System* [2000] EPOR 156 with *IBM/Document Abstracting and Retrieving* [1990] EPOR 98, *IBM/Text Clarity Processing* [1990] EPOR 606, and *IBM/Semantically-Related Expressions* [1989] EPOR 454; *KEARNEY/Computer-Related Invention* [1997] EPOR 236 with T_167/92 (*IBM/Document Composition*) [2003] EPOR 20 ; *SOHEI/General-Purpose Management System* [1996] EPO 253 with *IBM/External Interface Simulation* [1998] EPOR 431, *IBM/Document Composition* [2003] EPOR 20 and *IBM/Editing Business Charts* (TBA, 29 October 1993); *IBM/Rotating Displayed Objects* (TBA, 20 April 1994) with *IBM/Homophone Checker* [1990] EPOR 181, *IBM/Document Abstracting and Retrieving* [1990] EPOR 98, *IBM/Text Clarity Processing* [1990] EPOR 606, and T_186/86 (*IBM/Spatially Related Data*) (TBA, 5 December 1989).

[69] T_603/89 (*BEATTIE/Marker*) [1992] EPOR 221.

[70] T_119/88 (*FUJI/Coloured Disk Jacket*) [1990] EPOR 615.

[71] T_77/92 (*BROSELOW/Measuring Tape* [1998] EPOR 266.

[72] T_163/85 (*BBC/Colour Television Signal*) [1990] EPOR 606.

of a non-excluded field. Hence its findings in *Beattie* and *Fuji* that a musical marker and coloured disk jacket were inherently unpatentable on the grounds, respectively, that the marker consisted in substance of an arrangement of musical notation that had no effect on the device's physical properties or mechanical functionality,[73] and that the jacket was conventional but for its colour, the novel purposes of which were non-technical.[74] Hence too its finding in *Broselow* that a method of selecting correct patient treatment values in emergency situations by measuring the body length of a patient, and then reading the treatment values corresponding to that length directly from the measuring tape, was inherently patentable. According to the Board, while the tape appeared on its face to differ from previously known methods in its informational content alone, its true contribution lay in the co-relation of that content and the patient's measured heel-to-crown length, which co-relation represented 'a new gauge' for directly measuring emergency patient values that was technical in nature. It followed that the distinguishing informational content of the tape had a technical significance that advanced the prior art of a non-excluded field—presumably the field of medical treatment—and was inherently patentable on that basis.[75]

Undermining if not contradicting this approach, however, are the remainder of the Boards' decisions in *Broselow* and *BBC*. Having already found the method to be inherently patentable, the Board in *Broselow* went on to consider the case on the assumption that the subject matter's technical significance was already known—ie that the co-relation of patient length and treatment values on which the method depended were known to the relevantly skilled practitioner so as not to represent any contribution to the prior art. According to the Board, the method was inherently patentable even on this assumption, for while previously known subject matter may have employed the same method of reading treatment values from patient length, none enabled those values to be read directly from the measuring tape itself, requiring users instead to obtain them from a separate table or chart. In the Board's opinion, the method's elimination of this intermediate step of consulting an independent source 'save[d] time and also eliminate[d] a possible source of error', which effects, it held, were 'technical in nature'.[76] This underlines the ultimate significance of *Broselow*, which lies in its suggestion that a subject matter distinguished from the prior art by its informational content alone will nonetheless avoid exclusion under Article 52(2)(d) if its presentation of that content improves the efficient performance of a known method by, for example, eliminating one of the latter's steps. By this reasoning the contribution theory's requirement for a (secondary) technical contribution to the prior art of a non-excluded field was affirmed by the Board, but with a weakened understanding of

[73] See [1992] EPOR 221, [2.5]–[2.9].

[74] [1990] EPOR 615, [4].

[75] *BROSELOW/Measuring Tape* [1998] EPOR 266 [3.5].

[76] ibid, [3.7].

'technical contribution' that embraced new, if manifestly obvious, improvements in efficiency brought about by a method's increased automation. This understanding and aspect of the Board's approach in *Broselow* generally were consistent also with the string of whole contents cases decided during the 1990s, in which the fact of a method's automation by computer means was treated as sufficient to confer the technical character required to overcome an Article 52(2) objection, regardless of the nature of the automated method itself. This approach reflects the counter-intuitive logic of *Vicom* that while a mental method and a computer program are individually excluded by Article 52(2), put them together and one has an invention.

A further weakening of the criterion for inherent patentability in the context of presentations of information during this period was the result of a narrow interpretation of the word 'information' in Article 52(2)(d) itself. That interpretation, already apparent in *Broselow*, was seen most clearly in *BBC*, where the TBA rejected an opposition to the patentability of a television signal under Article 52(2)(d) on the basis of its technical character overall. In reasoning its decision the TBA emphasized the technical significance of the signal's informational content in exhibiting the technical features, and thereby determining the technical functionality of the system in which it occurred, without considering whether the signal thus conceived also represented an advance on the prior art of a non-excluded field.[77] Interestingly, however, the TBA in *BBC* seems to have been uncertain as to the precise doctrinal basis for its decision, stating that even if Article 52(2) were construed as resolving to a general exclusion in respect of 'essentially abstract' subject matter—which it understood to mean subject matter that is 'non-physical' and thus 'not characterised by technical features in the sense of Rule 29(1) [of the EPC Regulations]' —the signal before it would still have been an invention in the form of 'a physical [albeit transient] reality which can directly be detected by technological means'.[78] Thus, where *Broselow* signalled an expansive conception of technical contribution to cover improvements in the efficiency of a method's performance, *BBC* signalled an expansive conception of physical (cf abstract) phenomena to cover electronic manifestations of information, consistent with the treatment of the digital objects in *Vicom*. In this way both cases underscored, in combination with *Beattie* and *Fuji*, the failure of the TBA during the 1990s to resolve the substantive and interpretive issues raised in its earlier decisions for inherent patentability, including particularly the relationship of both novelty and inventiveness, and the importance of the subject matter's physicality, to conceptions of inherent patentability. The same uncertainty was also reflected in the cases during this period involving computer-implemented transactional or abstract schemes of the type at issue in *Christian Franceries* and *Stockburger*,[79] which showed in addition

[77] [1990] EPOR 606, [2].

[78] ibid.

[79] See also T_636/88 (*NAT/Bagging Plant*) [1993] EPOR 517; *IBM/Rotating Displayed Objects* (TBA, 20 April 1994) [3.3]; *IBM/Editing Business Charts* (TBA, 29 October 1993).

the seeming arbitrariness of the TBA's approach to inherent patentability, and the extent to which that arbitrariness was lost in the amorphous language of technical character and the minutiae of technical description.

2. Inconsistent Conceptions of 'Technical Character' in the Context of Automated Transactional Schemes

The first time the TBA considered the inherent patentability of an automated transactional scheme during the 1990s was in *IBM/Card Reader*.[80] That case involved an appeal from the Examining Division's rejection of an application directed to a method for operating an automatic self-service machine. As defined in the patent claims, the method involved the insertion, by a person, of a card pre-recorded with identification information into a machine that had been pre-programmed to read the information and verify the user's identity against information contained in its own pre-recorded database.[81]

In its decision, the TBA refused the appeal and affirmed the Examining Division's finding that the method thus defined was an inherently unpatentable business or other abstract method as such within the meaning of Article 52(2)(c). In reaching that decision the TBA relied on the distinction drawn in *Beattie*[82] between the (inherently unpatentable) use of technical means to support a non-technical activity on one hand, and the (inherently patentable) use of a combination of technical and non-technical components to solve a technical problem on the other.[83] Before declaring the method to fall within the first of those categories the TBA stated as follows:[84]

> [W]hen taken as a whole the claimed method is . . . essentially a method of deciding whether a card which is presented for the first time to a machine should thereafter be recognised as an authorised card by that machine. Such a method is part of a business operation. Of course, the claimed method does include steps which include a technical component in them (for example using the machine to store and read information). But the presence of such technical components does not alter the fact that the claimed method is a business method as such, rather than a technical method (just as the use of a typewriter to perform a business activity would not change such an activity into a technical method).
>
> As was stated in [*IBM/Document Abstracting and Retrieving*], '[t]he contribution to the art and the effects obtained are only in the area of an excluded activity and the true nature of the invention remains the same, whether or not a technical terminology is used in expressing it. In the present case, the true nature of the claimed subject matter remains the same, even though some technical means are used to perform it'.

[80] T_854/90 (*IBM/Card reader*) [1994] EPOR 89.

[81] See ibid, [VI].

[82] See [1992] EPOR 221, [2.5].

[83] See [1994] EPOR 89, [2.3].

[84] ibid.

Diagram A Cases involving the application of the contribution theory to deny patentability to general-purpose computer systems from 1988 to 1999

Case	Subject Matter and Decision	Précis of Reasoning
T_22/85 (*IBM/Document Abstracting and Retrieving*) [1990] EPOR 98	A system for automatically abstracting, storing, and retrieving electronic documents, and a system for automatically detecting and replacing incomprehensible linguistic expressions.	The only new concept disclosed in the application derives from its rules, which are of an intellectual nature and which do not define a new way of operating the computer in a technical sense.
T_52/85 (*IBM/Semantically-Related Expressions*) [1989] EPOR 454	A system for automatically generating a list of expressions. The system represents the conventional use of a general-purpose computer to solve a non-technical (linguistic) problem and produce a non-technical (non-physical) effect, and is therefore inherently unpatentable.	The functionality of the system depends on the meaning or 'abstract linguistic information content' of the expressions, and does not involve the processing of data or signals that represent physical entities in the manner of *Vicom* and *Koch*. Its contribution to the art therefore resides exclusively in the fields of linguistics and programming.
T_38/86 (*IBM/Text Clarity Processing*) [1990] EPOR 606	The systems' contribution to the art resides in their provision of a scheme governed by rules of a purely intellectual nature, which makes them mental methods excluded by Article 52(2)(c).	As for *IBM/Document Abstracting and Retrieving*.
T_65/86 (*IBM/Homophone Checker*) [1990] EPOR 181	A system for automatically detecting and correcting contextual homophone errors in a text document. The system's contribution to the art resides in its performance of a mental method within the meaning of Article 52(2)(c).	The system's inventiveness derives from its performance of steps capable of being performed by a person using skill and judgment with the purpose of producing a purely linguistic effect, which is an effect in a field excluded by Article 52(2)(c).
T_186/86 (*IBM/Spatially Related Data*) (TBA, 5 December 1989)	A system for editing digital text. The system is a technically conventional method of creating an object of informational and aesthetic content, which is a mental method excluded by Article 52(2)(c).	Relying on the principles of *IBM/Document Abstracting and Retrieving*.

T_158/88 (*SIEMENS/Character Form*) [1992] EPOR 69	A system for displaying characters on a visual display unit in a pre-determined form on the basis of certain pre-defined rules. The system solves no technical problem in the sense of making a technical contribution to the prior art, and is therefore inherently unpatentable.	The system does not alter the physical or technical functionality of the unit in any non-conventional manner, and solves non-technical problems aimed at facilitating the user's mental assimilation of the displayed characters.
T_833/91 (*IBM/External Interface Simulation*) [1998] EPOR 431	An interactive system for designing external interfaces for a simulated computer application. The system does not present a contribution in a non-excluded field, and is therefore inherently unpatentable.	The system's technical features are conventional, and do not improve the internal functioning of the computer.
T_167/92 (*IBM/Document Composition*) [2003] EPOR 20	A system for enabling a user to create a document by selecting from a screen suitable text parts previously defined by the user and stored as data files. The system does not make a contribution to the art in a field not excluded by Article 52(2). Its features fall under Article 52(2) and do not contribute to a technical effect or the solution of a technical problem.	The system's novelty lies in its display of information on a screen—which is a presentation of information within the meaning of Article 52(2)(d)—for the sole purpose of conveying information to the operator. The effect of that presentation is to enable the operator to take certain actions according to his intentions, which involves the judgment of the operator and cannot therefore constitute a technical effect.
T_790/92 (*IBM/Editing Business Charts*) (TBA, 29 October 1993)	An interactive method of graphically editing a chart in a data processing system controlled by a program and having a display on which the graphic objects of the chart can be selected by a cursor. The method makes no contribution to the art outside the field of matters excluded by Article 52(2), and is therefore inherently unpatentable.	The method's features depend on the purely mental considerations of the operator, and on the display of business data and related information by means of a chart, neither of which confers the contribution required to show the technical character required of inventions.
T_204/93 (*AT&T/Computer System*) [2001] EPOR 300	A system and method for generating (unspecified) software programs from generic specifications. The system automates a mental act in a technically conventional manner without producing any new technical effect, and is therefore inherently unpatentable.	The system's features are either technically conventional or represent the automation of an essentially mental act (namely, the writing of a computer program), and do not improve the technical functionality of a computer as a computer in contrast to a computer as a programming tool.

By this reasoning the TBA applied the contribution theory of patentability to support a decision reminiscent of that in *Christian Franceries* seven years earlier, emphasizing that the substance of a business method cannot be altered by its description in technical terms, and that a subject matter aimed at implementing a business method cannot be saved from exclusion under Article 52(2)(c) by the mere fact of its dependence on such means.

A quite different perspective on the same issue is reflected in the second of the cases in which the inherent patentability of a transactional scheme was considered during this period. That case is *PETTERSSON/Queuing System*,[85] decided in July 1994, and involving a system for determining the queue sequence for serving customers that was described, in the claims, in terms of the following five functionally defined components: a number allocation unit by which numbers were allocated to customers waiting to be served, terminals for dispensing the numbers to waiting customers, an information unit that received signals identifying the next customer to be served, a free service point for indicating the signals to customers, and computing means for executing the system thus described.[86]

The appellant's objection to the application in respect of this system was that it sought to patent a scheme, rule, or method for doing business as such within the meaning of Article 52(2)(c). In support of that objection it relied on two principal arguments premised on the contribution approach above. The first was that the system's individual features each corresponded to the steps of an inherently unpatentable method, which method thus represented the extent of the system's contribution to the art.[87] And the second was that the system's only novel feature was of a non-technical nature, confirming the non-technical character of the system's teaching overall.[88]

The TBA rejected those arguments and the conclusion for which they were advanced. In its view, the system was a technical apparatus in the form of a three-dimensional object having specific functional capacities and practical application in the service of customers.[89] On this basis, it was an invention within the meaning of Article 52(1). The fact that the apparatus was partly defined and wholly distinguished from the prior art by its basic working principle of 'deciding which particular turn-number [was] to be served at the particular free service point,'[90] and that such principle represented a step in a method for doing business capable of being performed by cerebral means, did not, it was held, detract from this result, for not only were prior art considerations irrelevant to inherent patentability, the principle itself was an integral part of the claim overall that could not be separated out and examined as an isolated intellectual concept. On the contrary, when read

[85] T_1002/92 (*PETTERSSON/Queuing System*) [1996] EPOR 1.
[86] ibid, [I].
[87] See ibid, [VII(a)].
[88] See ibid, [VII(b)].
[89] See ibid, [2.1].
[90] ibid, [I], referred to at [2.4].

in the context of the claim as a whole, the functional term representing the relevant working principle was properly to be understood as 'limited to a hardware property', and as thus constituting 'a computer program according to which the claimed hardware operates'.[91] Hence the TBA's conclusion, that the system involved a combination of technical and non-technical components directed to the solution of a technical problem consistent with the distinction of *Beattie*. In addition, while the TBA did not explicitly identify the 'technical problem' being solved, it seems clear from its reasoning to have been the elimination of human involvement from the apparatus's method by means of the latter's automation.[92]

Thus, in *Pettersson*, the TBA adopted an approach reminiscent of that in *Stockburger*, by treating the system's dependence on technical means as sufficient to distinguish it from a mental or business method as such within the meaning of Article 52(2)(c), and produced a result that is difficult to reconcile in reasoning or conclusion with *IBM/Card Reader* or *Christian Franceries* despite its purported application of the same legal principles. Indeed, the main difference between *Pettersson* and those cases, and thus between the *Beattie* alternatives themselves, seems on first reading to have derived from the terms in which the relevant subject matter was characterized and the question of its inherent patentability consequentially framed. Specifically, asking whether a device was excluded from patentability on the ground that its idea resided in a business method seems to have ensured a finding of inherent patentability, whereas asking whether a business method was entitled to a patent on the ground that it used technical means for its implementation seems to have ensured a finding of inherent unpatentability.

Closer analysis, however, reveals a deeper difference. Specifically, it shows the TBA's approach to inherent patentability in *Pettersson* to have been driven by its findings on three specific points, which put it in conflict with past decisions. The first, which distinguishes it from *Christian Franceries*, was that a product defined in a claim by its function and not its form is not for that reason precluded from being characterized as a device for the purpose of determining its exclusion under Article 52(2).[93] The significance of this finding has already been alluded to, and derives from its effect in underscoring both the importance of the terms in which an individual subject matter is conceived for its inherent patentability and, following *Pettersson*, the role of decision makers' discretion in undertaking that conception. The second, which also distinguishes *Pettersson* from *Christian Franceries*, was that an apparatus that has practical application in the service of customers cannot for that reason alone be excluded as an abstract subject matter in reliance on Article 52(2).[94] Thus, in *Pettersson*, the TBA cast 'practical applicability' as an indicator of technical character to support the patenting of a service industry apparatus, just as

[91] See ibid, [2.4].
[92] ibid, [2.5]–[2.6].
[93] See ibid, [2.2].
[94] See ibid, [2.1].

in *Christian Franceries* it cast (lack of) 'industrial applicability' as an indicator of (lack of) technical character to preclude the patenting of an industry service method;[95] the difference in approach arising from the Boards' contrasting treatment of functionally characterized products in each case. And the third finding, which distinguishes *Pettersson* from *IBM/Card Reader* and the majority of other 'contribution' decisions after 1988 involving computer systems, was that arguments directed to a subject matter's failure or otherwise to contribute to the technical arts relate solely to the questions of inventiveness and novelty, and are thus not relevant to the question of inherent patentability itself.[96] By this last finding in particular, the TBA rejected the contribution theory of technical character with the following three main consequences for a law. First, for a subject matter to be an invention it requires technical character (ie a technical effect or solution to a technical problem), but need not reflect any technical advance (ie a contribution to the prior art of a technical field). Second, in determining an objection to inherent patentability the exclusions of Article 52(2) are to be read and treated independently of each other, and not extrapolated beyond the requirement for technical character itself to support, for example, any general exclusion pegged to that Article's 'excluded fields of activity'. And third, in determining whether a subject matter possesses the technical character required of an invention, the subject matter needs to be construed in a holistic fashion, and its individual elements treated as cooperative or interdependent aspects of a single method or object, and not deconstructed with a view to identifying its distinguishing, defining, or 'essential' features.

In doctrinal terms, the effect of *Pettersson*, was to support a view of inherent patentability as at once discrete, in the *Sternheimer* sense of existing independent of the other conditions of patentability imposed by Article 52(1), and dilute, in the *Stockburger* and *IBM* sense of protecting subject matter that depends on technical means, including any computer apparatus cast as solving a technical problem. It also further exemplifies, in combination with *IBM/Card Reader* and the TBA's pre-1988 decisions, the ease with which different Boards of Appeal have been able to employ the same language of inherent patentability to arrive at diametrically opposed conclusions in respect of similar classes of subject matter, and to obscure in doing so the different legal principles driving those conclusions by embedding their reasoning in factual analysis.

3. Biotechnology as a Threat to Legalistic Conceptions of the Requirement for an Invention

Until now the analysis of Article 52(2) and (3) has focused entirely on non-naturally occurring products and processes, reflecting the small number of EPO

[95] cf the reasoning in *Pettersson* [1996] EPOR 1 with the reasoning in *Vicom* [1987] EPOR 74 on this point.

[96] See [1996] EPOR 1, [1].

decisions concerning the inherent patentability of biological subject matter.[97] While on its face this seems surprising in an era of biotech patenting, by the time of the EPC's introduction in 1977 the capacity of living and natural phenomena to support a patent had been largely accepted, both among Contracting States of the EPC and in other common law jurisdictions.[98] That acceptance explains the early development of a permissive granting practice within the EPO covering all manner of biotechnological products,[99] from recombinant DNA sequences to plants and animals.[100] That practice was tested, however, in a challenge to an EPC application directed to a naturally occurring phenomenon before the EPO's Opposition Division in 1994, in *HOWARD FLOREY/Relaxin*.[101]

The 1994 challenge involved an application seeking the refusal of a patent to the Howard Florey Institute of Experimental Physiology and Medicine for a DNA sequence that had been isolated from the human body for the first time using known recombinant techniques.[102] The central argument put by the opponents was that the sequence comprised a discovery as such within the meaning of Article 52(2)(a), such that granting the patent would confer excessive rights on the applicant, as well as concede the patentability of other discoveries, such as 'the moon (after the Americans landed on it in 1969), "Otzi" (a mummified, around 5,000-year-old man found in ice in the Italian/Australian Alps), or a new animal found in some remote area'.[103]

The Opposition Division rejected this argument in a decisive opinion built on two central findings.[104] The first was that it ignored EPO patent granting practices and the distinction they recognized between the (inherently unpatentable) act of finding a naturally occurring substance on one hand, and the (inherently patentable) act of isolating the same substance from its natural environment on the other. And the second was that the inventions produced by such acts of isolation included

[97] Most of the cases in this area focus on EPC, Art 53(a) and/or (b).

[98] See, eg *Ex p Hibberd*, 227 USPQ (BNA) 443, BPAI (1985); *Ex p Allen*, 2 USPQ 2d (BNA) 1425, 1427, BPAI (1987); aff'd 846 F 2d 77, Fed Cir (1988); Australian Patent Office, *Patent Applications Concerned with Living Organisms* (1980) 50 AOJP 1162.

[99] See Application No 83 307 553.4 (*HOWARD FLOREY/Relaxin*) [1995] EPOR 541, [5.1]; T_49/83 (*CIBA-GEIGY/Propagating Material*) [1979–85] EPOR vol C758, [2], cited with approval in T_356/93 (*PLANT GENETIC SYSTEMS/Glutamine Synthetase Inhibitors*) [1995] EPOR 357, 367. See also J Straus, 'Patenting Human Genes in Europe—Past Developments and Prospects for the Future' (1995) 26 IIC 920, 925.

[100] The inherent patentability of genetically engineered life forms has been supported by the TBA's restrictive interpretation of EPC, Art 53(a) and (b). See, eg T_320/87 (*LUBRIZOL/Hybrid Plants*) [1990] EPOR 173, 179–180; G01/98 (*NOVARTIS/Transgenic Plant*) [2000] EPOR 303, 312–313; T_19/90 (*HARVARD/Onco-Mouse*) [1990] EPOR 501, 527; Application No 88 312 222.8 (*LELAND STANFORD/Modified Animal*) [2002] EPOR 2, [42]; *HOWARD FLOREY/Relaxin* [1995] EPOR 541. cf *PLANT GENETIC SYSTEMS/Glutamine Synthetase Inhibitors* [1995] EPOR 357.

[101] *HOWARD FLOREY/Relaxin* [1995] EPOR 541.

[102] See ibid.

[103] ibid, [5.4].

[104] See ibid, [5.1].

not only the process of isolation, but also the substance itself, provided it could be 'properly characterised by its structure' and was 'new in the absolute sense of having no previously recognised existence'. In reaching this conclusion the Division emphasized that it was not deterred by the prospect of broad protection for the sequence thus produced however it might be created, stating that it would be 'perfectly justified to grant broad protection in view of the fact that H2-relaxin has been made available to the public for the first time'.[105]

This latter statement reveals the central significance of *Howard Florey* for inherent patentability, which concerns the basis for the Division's distinction between the act of first isolating a DNA sequence from its natural state on one hand and the isolated DNA sequence on the other. According to the Division, the justification for treating the isolated sequence as an invention, independent of the method by which it was isolated, lay in the value of that method in giving the public something of industrial applicability that was not previously known to exist.[106] In so doing, the Opposition Division allowed considerations of social benefit to affect its conception of the subject matter *qua* invention, just as Lord Diplock had done in the context of a similar subject matter in *American Cyanamid Co (Dann's) Patent.*[107]

The Opposition Division's treatment of a newly isolated biological material having no previously recognized existence as an invention is not the only reason for the interest of its decision. A further reason derives from the separate submission of the opponents that the DNA sequence should have been refused a patent on the basis that its publication or exploitation was contrary to *ordre public* or morality within the meaning of Article 53(a) of the EPC. In support of that submission the opponents argued (among other things) that granting a patent for the DNA sequence amounted to granting a patent for 'life', which was intrinsically immoral and thus prohibited by Article 53(a).[108] The significance of that argument for inherent patentability lay in the attempt it reflected to bolster the inherent patentability exclusion of Article 52(2)(a) by introducing, through the door of Article 53(a), a per se exclusion in respect of all human biological material on explicitly ethical grounds.

The Opposition Division's response to that attempt was unequivocal. It dismissed the opponents' concerns regarding DNA patenting as ill-conceived, stating that patents confer rights of exclusion rather than of use,[109] and suggesting as their

[105] [1995] EPOR 541, [5.3].

[106] ibid, [5.4].

[107] [1971] RPC 425, HL, discussed at pages 95–97.

[108] See [1995] EPOR 541, [6.1].

[109] See ibid, [6.3.3]; also Application No EP85304490 (*HARVARD/Onco-Mouse*) [1991] EPOR 525, 526; *LELAND STANFORD/Modified Animal* [2002] EPOR 2, [48]. cf *PLANT GENETIC SYSTEMS/Glutamine Synthetase Inhibitors* [1995] EPOR 357, 371 ('A patent confers on its owner(s) for a specified time an excusive right to exploit the subject matter of the claims, that is, to manufacture, use and market it, and to prevent others from doing the same.')

bases the conditions of novelty, inventiveness, and industrial applicability rather than ethics.[110] In addition, to the extent that those concerns reflected a view of DNA as 'life' they were answered by the assertion, consistent with the Division's reasoning in respect of Article 52(2)(a), that DNA is not 'life', but rather 'a chemical substance which carries genetic information and can be used as an intermediate in the production of proteins which may be medically useful'.[111] On the question of the intrinsic immorality and consequential inappropriateness of granting patents for gene sequences more generally, the Opposition Division declined to comment, suggesting that to do so would involve an inappropriate incursion into the domain of national legislatures to undertake an enquiry not permitted by Article 53(a).[112] Indeed, and contradicting the only prior EPO authority on this point,[113] the Opposition Division in *Howard Florey* construed Article 53(a) as having the limited function of ensuring 'that patents . . . not be granted for inventions which would universally be regarded as outrageous',[114] and as applicable only in respect of subject matter 'that the public in general would regard . . . as so abhorrent that the grant of patent rights would be inconceivable'.[115] The consequence of that construction for the opponents' case was put as follows:[116]

> As for the opponents' general assertions concerning the alleged intrinsic immorality of patenting human genes, these are founded on the premise that there is an overwhelming consensus among the Contracting States that the patenting of human genes is abhorrent and hence prohibited under Article 53(a). This assumption is false.

Speaking generally, the significance of *Howard Florey* for inherent patentability derives from its exemplification of two things about the EPO's treatment of Article 52(2) in the first twenty years of its existence. The first is the uneasy relationship between that understanding and the non-threshold conditions of patentability, and the second is the uneasy relationship between inherent patentability and policy considerations. On one hand, the Division was clearly influenced by considerations of social benefit in its conception of the invention as the isolated sequence. On the other hand, it rejected the relevance of such considerations to patentability, underlining in doing so the policy vacuum in which Article 52 tended to be construed. That this was challenged in the ethically charged context of an isolated human DNA sequence is not surprising, particularly given the wider

110 See [1995] EPOR 541, [6.2.2]. cf *PLANT GENETIC SYSTEMS/Glutamine Synthetase Inhibitors* [1995] EPOR 357, 366–367, 371, where the TBA described EPC, Art 53(a) as positioning the EPO 'at the crossroads between science and public policy' and thereby entrusting it with responsibility for ensuring that 'inventions the exploitation of which is not in conformity with the conventionally-accepted standards of conduct pertaining to . . . the culture inherent in European society and civilisation . . . be excluded from patentability'.

111 [1995] EPOR 541, [6.3.4].

112 See ibid, [6.4]–[6.5]. See also *LELAND STANFORD/Modified Animal* [2002] EPOR 2, [47].

113 See *PLANT GENETIC SYSTEMS/Glutamine Synthetase Inhibitors* [1995] EPOR 357.

114 See [1995] EPOR 541, [6.2.1].

115 ibid. The example cited by the Opposition Division was a letter bomb.

116 [1995] EPOR 541, [6.4.3].

developments that were occurring at the time of *Howard Florey* within the European Community regarding DNA patenting. What is surprising is that EPC jurisprudence in relation to inherent patentability remained as impervious to those developments, and the public policy debates that were driving them, for as long as it did. By 1999, however, the issue of biotech patenting and the EC's response to it had reached the point where the EPO's continued treatment of Article 52(2) as requiring a legalistic enquiry into the technical character of individual subject matter was becoming increasingly strained, particularly in light of the failure of that treatment to have delivered clarity and certainty even of doctrinal law.

D. Inherent Patentability in an Era of Harmonization: 1999 to Present

1. The EC Biotech Directive 1998

One of the arguments put by the opponents in *Howard Florey* to support their submission that the DNA sequence should have been denied a patent on *ordre public*/morality grounds was that there existed a public consensus in Europe that human genes should not be patented.[117] In advancing that argument the opponents relied on the then draft EC Directive on the legal protection of biotechnological inventions, and its proposed prohibition against the granting of patents for elements of the human body.[118] In the end, however, that tactic backfired, with the Opposition Division referring to the EC's disagreement over the terms of the draft as evidence not only of a lack of public consensus, but also of the inherently subjective and arbitrary nature of the issues raised by the opponents, which, in the Division's opinion, merely confirmed the view that the EPO was neither an appropriate nor a competent forum for considering the ethics of DNA patenting.[119]

In 1998, however, the landscape changed with the EC's finalization of its draft and consequential adoption of the Biotech Directive.[120] Concluded after a ten-year legislative passage mired in controversy, the purpose of the Directive was stated as being to establish[121]

> harmonised, clear and improved standards for the protection of biotechnological inventions with a view to furthering the innovative potential and the competitivity [*sic*] of Community science and industry in this important area of modern technology.

[117] See ibid, 541, [6.4] (excluding, on the opponents' arguments, industry and the EPO).

[118] Common Proposal of the European Parliament and the Council of Ministers for a Directive on the legal protection of biotechnological inventions, COM(88) 496 final—SYN 159, 17 October 1988, OJ C10/3 (13 January 1989). See ibid, [6.4].

[119] See [1995] EPOR 541, [6.4.1]–[6.4.5].

[120] Council Directive (EC) 98/44 of the European Parliament and of the Council of 6 July 1998 on the legal protection of biotechnological inventions (Biotech Directive) [1998] OJ L213/13.

[121] Common Position (EC) 19/98 [1998] OJ C110/26, [4].

While the Biotech Directive did not apply of its own accord to the EPO,[122] on 16 June 1999 the European Patent Organisation's Administrative Council incorporated one of the Directive's two sets of key provisions into a new chapter of the EPC's Implementing Regulations with effect from 1 September 1999.[123] The provisions in question relate to the inherent patentability of genetic material and other elements of the human body, and relevantly provide as follows:

Rule 27 (formerly Rule 23c)

Patentable biotechnological inventions

Biotechnological inventions shall also be patentable if they concern:
(a) biological material which is isolated from its natural environment or produced by means of a technical process even if it previously occurred in nature; . . .

. . . .

Rule 29 (formerly Rule 23e)

The human body and its elements
(1) The human body, at the various stages of its formation and development, and the simple discovery of one of its elements, including the sequence or partial sequence of a gene, cannot constitute patentable inventions.
(2) An element isolated from the human body or otherwise produced by means of a technical process, including the sequence or partial sequence of a gene, may constitute a patentable invention, even if the structure of that element is identical to that of a natural element.
(3) The industrial application of a sequence or a partial sequence of a gene must be disclosed in the patent application.

The effect of these Rules was to entrench the principle of *Howard Florey* that human DNA sequences and other biological materials are capable of supporting a patent, provided they are derived by an act of isolation or other 'technical' process, and notwithstanding that the protection conferred by the patent will extend beyond that act or process itself. For this reason, it could be said that the transposition of the Biotech Directive introduced no change to substantive EPC law and thus had no direct significance for Article 52(2). However, even if the doctrinal impact of that transposition was minimal at the time, its normative significance was substantial for two reasons. First, it restricted the discretion of decision-makers in conceiving individual subject matter *qua* invention with reference to industrial or other policy considerations. And second, and in combination with the Directive's other provisions, it supported the trend of EPO jurisprudence towards a field of

[122] The EPC is not a legal instrument of the European Community, and is therefore not subject to subsequent EC laws (see Report from the Commission to the European Parliament and the Council—Development and implications of patent law in the field of biotechnology and genetic engineering, COM(2002) 545 final (7 October 2002) [2.1]).

[123] See Decision of the Administrative Council of 16 June 1999 Amending the Implementing Regulations to the European Patent Convention [1999] OJ EPO 437.

patentability constrained by technical character alone.[124] On the other hand, and paradoxically, the Rules created a means for challenging that trend. Specifically, by entrenching the Directive's provisions on inherent patentability as a supplementary aid to the EPC, Rules 27 and 29 import the various policy objectives underlying those provisions into EPC jurisprudence.[125] Those objectives are given explicit form in the Directive's 56 recitals, from which the following central propositions can be drawn:

1 Modern biotechnology is important to European economic and social well-being, and depends for its existence on strong patent protection.[126]

2 The scope and availability of patent protection for modern biotechnology is uncertain, and is at risk of becoming more uncertain if the 'uncoordinated development of national laws' in respect of biotech patenting is allowed to continue.[127]

3 Uncertainty in the law of biotech patenting harms European biotechnology industry and trade.[128]

4 Consistent with Article 27.1 of the TRIPS Agreement, the patent system should be applied in a technologically neutral manner so as to protect inventions across all technological fields, even controversial fields such as modern biotechnology.[129]

5 It is neither the purpose nor the function of the patent system to regulate biotechnological research and development.[130]

6 While there is no general prohibition or exclusion in the EPC that precludes the patenting of biological material, there is the requirement for an invention, which must be maintained in order to safeguard the dignity and integrity of the person consistent with Opinion No 8 of the Group of Advisers on the Ethical Implications of Biotechnology to the European Commission.[131]

[124] On the effect of the Biotech Directive's provisions regarding the exclusion of EPC, Art 53(b) in this regard, see J Pila, 'Article 53(b) EPC: A Challenge to the *Novartis* Theory of European Patent History' (2009) 72 MLR 436–462.

[125] Rule 23b(1) provides: 'For European patent applications and patents concerning biotechnological inventions, the relevant provisions of the Convention shall be applied and interpreted in accordance with the provisions of this chapter. [The Biotech Directive] shall be used as a supplementary means of interpretation.'

[126] See Biotech Directive, Arts 1, 2, 3, 10, 11.

[127] See ibid, Arts 4, 5, 6, 7.

[128] See ibid, Art 14.

[129] See ibid, Arts 12, 22.

[130] See ibid, Arts 1, 2, 3, 10, 11.

[131] Group of Advisers on the Ethical Implications of Biotechnology to the European Commission, Opinion No 8 (Ethical Aspects of Patenting Inventions Involving Elements of Human Origin) 25 September 1996, available at <http://ec.europa.eu/european_group_ethics/docs/opinion8_en.pdf>. See Biotech Directive, Arts 15, 16, 19.

As these propositions reflect, the Biotech Directive is premised on the basic policy assertion, consistent with its stated objective above, that because society needs modern biotechnology, and modern biotechnology needs patent protection, society needs patent protection for modern biotechnology. However, its recitals also go beyond that assertion by giving the Directive's premise a doctrinal foundation drawn from the TRIPS Agreement's principle of non-discrimination (proposition (d)),[132] and the non-regulatory function of patent law generally (proposition (e)).[133] In so doing they recall the reasoning of the Division in *Howard Florey*,[134] while also striking a note that contrasts with that case by describing inherent patentability as one means by which the application of patent law is restricted 'so as to respect fundamental principles safeguarding the dignity and integrity of the person'.[135] That description is consistent with the reference in recital 19 of the Directive to Opinion No 8 of the Ethics Committee, in which the implications of inherent patentability for biotech patenting were described as follows:[136]

> The traditional distinction between discovery (not patentable) and invention (patentable) involves, in the field of biotechnology, a particular ethical dimension. It follows from this distinction that the knowledge related to the human body or its elements is relevant to scientific discovery and cannot be patented. It has to be clearly specified that the simple knowledge of the complete or partial structure of a gene cannot be patented.
>
> The human body, at different stages of its constitution and development, as well as its elements, do not constitute patentable inventions. Such an exclusion does not come only from the usual conditions of patentability, but it is also inspired by the ethical principle of non-commercialisation of the human body. Therefore no patent can be given on the human body or on its elements.

From this analysis the following conclusion might be drawn. By importing the Directive's provisions and recitals on inherent patentability as a supplementary aid to the EPC, Rule 23b(1) of the Regulations supports a reading of Article 52(2) and (3) as requiring a different approach to inherent patentability having regard to the nature of the subject matter at issue. In the context of biotech subject matter, the invention properly conceived includes 'a particular ethical dimension'. Given this, one might say that rather than supporting the trend of EPO jurisprudence towards a field of patentability constrained by technical character alone, the introduction of Biotech Directive jurisprudence into the previously closed world

[132] See Biotech Directive, recital 12.

[133] ibid, recital 14.

[134] See also Report from the Commission to the European Parliament and the Council—Development and implications of patent law in the field of biotechnology and genetic engineering, COM(2002) 545 final (7 October 2002) [49].

[135] See Biotech Directive, recital 16. On the Biotech Directive's marriage of commercial and ethical concerns, see generally DM Gitter, 'Led Astray by the Moral Compass: Incorporating Morality into European Union Biotechnology Patent Law' (2001) 19 Berkley J Int'l L 1–43.

[136] Ethical Aspects of Patenting Inventions Involving Elements of Human Origin, [2.2]–[2.3].

of the EPC system threatened to undermine it, by opening the way for a range of new policy considerations in the context of Article 52(2) around the question '*what constitutes an invention in the field or art in question, and in the context of the biotech field, what is the ethical dimension which an invention includes?*'

2. The TRIPS Agreement 1994

As has been seen, after the decision in *IBM* the TBA's divergent approaches to inherent patentability resolved to a competition between the whole contents and contribution theories of inherent patentability, with a majority of decisions reflecting the latter in their exclusion of technically conventional subject matter for its failure to contribute to the prior art of a technical field. Since 1999, however, that trend has been reversed as a result of the TBA's rejection of the contribution approach in favour of a view of inherent patentability premised entirely on the subject matter's technical character. The result has been the clearest statements to date of the EPO's commitment to an expansive conception of inherent patentability.

Of interest in relation to that commitment is the context from which it originally emerged and the means by which it was originally realized. The context was an increasing gap between the EPC and the patent systems of other major industrial jurisdictions as a result of the continually expanding understandings of inherent patentability that the latter supported,[137] and the means were the two central aspects of the public policy agenda of the TRIPS Agreement: technological neutrality and harmonization.

(a) *Revisiting the Computer Programs Exclusion in Light of Harmonization:* IBM/Computer Programs

IBM/Computer Programs[138] involved a patent application that was directed to a computer program recorded on a computer readable medium without the conjunction of any implementing apparatus.[139] The application was refused by the Examining Division for claiming a computer program as such within the meaning of Article 52(2)(c) on the basis, consistent with the TBA's assumption since *Koch*, that a computer program is an informational product the presentation of which, whether 'by itself or as a record on a carrier', is excluded from patentability

[137] In the context of business methods, see, eg *State Street Bank & Trust Co v Signature Financial Group Inc* 149 F 3d 1368 (Fed Cir, 1998); *Welcome Real-Time SA v Catuity Inc* [2001] FCA 445. *State Street Bank* has since been disproved (see *Re Bilski* 545 F 3d 943 (Fed Cir, 2008)) and *Welcome Real-Time SA* is under review as part of the Australian Government's review of 'patentable subject matter' (see http://www.acip.gov.au/reviews.html#%23subject).

[138] [1999] EPOR 301. See also T_1173/97 (*IBM/Computer Programs*) [2000] EPOR 219.

[139] See [1999] EPOR 301, [IV].

under Article 52(2)(c) and (d).[140] In an unusually drafted appeal the TBA was asked to consider solely whether that finding was correct in law; the aim presumably having been to elicit a decision of principle uncluttered by judicial analysis of fact.

In arguing against the Examining Division's decision, the appellant in *IBM/Computer Programs* relied heavily on two policy submissions, which it suggested to demand a narrow construction of the computer programs exclusion. The first was the need to ensure complementarity between the patent and copyright systems, the point being that as the ideas and principles underlying a computer program are not capable of protection by copyright law, they should be capable of protection by patent law.[141] And the second was the need to ensure the EPC's compatibility with both the provisions of the TRIPS Agreement, and the patent-granting practices of Japan and the United States.[142]

In its reasons for decision the TBA gave considered attention to the second of these submissions in a rare deviation from its usual approach. The conclusion it reached was that both the TRIPS Agreement and United States/Japanese practice were relevant and legitimate interpretive aids to the EPC, as an indicator of current jurisprudential trends the EPO's observance of which 'contribute[s] to the further highly desirable (worldwide) harmonization of patent law'.[143] The effect of that conclusion was to require the TBA's determination of the appeal in light of the fact, which it effectively accepted, that per se exclusions for computer programs were out of step with international practice and jurisprudential trends.[144] It was in that context that the TBA turned to the text of Article 52, which it interpreted as follows.[145]

1. A computer program as such within the meaning of Article 52(2)(c) is an abstract creation lacking technical character. It follows that any computer program having technical character is an invention, and that the scope of the computer programs exclusion depends on the meaning of 'technical character' in the specific context of computer programs.
2. A computer program cannot be regarded as technical by virtue of its inclusion of features common to all such programs.[146]

[140] ibid, [I].

[141] ibid, [III].

[142] ibid, [IV].

[143] ibid, [2.3], [2.6]. The TBA noted that the TRIPS Agreement has no direct application to the EPC, by direct effect or application of the Vienna Convention on the Law of Treaties (Vienna Convention) (Vienna, 22 May 1969, 1155 UNTS 331, 8 ILM 679), but found it appropriate to take it into account nevertheless 'since it is aimed at setting common standards and principles concerning the availability, scope and use of trade-related intellectual property rights, and therefore of patent rights [and thus] gives a clear indication of current trends'. The same reason was given by the TBA for its reference to US and Japanese Patent Office practice.

[144] See [1999] EPOR 301, [2.3].

[145] See ibid, [5].

[146] See ibid, [6].

3. As a result of 2., the required technical character of a computer program cannot derive from the program's modification of its implementing apparatus. Rather, its technical character must be found in the further effects created by its implementation—either directly from the technical character of those effects themselves, or indirectly from their effect in causing the hardware to solve a technical problem. This means that a computer program to be an invention must either manage, by means of a computer, an industrial or mechanical process, or comprise a necessary (or the only) means of obtaining a technical effect.
4. The mere fact that the basic idea underlying a subject matter derived from a computer program was of itself no bar to the subject matter's inherent patentability,[147] and nor was the fact that the 'further effects' of a program (from which its technical character derived) were known in the prior art. The question of a computer program's technical contribution to the art was relevant to its novelty and inventiveness, but not to its inherent patentability.[148]

By this reasoning the TBA relied on the 'as such' proviso of Article 52(3) to extract the following central principle consistent with *Koch*: a computer program is a set of instructions the inherent patentability of which depends on its production, when implemented, of a predetermined technical effect beyond that inherent in its conventional physical interaction with an implementing apparatus. Contrary to *Koch*, however, the TBA held that the effect in question does not need to be disclosed in physical reality, with the result that a computer program claimed in isolation could still support a patent on the basis of its 'potential' to produce a technical effect when executed by a computer.[149] Also contrary to *Koch* was the TBA's finding in *IBM/Computer Programs* that the technical character of a computer program, while necessarily deriving from the program's further effects, need not reflect any change in the technical functionality of a physical apparatus itself, and might therefore derive from the program's instructional content.[150] In reaching that decision it referred to the general interpretive rule of the Vienna Convention,[151] that '[a] treaty shall be interpreted in good faith in accordance with the ordinary meaning to be given to the terms of the treaty in their context and in light of its object and purpose',[152] which it held to be in full agreement with its construction of the proviso, as follows:[153]

> [T]he object and purpose of the EPC is the grant of patents for inventions and thus to promote technical progress by giving proper protection to these inventions. With this in mind, the Board has arrived at its interpretation in the light of developments in

[147] See [1999] EPOR 301, [7.4].
[148] See ibid, [8].
[149] See ibid, [9.4].
[150] See ibid, [11].
[151] See at pages 110–111.
[152] Vienna Convention, Art 31.
[153] See [1999] EPOR 301, [10.2].

information technology. This technology tends to penetrate most branches of society and leads to very valuable inventions. In its interpretation the Board has in its view not gone beyond the ordinary meaning given to the terms of the EPC. The meaning it has attributed to the expression 'as such' in Article 52(3) EPC is, in its opinion, not a special meaning within the meaning of Article 31(4) Vienna Convention, which would have required the consent of the parties to the EPC.

Thus, the TBA concluded its decision by referring, for the first time in the history of Article 52(2) and (3), to the interpretive context of that provision in order to legitimate its expansive conception of the 'as such' proviso under the Vienna Convention, and thereby avoid the need to submit its decision to the Contracting States for approval. In so doing it completed its circle of reasoning by ending from where it began, namely, with an appeal to extraneous sources to justify its conclusion that technical programs in any form were inherently patentable. It also contradicted the view of previous Boards since *Vicom*, that however expansive the meaning of 'technical' (which the Board in *IBM/Computer Programs* did not define), it would not permit the patenting of programs independent of their implementing apparatus.[154]

(b) Revising Article 52 in Light of the TRIPS Agreement

The history of the development of the law of inherent patentability under the EPC to 2000 is largely a history of the computer programs exclusion of Article 52(2)(c), and the TBA's attempt to reconcile that exclusion with a positive conception of the invention as any technical subject matter. In the years immediately following *IBM/Computer Programs*, that struggle was obviated by the TBA's declaration that on a proper construction, Article 52(2) and (3) *do* permit the patenting of technical subject matter, in the form of technical programs, technical information, and useful products, and that the EPC as a whole requires such patenting. In reaching that position the Board drew support from a legalistic interpretation of the 'as such' proviso and considerations of European and international harmonization. Those considerations included compliance with the EC Biotech Directive and, after 2000, compliance with the revision of Article 52 itself. That revision took place at the Diplomatic Conference that was held by the European Patent Organisation in November 2000 for the express purposes of adapting the EPC to comply with the TRIPS Agreement and other international patent treaties, and of creating the conditions necessary to ensure the EPC's future capacity to respond to changes in technology and law.[155] Among the proposals contained in the Conference's

[154] See generally ibid, [11].

[155] See European Patent Organisation, *Notice from the President of the EPO*, 24 March 2000, (2000) 4 OJ EPO 173; Act Revising the Convention on the Grant of European Patents (European Patent Convention) of 5 October 1973, last revised on 17 December 1991 (Munich, 29 November 2000).

official discussion papers were two of relevance to inherent patentability.[156] The first was a proposal to bring Article 52(1) into line with Article 27.1 of the TRIPS Agreement by[157]

> enshrining 'technology' in the basic provision of substantive European patent law, clearly defining the scope of the EPC, and making it plain that patent protection is available to technical inventions of all kinds.

And the second was a proposal to delete the computer programs exclusion of Article 52(2)(c).[158] The reason for the second of those recommendations was described in the Conference documents as formal rather than substantive; the Administrative Council expressly noting that[159]

> [t]he EPO and the Boards of Appeal have always interpreted and applied the EPC in such a way that this exception in no way excludes appropriate protection for software-related inventions, ie inventions whose subject matter consists of or includes a computer program. . . . Nevertheless, the point must be made that patent protection is reserved for creations in the technical field.

It is unsurprising in light of this statement that the Conference decided to amend Article 52(1) in the manner suggested but to retain the computer programs exclusion of Article 52(2)(c). Thus amended, Article 52(1) provides that 'European patents shall be granted for any inventions, in all fields of technology',[160] rather than (as previously) 'for any inventions'.

In the Explanatory Remarks accompanying the EPC 2000 Revision Act, this amendment is described as 'not chang[ing] the actual legal position',[161] but rather as expressing, in the 'internationally binding terminology' of Article 27.1 of the TRIPS Agreement, the 'long-standing principle of European patent law that patent protection is available only for creations in a technological field'.[162] However, while the principle that 'patent protection is available only for creations in a technological field' requires the exclusion from patentability of creations outside that field, it does not require the inclusion of 'inventions, in all fields of technology'. Further, given that Article 52(2) and (3) define 'inventions', the requirement for which is expressed in Article 52(1) as being separate from the requirement for the availability of protection 'in all fields of technology', the latter could not strictly speaking determine the former. It is unclear what conclusions to draw from this. While Article 52(1) and the TBA's interpretation of Article 52(2)

[156] See Administrative Council, Basic Proposal for the Revision of the European Patent Convention, EPO Doc MR/2/00 (2000) 43ff.

[157] ibid, 43, [1].

[158] ibid, 43, [3].

[159] ibid, 43, [3]–[4].

[160] Act Revising the Convention on the Grant of European Patents (European Patent Convention) of 5 October 1973, last revised on 17 December 1991, Special Edition No 4 OJ EPO (2001) 3 ff.

[161] See ibid, 135, [6].

[162] ibid.

and (3) are undoubtedly sympathetic in their focus on technology, it seems a stretch to attribute the latter to the terms of the former. In addition, whether Article 27.1 itself requires a reading of Article 52(2) and (3) as resolving to a requirement for a subject matter having technical character depends on whether one infers such a requirement from its support for the availability of patents in all fields of technology, or its prohibition against discrimination as to technological field. [163]

3. Technical Information and Automated Transactional Schemes after *IBM/Computer Programs*

The doctrinal significance of the TBA's decision in *IBM/Computer Programs* derives from its use of the 'as such' proviso of Article 52(3) to reconcile the EPO requirement for technical character with the computer programs exclusion of Article 52(2)(c) as follows: in the context of Article 52(2)(c), the proviso creates a distinction between computer programs (which are technical and thus inherently patentable) on one hand and computer programs as such (which are non-technical and thus inherently unpatentable) on the other, such that any computer program of a technical nature is an invention within the meaning of Article 52(1). On its face this distinction begs two essential and related questions. First, what in substance differentiates a computer program from a computer program as such? And second, to what extent can the use of a computer program to automate an otherwise unpatentable method confer the technical character required to show an invention?

In considering the first of these questions, the TBA in *IBM/Computer Programs* emphasized the need for a program to derive its technical character from features not common to all computer programs. It did not require, however, that such features be unique to the program itself, nor that the secondary effect they create (or problem they solve) be actually or allegedly new or inventive. On the contrary, the TBA in *IBM/Computer Programs* explicitly rejected the relevance of such considerations in the context of Article 52(2), which accounts for its rejection of the contribution theory itself.

The implications of *IBM/Computer Programs* for the second question are less clear, and depend on the meaning of 'industrial or mechanical process' and 'technical effect'.[164] Assuming, however, that the *Pettersson* understanding of 'industry' is correct, then *IBM/Computer Programs* effectively establishes the inherent patentability of automated business and other methods of use in a service industry. That result underscores the expansive tendencies of *IBM/Computer Programs* and the potential it creates for a further restriction of Article 52(2) with reference to notions

[163] On the consistency of TRIPS, Art 27.1 with the EPC in general, see D Schiuma, 'TRIPS and Exclusion of Software "as Such" from Patentability' (2000) 31 IIC 36 (arguing that the computer programs exclusion of EPC, Art 52(2)(c) is in breach of TRIPS, Art 27.1).

[164] See [1999] EPOR 301, [6.5]; n 146.

of industrial applicability. Paradoxically, it also vindicates the EPC drafters' original arguments for and against the relocation of the computer programs exclusion from its own paragraph of Article 52(2) to Article 52(2)(c), by showing the elasticity of the exclusion thus placed, and the potential it creates for overriding the computer programs and other Article 52(2) exclusions by 'indirect means'.[165] That potential has been confirmed in the most recent TBA decisions in this area.

(a) Presentations of Information

The first of those cases is *PHILIPS/Record Carrier*,[166] which involved an appeal from a decision of the Examining Division to refuse an application for a patent directed to a record carrier of use in a picture retrieval system in reliance on Article 52(2)(d).[167] The TBA allowed the appeal and found the carrier to be an invention on the same principles as the television signal of *BBC*,[168] which it affirmed as correct following the decision in *IBM/Computer Programs*.[169] In doing so it adopted an interpretive approach reminiscent of that decision by referring to the drafting history of the EPC to support a narrow reading of Article 52(2)(d) as restricted to information of a purely cognitive nature.[170] One result, as it acknowledged, was to cast computer programs outside Article 52(2)(d) and thereby to restrict the doctrinal basis for their exclusion to Article 52(2)(c).[171] Another result, which it did not acknowledge, was to undermine *IBM/Computer Programs* by supporting the inherent patentability of all technical information including (on the TBA's definition) all computer programs 'as such' within the meaning of *IBM/Computer Programs*.[172] This result flowed from its distinction between (inherently unpatentable) cognitive information and (inherently patentable) technical information, and from its definition of the latter as including information that supports physical interactions within and between machines but that does not convey any humanly understandable meaning. According to the TBA, such information falls outside the ordinary meaning of the term 'information' as it existed at the time of the EPC's introduction, and is therefore not 'reasonably interpretable' as coming

[165] See Report on the 11th meeting of Working Party I held in Luxembourg from 28th February to 3rd March 1972, Luxembourg Conference (Brussels 13 April 1972), Doc BR/177/72 (1972), [8], discussed at page 154.

[166] T_1194/97 (*PHILIPS/Record Carrier*) [2001] EPOR 25.

[167] See ibid, [III].

[168] [1990] EPOR 606, discussed at pages 213–214.

[169] See ibid, [3.6].

[170] ibid, [3.7.3].

[171] ibid, [3.7.4].

[172] ibid, [3.7.2]. Note in this context the proposal by the UK delegation at the Luxembourg Conference to clarify the scope of (what become) Art 52(2)(c) on the basis of its uncertainty, and the rejection of that proposal in order, among other things, to avoid conveying the false impression that that Article was intended for more restrictive interpretation than the corresponding provision of the PCT (Washington, 19 June 1970, 9 ILM 978); see the discussion at pages 153–155.

within the intended exclusions of Article 52(2)(d) and (3); a view more recently affirmed and applied in a separate decision to support the patenting of an 'electronic message'.[173] To the extent that the Board's reasoning in this respect undermines the separate exclusion of computer programs contained in Article 52(2)(c), it is consistent also with the decision in *Broselow* that a method distinguished by its informational content alone and opposed on that basis under Article 52(2)(d) will avoid exclusion if its effect is to render performance of a previously known method more efficient, whether by removing a step from the method itself or by reducing the degree of human involvement (and thus risk of human error) involved in its performance; an effect one would expect to be easily demonstrated in cases involving a computer program.

(b) *Automated Business Methods*

(i) *PBS PARTNERSHIP/Controlling Pension Benefits System*

To similar effect was another decision involving Article 52(2), which achieved a similar result to *Philips* but by different interpretive means and with wider ramifications for exclusions other than the computer programs exclusion itself. That case is *PBS PARTNERSHIP/Controlling Pension Benefits System*,[174] which involved an appeal from a decision of the Examining Division to refuse a patent application directed to an actuarial method for controlling a pension benefits program, and to the programmed general-purpose computer by which the method was implemented,[175] as a business method and computer program respectively under Article 52(2)(c). In relation to the method, the TBA agreed with the Examiner's finding of inherent unpatentability, describing it as a use of data processing means to perform a series of steps having a purely economic purpose, the technical features of which did not change its character as a business method as such.[176] Indeed, according to the TBA, any other conclusion 'would be too formalistic and would not take due account of the term "invention"'.[177]

By contrast is the TBA's finding in relation to the computer itself, which it held to be inherently patentable on the basis that any[178]

> computer system suitably programmed for use in a particular field, even if that is the field of business and economy, has the character of a concrete apparatus in the sense of a physical entity, man-made for a utilitarian purpose and is thus an invention within the meaning of Article 52(1) EPC.

[173] See T_858/02 (*LUCENT TECHNOLOGIES/Structured Voicemail Messages*) [2006] EPOR 2.
[174] [2002] EPOR 52.
[175] See ibid, [II].
[176] See ibid, [3].
[177] ibid, [3].
[178] ibid, [5].

The TBA, therefore, invoked the same formalistic reasoning it had eschewed in the context of the method to permit the patenting of the same method when characterized in terms of its implementing computer. In justifying that approach the TBA relied explicitly on the failure of Article 52(2)(c) to refer to physical entities, which it construed as meaning that[179]

> if a claim is directed to such an entity, the formal category of such a claim does in fact imply physical features of the claimed subject matter which may qualify as technical features of the invention concerned and thus be relevant for its patentability.

The effect of this finding was to reduce the computer programs exclusion to *de minimis* status by enabling the grant of a patent for any program characterized in an application by reference to its implementing apparatus and not its instructional content per se, on the ground that an 'invention' within the meaning of Article 52(1) includes any 'concrete apparatus in the sense of a physical entity, man-made for a utilitarian purpose'.[180] By this reasoning the TBA took the *Vicom* logic regarding automated methods to an extreme. As has been seen, in *Vicom* the implementation of a method by computer means was held to facilitate the method's claim to inherent patentability, which claim, once established, precluded its exclusion as a computer program. In *PBS Partnership*, by contrast, the implementation of the method by computer means not only facilitated the method's entitlement to inherent patentability, it established that entitlement, despite the fact that the method itself was explicitly regarded—even in combination with the computer means—as inherently unpatentable. On this logic, the conjunction of one (inherently unpatentable) business method and one (inherently unpatentable) computer program produces, when claimed by means of the latter's implementing hardware, one (inherently patentable) invention in the form of a physical apparatus of practical utility. The result is a reading of Article 52(2) that achieves by indirect means the effect of both *Philips* and *IBM/Computer Programs* in conjunction with *Pettersson,* namely, the recognition respectively that computer programs and automated business methods are technical and therefore inherently patentable. Paradoxically, however, *PBS Partnership* also undermines *IBM/Computer Programs* by conflating the essence of a computer program with its material forms of embodiment in contradiction of the TBA's reasoning in that case.[181]

(ii) *PITNEY BOWES/A System for Processing Mail and HITACHI/Auction Method*

Since PBS Partnership, the TBA's reading of Article 52(2) and (3) has been further extended in three important cases: *PITNEY BOWES/A System for Processing*

[179] [2002] EPOR 52, [5].
[180] ibid.
[181] See [1999] EPOR 301, [9.8].

Mail,[182] *HITACHI/Auction Method*,[183] and *DUNS LICENSING ASSOCIATES/ Estimating Sales Activity*.[184] In *Pitney Bowes*, a method for shipping mail that involved a series of steps including the sorting of mail by a van driver according to a predetermined system was held not to be a method for doing business as such on the grounds that it did not depend on 'the driver exercising a judgement of the kind usually involved in doing business'.[185] Further, the system for sorting mail itself, while described by the TBA as 'inspired by that mathematically based approach to planning and resource allocation known as operational research or logistics which is nowadays a typical management "tool" used in running a business',[186] was held not to be a method for doing business as such, but rather an 'invention' within the meaning of Article 52(1). The reason was its 'practical application to mail processing, which is itself, in essence, a particular kind of mechanical handling and selective conveying of articles to respective destinations under given time constraints'.[187] According to the Board[188]

> [t]he established jurisprudence of the EPO Boards of Appeal has construed Articles 52(2) and (3) EPC to mean that the fact that a measure may have been derived from or inspired by an insight originating in an activity which is *per se* excluded from protection – be it a discovery, a mathematical method, a mental act or a method of doing business – does not imply that a claim including the material expression or embodiment of such a measure in its specific practical application in the solution of a technical problem is a claim to the excluded activity as such.

In the second case, *Hitachi*, the TBA went even further than this, overturning a decision of the Examining Division that an automated method of conducting an auction was not an invention within the meaning of Article 52(1). In so doing it rejected the formalism of the *PBS Partnership* approach to support a view of inherent patentability as encompassing any subject matter having technical features regardless of 'the category of the claim' in issue.[189] Emphasizing the technical (computer) features of the auction method in question, the TBA affirmed that the contribution approach of earlier EPO jurisprudence was wrong, citing the independence of inherent patentability from novelty and inventive step,[190] and the longstanding recognition of the TBA (in reliance on Article 52(3)) that

182 T_388/04 (*PITNEY BOWES/A System for Processing Mail*) [2007] OJ EPO 1/16.

183 T_258/03 (*HITACHI/Auction Method*) [2004] OJ EPO 12/575.

184 T_154/04 (*DUNS LICENSING ASSOCIATES/Estimating Sales Activity*) [2008] OJ EPOR 38.

185 ibid, [2.3].

186 ibid.

187 ibid, [2.4].

188 ibid.

189 ibid, [3.7] (finding that the Examining Division's assessment of a computer program carrying out a method of conducting an auction was an invention within the meaning of EPC, Art 52(1) 'since it comprises clearly technical features such as a "server computer", "client computers" and a "network"').

190 ibid, [3.2]–[3.3].

'a mix of technical and non-technical features may be patentable'.[191] In its judgment, the first of these reasons was inconsistent with the finding in *PBS Partnership* that '[a] feature of a method which concerns the use of technical means for a purely non-technical purpose and/or for processing purely non-technical information does not necessarily confer a technical character to such a method',[192] confirming (with reference to the decision in *Koch*) that the mere presence in a claim of any technical means would confer the one essential (constitutive) feature of an invention.[193]

What matters having regard to the concept of 'invention' . . . is the presence of technical character which may be implied by the physical features of an entity or the nature of an activity, or may be conferred to a non-technical activity by the use of technical means. In particular, the Board holds that the latter cannot be considered to be a non-invention 'as such' within the meaning of Articles 52(2) and (3). Hence, in the Board's view, activities falling within the notion of a non-invention 'as such' would typically represent purely abstract concepts devoid of any technical implications. . . . The Board is aware that its comparatively broad interpretation of the term 'invention' in Article 52(1) will include activities which are so familiar that their technical character tends to be overlooked, such as the act of writing using pen and paper. Needless to say, however, this does not imply that all methods involving the use of technical means are patentable. They still have to be new, represent a non-obvious technical solution to a technical problem, and be susceptible of industrial application. . . . It is therefore concluded that, in general, a method involving technical means is an invention within the meaning of Article 52(1).

(iii) *DUNS LICENSING ASSOCIATES/Estimating Sales Activity*

This brings us to the last of the EPO cases for detailed discussion, *Duns Licensing Associates*. According to the TBA in that case, writing on 15 November 2006, 'the constant jurisprudence of the Boards of Appeal' is capable of succinct summary in the following three propositions of relevance.[194] First, Article 52(1) sets out four 'essentially separate and independent criteria of patentability': the existence of an invention, and establishment of its novelty, inventive step, and industrial applicability. Second, Article 52(2) and (3) define an invention as any subject matter having technical character. And third, an invention may include any mix of

[191] ibid, [3.4]. See also ibid, [3.6] (supporting the second of these factors with reference to the practical difficulty of separating a claim into technical and non-technical features, and the possibility of an invention having 'technical aspects which are hidden in a largely non-technical context').

[192] [2008] OJ EPO 2/46, [4.3].

[193] ibid, [4.4], [4.5]–[4.7].

[194] See ibid, [5]. A fourth proposition identified by the TBA was that in defining the problem–solution approach, the problem must be a technical problem which the relevant skilled addressee might be asked to solve at the priority date, though it may be formulated using an aim to be achieved in a non-technical field, which aim is thus not part of the technical contribution which the invention makes. The result in *Duns Licensing Associates* was a finding that a claim in respect of a system for estimating sales activity was invalid as being for a business method as such. The decision was recently applied to reach a similar result in T_1449/06 (*IBM/Database Back-Solving*) [2010] EPOR 3, [34]–[35].

technical and non-technical features, including one in which the non-technical features dominate. How these propositions sit with the TBA's further finding that only the *technical* features of an invention can be counted in determining its novelty and inventive step, is unclear, unless by 'dominate' the TBA meant something other than constitutive of the subject matter *qua* invention.

With respect to the requirement for an invention itself, the Board went to some effort to justify its three propositions above, relying to that end on the following factors:

1 The case law of the Boards of Appeal and the Enlarged Board;[195]

2 The expression in Article 52(1) of 'the fundamental maxim of the general entitlement to patent protection for any inventions in all technical fields';[196]

3 The absence of any accepted definition of 'invention' in the history or case law of the EPC, and the evidence in the *travaux préparatoires* of an intention on the part of the drafters that Article 52(2) and (3) reflect conventional patentability criteria, and 'not be given a too broad scope of application';[197]

4 The language of Article 52(2) itself, which 'covers subjects whose common feature is a substantial lack of technical character', and which 'ultimately derives from the classical notion of invention . . . which distinguishes between practical scientific applications and intellectual achievements in general';[198]

5 The support given for the TBA's view of the law by the EPC 2000 Revision Conference;[199]

6 The expression of the connection of the notions of invention and technical character in other parts of the EPC, including particularly Articles 18 and 56, and Rules 27(1) and 29(1);[200]

7 The logical and practical problems that would result from mixing the 'absolute requirement' for an invention—which depends on 'the inherent character of the claimed subject matter'—with the 'relative criteria' of novelty and inventive step, notwithstanding lay understandings of the two as interdependent.[201]

In response to these factors one can say as follows.

First, the case law of the Boards of Appeal lacks the consistency and clarity even of persuasive judicial authority. At most, it can be said to support the second proposition, that a subject matter requires technical character in order to be an invention

195 See ibid, [6].
196 See ibid.
197 See ibid, [6]–[7].
198 See ibid, [8].
199 See ibid.
200 See ibid.
201 See ibid, [10].

at law, which still leaves the questions, what is technical character, and how is the subject matter properly conceived *qua* invention? Second, Article 52(1) does not express, whether as a 'fundamental maxim' or otherwise, a 'general entitlement to patent protection for any inventions in all technical fields'; much less, as the TBA implies, a general entitlement to patent protection for any subject matter in a technical field. Rather, it expresses a prima facie entitlement to patent protection for *an invention* that is 'patentable' (in the sense defined), and a requirement that the patent system apply *across all fields of technology*. This is more than a semantic disagreement with the TBA's reasoning: it is calling attention again to the point above that the principles of Article 27.1 of the TRIPS Agreement, as expressed in Article 52(1) of the EPC, do not support the reading advanced by the Board. Indeed, it is submitted that they do not restrain the permissible reading of Article 52(2) and (3) beyond prohibiting a construction of those paragraphs as supporting a threshold exclusion covering an entire field of technology. Third, while it is true that neither the history nor case law of the EPC contains an accepted definition of the 'invention', this does not mean either that such a definition cannot be derived, nor, even if one cannot be derived, that it is not important to have some understanding of what constitutes a subject matter *qua* invention. As Lord Mustill said in response to a similar point when advanced by Lord Hoffmann in *Biogen Inc v Medeva plc*, 'in some instances a close conceptual analysis of the nature of patentability will not be a waste of time', particularly in the context of 'subject[s] far distant from the mechanical and chemical inventions to which so much of traditional patent law relates'.[202] Indeed, this view is consistent with the TBA's own suggestion in *Duns Licensing Associates*, consistent with prior TBA jurisprudence,[203] that the patentability of an established invention depends on the contribution made by its technical features; a view which directly contradicts factor (g) by recognizing the relevance of the 'inherent [ie technical] character' of a subject matter to determinations of its novelty and inventive step. Further, the basis for suggesting that the drafters intended Article 52(2) and (3) to reflect 'conventional patentability criteria' is unclear, as is the consistency of that suggestion with the TBA's view of the invention as 'separate and independent' of novelty, inventive step, and industrial applicability, which were presumably the conventional criteria in question. And finally, and as a matter of logic, the absence of an accepted definition of the 'invention' does not support, as the TBA's comments imply, a *de minimis* reading of the statutory requirement for the same.

[202] See [1997] RPC 1, HL, 31–32 per Lord Mustill, 41 per Lord Hoffmann, discussed at pages 298–299.

[203] See, eg T_939/92 (*AGREVO/Triazole Herbicides*) [1996] EPOR 171 (described by Karet (in I Karet, 'Over-Broad Patent Claims: An Inventive Step by the EPO' (1996) 18 EIPR 561–563) as the EPO's response to the problem of broad claims, equivalent to Lord Hoffmann's response in *Biogen v Medeva* itself, on which see pages 298–300); T_641/00 (*COMVIK/Two Identities*) [2004] EPOR 10. See also, since *Duns Licensing Associates*, T_368/05 (*CITIBANK/Integrated Account*) [2007] EPOR 54, [18]–[20]; T_1023/06 (*IGT/Computer Implemented Game Process*) [2007] EPOR 36, [10]–[12].

This leaves two substantive matters raised by the Board. The first concerns the EPC 2000 Revision Conference, and the implication by the Board that it evidences agreement among EPC states with respect to the Board's three propositions. And the second concerns the language of Article 52(2) itself, and its purported derivation 'from the classical notion of invention . . . which distinguishes between practical scientific applications and intellectual achievements in general'.[204]

a. EC Software Patenting Directive Proposal The suggestion of the TBA that there exists even broad agreement among EPC states for its three propositions, including particularly its theory of Article 52(2) and (3), is undermined by the history of another legislative reform initiative in this area; the European Commission's 2002 proposal for a directive on the patenting of computer programs.[205]

This proposal followed the EC's earlier identification of software patenting as an issue requiring urgent action at the Community level, on account of the harm allegedly being caused to the Internal Market by divergent applications in EU states of Article 52(2)(c) and (3) of the EPC.[206] The action it envisaged was two-fold: first, the introduction of a Community directive requiring the availability of software patents throughout the EU, consistent with the position in Japan and the United States; and second, the amendment of Article 52(2) of the EPC to remove the computer programs exclusion in sub-paragraph (c).[207] As has been seen, the second of these actions was considered at the EPC 2000 Revision Conference, and rejected.[208] The first was the aim of the EC in proposing a software patenting directive.[209]

On its face, the aim of the Commission Proposal was to codify the principles of EPO jurisprudence reflected in *PBS Partnership*. That aim is reflected in the Proposal's requirement that Member States: (a) ensure the prima facie availability of patents for subject matter involving the use of a computer, computer network, or other programmable apparatus as a 'computer-implemented invention';[210] (b) require, as a condition of inventiveness, that such 'inventions' make a non-obvious contribution to a technical field;[211] and (c) define 'technical field' to include computer programming.[212] The result was a draft Directive that used the language of the contribution theory of Article 52(2) to achieve the result of the whole contents

204 See [2008] OJ EPO 2/46, [8].

205 See Proposal (EC) 2002/92 for a Directive of the European Parliament and of the Council on the patentability of computer-implemented inventions (Commission Proposal) [2002] OJ C151/129.

206 See United Kingdom, Promoting innovation through patents—The follow-up to the Green Paper on the Community patent and the patent system in Europe, COM(1999) 42 final (3 February 1999) 2, 12.

207 See ibid, 13–14.

208 See Administrative Council, Basic Proposal for the Revision of the European Patent Convention, EPO Doc MR/2/00 (2000) 43, [3]–[4].

209 See Commission Proposal, recitals 2, 3, 5, and 16.

210 ibid, Arts 4(1), 2(a).

211 ibid, Arts 4(2), 2(b).

212 ibid, Arts 4(2), 3.

theory, and thereby give the principles of post-1999 EPO case law the imprimatur of EU legislation.[213]

The first significant blow to the Proposal came in September 2002, when the European Economic and Social Committee published its Opinion on the Proposal in accordance with the mandatory EC legislative process. The Opinion was deeply critical of both the draft and the manner in which it was prepared. In particular, it described the Proposal as a partisan document[214] that lacked a sound economic basis, showed insufficient regard for public interest and democracy,[215] and put Europe at risk of increased legal uncertainty, and perhaps even 'legal chaos'.[216] The Proposal's substantive provisions were further criticized as offering 'a de facto acceptance and justification of the *a posteriori* drift of EPO jurisprudence',[217] and as opening the way to the abolition of Article 52(2) and (3) by envisaging the patenting of all methods implemented through software programs or on electronic networks, including the Internet.[218] According to the Committee, the doctrinal premise of the Commission Proposal—the TBA's distinction between 'computer software by itself' and 'software producing technical results'—was 'indefinable', and 'the product of legal casuistry'.[219]

In September 2004, the Committee was vindicated in these criticisms by the European Parliament's vote to amend the Proposal to reflect its comments.[220] While retaining the Proposal's basic structure, the Parliament's Draft effected three important changes. First, it required that the necessary technical contribution arise from something other than the digital representation, production, handling, processing, distribution, or presentation of information.[221] Second, it required that Member States exclude data processing from their definitions of 'field of technology', and that innovations in the field of data processing not be regarded as inventions for patent law purposes.[222] And third, it expressly precluded patent protection for any computer-implemented business or other method that failed to produce any technical effect beyond the usual physical interactions of a program and its implementing hardware,[223] or the improvement in the efficiency of a data processing system's use of resources.[224] In combination, these changes put the

[213] Just as the Biotech Directive had done some four years earlier; see the discussion at pages 224–228.

[214] Opinion of the Economic and Social Committee on the 'Proposal for a Directive of the European Parliament and of the Council on the patentability of computer-implemented inventions' [2003] OJ C61/154, [3.12].

[215] ibid.

[216] ibid, [3.16], [3.13].

[217] ibid, [3.1.1].

[218] ibid, [3.1.2].

[219] ibid, [3.1].

[220] Position of the European Parliament adopted at first reading on 24 September 2003 with a view to the adoption of Directive (EC) 2003/. . .EC of the European Parliament and of the Council on the patentability of computer-implemented inventions [2004] OJ C77 E/230.

[221] ibid, Art 2(b).

[222] ibid, Art 3.

[223] ibid, Art 5.

[224] ibid, Art 6.

Parliament's Draft *against* the trend of contemporary EPO authority, by supporting a theory of technical character consistent with the contribution theory, under which the patenting of even inventive uses of a technically conventional apparatus to implement a method otherwise excluded by Article 52(2) was prohibited.

In May 2004, however, the Parliament's Draft was rejected, and the Commission's original version largely reinstated following the EU Council's agreement to a common position proposal that incorporated that version's central provisions.[225] Political controversy ensued, amid strong lobbying from different groups and interests.[226] Tensions were further increased when, in 2004, the Dutch delegate to the Council was instructed by his national Parliament to change his 'yes' vote on the May 2004 text to 'neutral',[227] and the Polish Government announced its non-support for that text.[228] Joining the national opposition to the proposal was the European Parliament's Legal Affairs Committee, which voted on 2 February 2005 to request the Commission to withdraw the Directive and restart the legislative process. Despite this, in March the Council approved the May text,[229] and resubmitted it to the European Parliament under the EC Treaty's co-decision procedure as a 'common position with a view to the adoption of a directive'.[230] On 6 July 2005 the Parliament decisively rejected the position, ending the passage of the Commission Proposal.[231]

The history of the Commission Proposal is significant for many reasons, not least of which is its reflection of the EU's jealousy of the European Patent Organisation's jurisdiction.[232] Of greatest immediate relevance, however, is the

[225] See Proposal for a Directive of the European Parliament and of the Council on the patentability of computer-implemented inventions—Political agreement on the Council's common position, Council Doc No 9713/04, 2002/0047(COD) (24 May 2004); 'Council clashes with Parliament on software patents', *EurActiv* (19 May 2004), available at <http://www.euractiv.com/en/infosociety/council-clashes-parliament-software-patents/article-117930>.

[226] For a discussion of the various 'EU Actors' positions', see 'Tension grows as Council prepares to vote on software patents directive', *EurActiv* (11 May 2004), available at <http://www.euractiv.com/en/infosociety/tension-grows-council-prepares-vote-software-patents-directive/article-117825>.

[227] See 'Dutch Parliament causes EU software patents crisis', *The Inquirer* (2 July 2004), available at <http://www.theinquirer.net/inquirer/news/1024273/dutch-parliament-causes-eu-software-patents-crisis>.

[228] See 'Software patents law up in the air after Poland pull out', *EurActiv* (19 November 2004), available at <http://www.euractiv.com/en/infosociety/software-patents-law-air-poland-pull/article-132419>.

[229] See Common Position (EC) 20/2005 adopted by the Council on 7 March 2005 with a view to the adoption of a directive of the European Parliament and of the Council on the patentability of computer-implemented inventions [2005] OJ C144 E/9.

[230] See Treaty Establishing the European Community [2002] OJ C325, Art 251.

[231] See [2006] OJ C157 E/95; European Parliament legislative resolution on the Council common position with a view to the adoption of a directive of the European Parliament and of the Council on the patentability of computer-implemented inventions (11979/1/2004—C6-0058/2005—2002/0047 (COD)) [2006] OJ C157/265.

[232] This is apparent from the opening address made by Commissioner Frits Bolkestein to the Plenary Debate on the Commission Proposal on 23 September 2003, in which the Commissioner urged his EU colleagues to support the Proposal in order that the reforms could be achieved 'using the community method', rather than 'via the route of an intergovernmental treaty' in which the EC would 'take a back seat'. For a transcript of the Debate, including the Commissioner's address, see

controversy generated by its substantive provisions among EPC Contracting States in their capacity as EU Members. That controversy undermines the TBA's suggestion above that there exists even majority support among those states for its reading of Article 52(2) and (3), which was essentially the reading described by the European Economic and Social Committee as the 'product of legal casuistry', and rejected by the Parliament in 2005. It also calls into question the appropriateness of the TBA's trenchant support for that reading, notwithstanding the real European concerns regarding its doctrinal, social, and economic impact.

b. 'The Classical Notions of Invention'. The TBA's reliance on the 'classical notion of invention' and that notion's purported distinction 'between practical scientific applications and intellectual achievements in general' to support its theory of Article 52(2) and (3) is intriguing. According to its judgment, it is from this notion and distinction that those Articles 'ultimately derive[ ]', and which establish the criterion for technical character, as the 'exclusive criterion common' to the Article 52(2) subject matter.[233]

It was indeed always common ground that creations in engineering and technology were entitled to patent protection under the European Patent Convention. As the Board judged in T 930/05 . . . this criterion is reflected by the internal logic of Article 52(1) and (2) EPC. The mere fact that the Article 52(2) list of items not to be regarded as inventions is non-exhaustive . . . is indicative of the existence of an exclusion [*sic*] criterion common to all those items and allowing for additions to the list that were thought possible. The enumeration of typical non-inventions in Article 52(2) EPC covers subjects whose common feature is a substantial lack of technical character. The formulation of the law ultimately derives from the classical notion of invention adopted, which distinguishes between practical scientific applications and intellectual achievements in general. The connection of the notions of invention and technical character of the invention arises immediately, because the list of exclusions in Article 52(2) EPC, with its reference to Article 52(1) EPC, must be viewed as a negative definition of the notion of invention. This connection is also inherent in other provisions . . . which clearly express this underlying principle of patent law.

It is not clear what the Board meant by 'the classical notion of invention'. It seems unlikely to have been referring to a conception of invention from the classical antiquity period of history. This is because the notion of invention is of relatively recent origin, even more recent than the notion of technology itself, which did not emerge properly until the seventeenth century,[234] with Francis Bacon's idea

<http://swpat.ffii.org/papers/eubsa-swpat0202/plen0309/deba/index.en.html>. Similar arguments were made in 1976 and 1977 for the UK's ratification of the EPC (see Hansard HC (series 5) vol 910, col 803 (30 April 1976), vol 379, col 281 (24 January 1977).

[233] [2008] OJ EPO 2/46, [8].

[234] cf E Zilsel, 'The Sociological Roots of Science' (1942) 47 Am J of Sociology 544–562 (locating the origins of technology in the separation of the liberal and mechanical arts in around 1550).

of applying science to utilitarian ends, and Descartes' origination of the idea of physical laws and the mechanical theory of the natural world.[235] Similarly, if the Board's reference was to the notion of invention as a term of legal art, then its view conflicts with the nineteenth century observations of Daniell, cited in the Introduction, that the longstanding association of that term with the patent system was due to the ordinary meaning of the verb 'to invent', which was sufficiently wide to encompass both importation and original conception, and any other form of first introduction.[236] In light of this, it is not clear what the TBA understood as the source of Article 52(2) and (3).

Nevertheless, the distinction between practical scientific applications and intellectual achievements drawn by the Board is well-known, and might certainly be described as classical in this limited sense, even if not in the sense of being accepted or true. It is not a distinction central to original theories of technology—the Cartesian theory was likely too tethered to the machine analogy[237]—though it certainly pervades early twentieth century theories,[238] which support a broad conception of technology as 'the application of the method of science to practical problems'.[239] However, even this is a conception of *technology*, not of (the) invention, and cannot therefore in logic support a definition of the latter with reference to the former. Further, it is a notion that was accepted even by its proponents as

[235] See generally J Bury, *The Idea of Progress: An Inquiry into its Growth and Origin* (New York: Dover Publications, 1932), especially ch 2; WE Houghton, 'The History of Trades: Its Relation to Seventeenth-Century Thought as Seen in Bacon, Petty, Evelyn, and Boyle' (1941) 2 J History of Ideas 33–60. cf Zilsel, 'The Sociological Roots of Science', 558 (describing as 'the decisive event in the genesis of science' the 'rise of the methods of the manual workers to the ranks of academically trained scholars at the end of the sixteenth century'); L Edelstein, 'Recent Trends in the Interpretation of Ancient Science' (1952) 13 J History of Ideas 573, 579 (describing the view that ancient science failed to lead to technological application as 'one of those prejudices that die hard').

[236] A Daniell, 'Inventions and Invention' (1899) 11 Juridical Rev 151–172.

[237] See generally EW Strong, 'Mechanical Ideas in the Scientific Thought of the Seventeenth Century' (1938) 72 The American Naturalist 324–339; LS Feuer, 'Mechanism, Physicalism, and the Unity of Science' (1949) 9 Philosophy and Phenomenological Research 627–643.

[238] See, eg L White, 'Technology and Invention in the Middle Ages' (1940) 15 Speculum 141, 141 (defining technology as 'the way people do things', and adding '[i]n a certain sense there is even a technology of prayer'); RP Multhauf, 'The Scientist and the "Improver" of Technology' (1959) 1 Technology and Culture 38, 38 (citing Singer et al's definition of technology as 'concerned with systematic discourse about the (useful) arts', and noting '[o]ther definitions could be compiled indefinitely, not merely because of differences of viewpoint, but because the signification of the terms has evolved along with their subject matter'); M Kranzberg, 'At the Start' (1959) 1 Technology and Culture 1, 8–9 (criticizing the definition of 'technology' in Singer's classic work, *A History of Technology*, for dealing with technology 'only within the framework of its own subject matter, as if it were insulated from the rest of society', and describing that treatment as deriving 'inevitably' from his definition of 'technology' as 'how things are commonly done or made'); PF Drucker, 'Work and Tools' (1959) 1 Technology and Culture 28, 36 (arguing that technology 'must be considered as a system, that is, a collection of interrelated and intercommunicating units and activities').

[239] See M Bunge, 'Technology as Applied Science' (1996) 7 Technology and Culture 329, 331.

putting technology beyond the sphere of description, if not understanding; a fact which accounts for Singer's decision in his 'classic' *History of Technology*[240] not to attempt to give a comprehensive treatment of its subject after 1850 on the basis that technology, in the sense above, had by that time become so complex as to defy description, let alone understanding.[241]

E. Conclusion

From the outset, Article 52(2) of the EPC was accepted by the TBA as a definitional provision and its paragraphs resolved to a positive requirement, initially for a non-abstract subject matter, and then for a subject matter having technical character. In its earliest understandings, the TBA relied heavily on the individual exclusions of Article 52(2) to support a view reminiscent of early twentieth century UK law in its concern with the production or alteration of a physical product. By 1988, however, its focus was on the subject matter's use of technical means to solve a problem or create an altered state of affairs. The result was a definition closer to that supported by the High Court of Australia in *National Research Development Corp v Commissioner of Patents*,[242] albeit without the need for an end of commercial significance, and without the same assumption of mechanical or chemical ingenuity. Pursuant to that conception, the requirement for an invention was sufficiently dilute to ensure the prima facie patentability of most artefactual products and methods.

To the extent that this suggests a coherent trend towards liberalization, it is undermined by the cases decided in the eleven-year period after 1988, when the uncertainties of the TBA's early decisions re-emerged to support competing definitions of inherent patentability in the context, primarily, of programmed computers. Indeed, not only is a reversion to a more restrictive approach to Article 52(2) apparent during this period, it is that approach which predominates. This explains the denial of protection for many conventionally programmed general-purpose machines during the 1990s on the basis of their failure to advance the technical arts, and sometimes, their failure to advance the art of a non-excluded field. Since 1999, however, the predominance of the contribution theory of inherent patentability has ended, and much of the practical uncertainty surrounding the threshold status of computer-related subject matter and other informational products under the EPC has been resolved. That resolution has come with the TBA's confirmation of the inherent patentability of all technical objects, including computer programs and other forms of technical information, with reference to

[240] C Singer, EJ Holmyard, and AR Hall (assisted by E Jaffé, RHG Thomson, and JM Donaldson), *A History of Technology* (Oxford: Clarendon Press, 1967) 8 vols.

[241] Drucker, 'Work and Tools', 35.

[242] (1959) 102 CLR 252, HCA, discussed at pages 90–95.

the 'as such' proviso of Article 52(3) construed in light of extraneous legal and political developments. The result has been an expansive understanding of inherent patentability that, since *Re Bilski*[243] at least, creates a gap between the EPC and the laws of the United States, whatever practical benefits it may have for those charged with determining the validity of individual claims.[244] Nonetheless, even the most expansive reading of the threshold requirement for an invention cannot resolve the issues raised by that requirement, for there remains the question '*what is a computer program or other (inherently patentable) subject matter* qua *invention?*', the answer to which depends on an understanding of the constitutive elements of inventions, and of a method of perceiving individual subject matter *qua* invention informed by an ontology of the invention for patent law. To the extent that the TBA has purported to resolve these issues (in answer to the question) through the concept of 'technical character' itself, and the requirements of novelty and inventive step, its resolution has been acknowledged by the EPO President to be inadequate. That acknowledgement is implicit in her 2008 referral to the EBA, in which she described as an open question (not answered by the text or interpretive context of the EPC)[245] whether the exclusions of Article 52(2) and (3) are 'based on the fact that the items mentioned lack the necessary technical character or . . . on broader social or political considerations'?[246] Also important is her implicit recognition that the concept of 'technical character' may itself import 'social and political considerations', and that any ontology of the invention for patent law purposes must import the same.

The interpretation of Article 52(2) and (3) of the EPC since 1977 reflects many themes. Perhaps the most pervasive is the EPO's simultaneous reliance on the technical character requirement to avoid the terms of Article 52(2) and (3), and dependence on those terms to give life to that requirement. The circularity thus produced in many of its decisions underscores a further difficulty, which stems from their alternate use and eschewal of formalistic reasoning to support different approaches to similar subject matter according to their description in the patent documentation. For example, in *PBS Partnership* the Board confirmed simultaneously the inability of an applicant to rely on the technical character conferred by a business method's automation to avoid exclusion of the method under Article 52(2)(c), and the ability of the same applicant to rely on the physical form of the computer programmed to implement that method to enable protection of

[243] 545 F 3d 943 (Fed Cir, 2008) (rejecting a 'technological arts' test of inherent patentability on the grounds, amongst others, of the contention and ambiguity surrounding 'technological' and related concepts, and its lack of basis in US patent law). At the time of writing a decision of the US Supreme Court in *Re Bilski* is pending.

[244] See factor (g) of *Duns Licensing Associates* [2008] OJ EPO 2/46, [10], at page 237, emphasizing the practical benefits of its interpretation of Art 52(2) and (3).

[245] G_3/08 (*PRESIDENT'S REFERENCE/Computer Program Exclusion*) [2009] EPOR 9, [77]–[80].

[246] ibid, [79].

the computer thus programmed. Further, having initially employed formalistic reasoning to justify the patenting of computer programs when claimed in conjunction with a machine—as in *PBS Partnership*—the TBA in *IBM/Computer Programs* relied on the inappropriateness of formalistic reasoning to justify *extending* that principle to allow patents for the same computer programs when claimed in isolation.

Instances of the Boards' use of this kind of reasoning to overcome the threshold exclusions from patentability are numerous, and extend beyond the computer programs exclusion itself. For example, in *Howard Florey* the TBA relied on the chemical nature of isolated DNA sequences to justify their conception *qua* invention as products as such, capable of supporting expansive protection, and also of avoiding ethical objections to their patentability as forms of 'life'. Of more general applicability, there is the 'as such' proviso of Article 52(3), which the Boards have used to read down the Article 52(2) exclusions as exceptions to a primary inventor's right. In this way, what began as a non-exhaustive and seemingly diverse list of subject matter not capable of constituting an 'invention', properly defined, has become a limited exception from a presumptive statutory right.

If the TBA's legalistic approach has failed to put the law on a principled doctrinal basis, so too its occasional reference to policy and history has failed to put the law on a firm normative basis. For example, the TRIPS Agreement's principle of technological neutrality has been used by the TBA to support the threshold patentability of any subject matter having technical character. However, not only does this use strain the meaning of Article 27.1's terms, it also fails to engage with the purpose of those terms, and the significance of their prohibition against technological discrimination for the European system. Similarly limited has been the view taken by the TBA of the requirements for harmonization and international trade. In the reasoning of the TBA, those requirements support an alignment of the European threshold for patent protection with that of Europe's trading partners. As the European Economic and Social Committee's 2002 comments made clear, however, this aim has been arrived at without sufficient empirical or theoretical research regarding the impact of increased patentability on the European market and its industrial interests. Similarly, and notwithstanding the EPC drafters' apparent belief to the contrary, harmonization alone is not a valid basis for substantive legal reform. And finally, with respect to history, the TBA has noted the drafters' concern in implementing Article 52(3) that Article 52(2) not be given too wide an application, but failed to consider the reason for that concern, or what the drafters understood by *too* wide an application. Similarly, the absence of unequivocal evidence to support its view of the EPC Contracting States as having uniformly supported in their national laws, and agreed to support in a European law, the definition of 'invention' which it supports, gives some of its jurisprudence a tendentious feel. This is all the more so when one reads the EPC *travaux*, and the earlier records from the Council of Europe, which reflect an awareness among European experts that the concept of 'technology' was sufficiently elusive and differently understood among European States to be an unstable basis for harmonization.

Several ironies run through the EPO cases on Article 52(2) and (3). One concerns the 'as such' proviso on which the Boards' expansive reading of technical character has turned. As seen in chapter 4, that proviso originally applied solely to scientific discoveries, and was only extended, at the eleventh hour of the drafters' negotiations, so that its limited application would not be misconstrued by future decision-makers as implying that the Article 52(2) categories to which it did *not* apply were intended for broad construction. The effect of that extension, however, was to overcorrect the drafters' interpretive dilemma by turning a text they regarded as vulnerable to unduly broad interpretation into a text destined for unduly narrow interpretation. The TBA, it would seem, has fulfilled this destiny.

A second irony concerns the list of exclusions in Article 52(2) itself, independent of the Article 52(3) proviso. As has again been seen, that list was taken by the EPC drafters from the Rules of the PCT, less out of agreement with the Rules' content than as part of the drafters' commitment to facilitating harmonization by ensuring the EPC's literal harmony with the texts of other international instruments. Legal and technological developments, however, soon meant that the text originally formulated to ensure harmonization, began to be viewed as impeding it, by putting Europe in possible breach of the TRIPS Agreement and at odds with its major trading partners. The irony lies in the Boards' solution, which has been to invoke the objective of harmonization to support the reduction, and arguably the elimination, of the very provision to which the drafters' original pursuit of that objective gave life. By relying on harmonization to read down the 'conditions for the grant of patents' originally fixed in Article 52(2) in deference to the PCT and principles of certainty, the TBA has vindicated the UK's original opposition to that provision by showing inherent patentability to be, after all, a matter more suitably placed in the EPC's Implementing Regulations, from where it might more readily be amended with a view to keeping pace with technological change and international developments, the latter of which, as the decision in *Re Bilski*[247] (following *State Street Bank & Trust Co v Signature Financial Group Inc*[248]) demonstrates, are not destined to move in any clear or uniform direction.

This raises a third irony, which is the shift in the law of other developed nations, including the law of the United States, soon after the EPC 2000 revisions took effect. As a result of that shift, it is no longer possible for the EPO to claim consistency with the law of the US, which has (pending the decision of the Supreme Court in *Bilski v Kappos* at least) rejected a 'technological arts' conception of the invention. A natural question at this point is whether the historical problems associated with the technical character theory of Article 52(2) have been resolved by the amendment of Article 52(1) agreed at the 2000 EPC Revision, as again the

247 545 F 3d 943 (Fed Cir, 2008).
248 149 F 3d 1368 (Fed Cir, 1998).

TBA has implied. As has been seen, the result of that amendment is to specify the patentability of 'inventions . . . in all fields of technology', consistent with Article 27.1 of the TRIPS Agreement, and giving the theory a more explicit legal foundation. The conclusion to be drawn from the analysis above, however, is that rather than resolve the problems created by the technical character theory of inherent patentability, the 2000 EPC Revision has made them worse, precisely by entrenching the theory at a time of deep uncertainty over its meaning and basis. That the TBA has failed to engage with this possibility, and has deflected concerns in relation to its approach from a variety of legislative, judicial, academic, and other sources, is further evidence of its casuistical approach to this area of European patent law. The same was all but acknowledged by the President of the EPO in her 2008 EBA referral. In that referral, the President gave examples of divergent TBA authority on each of the core issues being referred, namely, 'the relevance of the category of the claim', 'where the line should be drawn between those aspects excluded from patentability and those contributing to the technical character of claimed subject matter', and how to conceive 'the activity (programming) which underlies the resulting product (computer program)'.[249] Further, while supporting the TBA's view of the EPC authors as having 'agreed that a computer program could only be patentable if it had technical character',[250] she did not take the TBA's further step of construing the requirement for an invention itself as satisfied by proof of technical character.[251] This is important, not only for conceptions of the invention itself, but also for conceptions of the role of the requirement, and the possibility that such role endures beyond the threshold (exclusion) stage to ensure the internal regulation and cohesion of the law.

[249] 149 F 3d 1368 (Fed Cir, 1998), [4].

[250] ibid, [78].

[251] See ibid, [78]–[80].

6

Inherent Patentability in UK Law Since 1977

A. Introduction

The drafting history of the Patents Act 1977 (PA) reveals the central objective in framing section 1 to have been the substantive harmonization of the UK and European Patent Office (EPO) principles of inherent patentability. Even without recourse to legislative history, that objective is apparent from the substance of the Act, including particularly the near-identical wording of section 1 of the PA and Article 52 of the European Patent Convention (EPC),[1] in addition to section 130(7), which has been construed to require that section 1 be interpreted in harmony with Article 52, consistent with the UK parliamentary intent.[2] However, to infer from this a substantive identity between the provisions at the time of their enactment, or a predestined development along parallel lines, would be to ignore the different institutional and jurisprudential contexts into which each provision was introduced. It would also be to ignore the different nature of the UK and EPC patent systems, and the fact that the former, unlike the latter, continues to regulate the life of a patent beyond the point of its grant. Finally, it would be to ignore the reality of the inherent patentability enquiry, and its dependence on the 'mindset' of individual judges.[3]

The significance of this for contemporary principles of inherent patentability is both substantive and methodological. Substantively, the rules of precedent subject UK decision-makers to an interpretive constraint that is largely anathema to the EPO, increasing immediately the potential for non-harmonious decision-making. That potential was further increased at the time of the PA's introduction by the belief among some that section 1 gave statutory form to pre-existing common law principles, and left intact the four centuries of jurisprudence on which those

[1] Convention on the Grant of European Patents (European Patent Convention, as amended) (Munich, 5 October 1973, 13 ILM 268).

[2] See, eg *Gale's Application* [1991] RPC 305, CA, 323 per Nicholls LJ; *Asahi Kasei Kogyo KK's Application* [1991] RPC 485, HL, 506 per Dillon LJ; *Merrell Dow Pharmaceuticals Inc v Norton & Co Ltd* [1996] RPC 76, HL, 82 per Lord Hoffmann; *Aerotel Ltd v Telco Holdings Ltd* [2006] EWCA Civ 1371, [6] per Jacob LJ; also Hansard HL (series 5) vol 381, cols 854–855 (29 March 1977). cf *Actavis UK Ltd v Merck & Co Inc* [2008] EWCA Civ 444, [48].

[3] See *CFPH's Application* [2005] EWHC 1589, Pat, [48]–[49] per Prescott QC, citing *Dyson Appliances Ltd v Hoover Ltd* [2001] EWCA Civ 1440 per Sedley LJ.

principles were based.[4] This raises a second consequence of the context into which section 1 was introduced, which derives from the content of that jurisprudence and its potential for influencing judicial constructions of section 1. As seen in chapter 3, the history of the requirement for inherent patentability in modern UK law was, at least until the 1960s, a conservative one, characterized by three specific and related attitudes. The first was an antagonism to non-manufacturing methods, reflected in conceptions of the invention as a subject matter of mechanical or chemical utility directed to the alteration or production of a vendible product. The second was a reluctance explicitly to acknowledge the relevance of ethics and other considerations of public policy to such conceptions, reflected in the treatment of medical methods. And the third was a disposition towards literal rather than purposive constructions of patent legislation, reflected in the emphasis on the meaning of the term 'invention', and the language of section 6 of the Statute of Monopolies 1623[5] with reference to which that term was statutorily defined. These factors offer one explanation for the length of time it took for the liberalizing decision of *National Research Development Corp v Commissioner of Patents*[6] to be taken up by the UK courts in the context of computer programs and other 'new' technologies during the 1960s and 1970s, and thus for the length of time it took to resolve the state of uncertainty that existed in relation to those technologies immediately prior to the PA's introduction. As seen in chapter 4, that uncertainty was a principal reason for the UK's concern to exclude computer programs from the EPC definition of the 'invention', and for its attempts to have Article 52(2) and (3) transferred from the main text of the EPC to the Implementing Regulations to permit future amendment by the European Patent Organisation's Administrative Council. It was also a reason for the inclusion of section 1(5) of the PA, and the proposal during the 1977 parliamentary debates to recognize the right of the Secretary of State under that section to prescribe different criteria for the grant of British and European patents.[7]

If modern patent jurisprudence gave the UK courts a predisposition towards a conservative interpretation of section 1 of the PA, a tendency towards a conservative methodology might be inferred from its common law traditions. The reason for this derives not only from the effect of precedent and Parliamentary sovereignty in encouraging a concern with legal doctrine and the law's internal coherence over theory and public policy itself, but also from their encouragement of a more parochial perspective on inherent patentability than might be expected of a supranational body. Again, this attitude can be seen in the position taken by the UK delegation during the European discussions of the 1950s and 1960s, and during the subsequent EPC debates, as well as in the parliamentary proposals regarding

[4] See at page 190.
[5] 21 Jac 1 c 3
[6] (*NRDC*) (1959) 102 CLR 252, HCA.
[7] See Hansard HL (series 5) vol 379, cols 1410–1411 (15 February 1977), discussed at page 192.

the PA.[8] The latter in particular underlie the tension that existed between the UK's commitment on ratifying the EPC to ensuring the future harmony of its patent laws with those of other EPC Member States on one hand, and the constitutional commitments of its courts on the other. They also underlie the tension that existed in 1977 between the UK's past and future patent traditions due to the different normative bases from which they proceeded. Informing the UK's pre-1977 case law was an understanding of the patent as a common law exception to a general prohibition against monopoly grants, conferred by the Crown at its discretion in recognition of a person's contribution to the industrial arts. Informing UK case law after 1977, by contrast, was a conception of the patent as a statutory right derived from an international treaty and conferred on the satisfaction of certain administrative requirements.

This leaves the nature of the UK patent system, and its ongoing concern with the lives of patents beyond the point of their grant. The significance of this for contemporary approaches to inherent patentability is also substantive and methodological, and derives from its effect in encouraging a holistic and contextualized approach to inherent patentability informed more directly by post-grant issues of infringement and scope than the EPC pre-grant system.

These themes are of more than academic interest. In an area of law notorious for the complexity and uncertainty of its decisions, it is only by understanding the peculiar position of the UK courts at the juncture of the national and international patent systems that the meaning of those decisions, and the reasons for their complexity and uncertainty, can be understood.

B. Different Approaches for Different Categories of Subject Matter: 1977 to 1989

Prior to 1990, inherent patentability under the PA was the focus of only two significant pieces of litigation in the UK, both of which were finally resolved by the Court of Appeal. The first involved a patent application by Merrill Lynch directed to a computerized financial method,[9] and the second involved a patent application by Genentech directed to a recombinant protein.[10] In combination, the decisions generated by these applications represent the first era of case law under section 1(2), and establish the legal basis from which all subsequent analyses of that section have proceeded.

[8] For examples see the discussion at page 192.

[9] See *Merrill Lynch Inc's Application* [1988] RPC 1, Pat, aff'd [1989] RPC 561, CA.

[10] See *Genentech Inc's Patent* [1987] RPC 553, Pat, aff'd [1989] RPC 147, CA.

1. Inherent Patentability in Light of the Business Methods Exclusion

Detailed consideration of inherent patentability under the PA was not undertaken until 1987, nearly ten years after its introduction, in the first of a series of important decisions involving the application by Merrill Lynch for a patent for a computerized method of analyzing buy and sell securities options.[11] The decision was of the Principal Examiner, who rejected the application on the ground that the method's sole contribution to the prior art lay in the performance of a method for doing business within the meaning of section 1(2)(c). In reaching his decision the Examiner relied principally on the terms of the 'as such' proviso to that paragraph, which he construed narrowly to support a restrictive view of inherent patentability premised on the following principle: while an invention can consist in part of subject matter expressly excluded from patentability by section 1(2), it can derive neither its status as such, nor its novelty or inventiveness, from the presence of that subject matter itself.[12] It followed that in determining whether an application for a patent related to an excluded item in the sense required by the proviso, regard would need to be had to whether the application's non-excluded features were 'already known or obvious'.[13] In addition, the absence of an established prior art meant that those issues would need to be judged against 'things which are common general knowledge or are acknowledged in the specification itself as being conventional'.[14]

In support of this reasoning the Examiner cited the provisions relating to inherent patentability contained in Chapter IV of the EPO's 1985 Guidelines for Examiners,[15] including in particular their instruction that the question of inherent patentability be resolved by identifying the subject matter's contribution to the art and determining whether that contribution was technical in nature.[16] Approving this instruction as consistent with his interpretation of the section 1(2) proviso above, the Examiner referred also to an article[17] citing the opinion of the EPO that

[11] See [1988] RPC 1, 8–9.

[12] ibid, 3–4.

[13] ibid, 4.

[14] ibid.

[15] [1985] OJ EPO 173, 177–179, discussed at page 210. The Examiner referred to the Guidelines on the ground that he knew of no decisions 'issued by competent authorities which would assist . . . in determining the effect of [EPC, Art 52(2)]' notwithstanding that his decisions followed T_16/83 (*CHRISTIAN FRANCERIES/Traffic Regulation*) [1988] EPOR 65, T_51/84 (*STOCKBURGER/Coded Distinctive Mark*) [1986] EPOR 229, and T_208/84 (*VICOM/Computer-Related Invention*) [1987] EPOR 74, discussed at pages 197–202.

[16] See [1988] RPC 1, PE, 5. See further MW Molyneaux, 'A History of Patenting Software-Related Inventions in the United Kingdom: Part II' [1996] Computer and Telecommunications L Rev 51, 52; J Richards and MW Molyneaux, 'Patenting Software-Related Inventions in the US and European Patent Offices' [1995] Computer and Telecommunications L Rev 117, 120; J Worthy, 'Software Patents in the UK after Merrill Lynch' [1989] EIPR 380, 381.

[17] The article was by Mr N Wallace, Vice-President of the EPO (see N Wallace, 'The Patentability of Software Related Inventions' (1985) 3 EPI Information 32, cited in [1988] RPC 1, 5).

'the list of exclusions in Article 52(2) EPC is of a declaratory nature and merely affirms what could be deduced from Article 52(1) EPC since the latter excludes from patentability any subject matter which is not of a technical nature'.[18] Commenting on this opinion the Examiner stated as follows:[19]

This accords with the view that I have taken that the word 'invention' as used in section 1(1) of the Patents Act 1977 should not be understood as being as broad as the phrase 'manner of manufacture'. Also the point is made in this Article that an invention may be technical by virtue of its structure, as is the case with a new chemical compound, or by virtue of its effect, and that in the case of computer program-related systems it is usually the effect which has to be technical. While agreeing with this I consider that it is necessary to take into account the possibility that an application which primarily relates to a computer system controlled by a program to perform a business method may nevertheless involve a novel technical structure which would not itself be excluded from patentability.

In this way, the Examiner reconciled his literal interpretation of section 1(2) with the EPO's requirement for technical character in terms of the following general proposition: to be an invention a computer system or apparatus must produce a technical effect, meaning an effect the actual or alleged contribution of which to the known art is not attributable to the system's implementation or incorporation of a section 1(2) item. In addition, while the Examiner left open the possibility that a computer system performing an excluded method might nonetheless constitute an invention on the basis of its incorporation of a novel technical structure, his failure to explain the kinds of systems for which that possibility was envisaged makes it unclear how, if at all, it could be reconciled doctrinally with his otherwise expansive reading of section 1(2) above.

Indeed, it seems clear from the Examiner's decision overall that his concern was with neither novel technical structures nor computer programs per se, but rather with the other method exclusions of section 1(2), and ensuring they could not be circumvented by the use of computer technology. Hence his exclusive reliance on the business methods exclusion of section 1(2)(c) in reasoning his decision, and his expansive interpretation of the same as covering any method of commercial transaction capable of performance by human means using conventional apparatus. According to the Examiner, not only was the method performed by the *Merrill Lynch* system a business method thus defined,[20] the technical means comprising that system consisted entirely of features either conventionally found in automated business systems, or specifically required for the performance of the *Merrill Lynch* method, preventing the system overall from being considered 'to constitute a new technical structure or to produce a technical effect in the [EPO] sense'.[21]

[18] [1988] RPC 1, 5.

[19] ibid.

[20] ibid, 6. The Examiner further found that the applicant's express exclusion of manual systems from the scope of its claims by means of its requirement that the system be automatic did not affect this finding, as it did not alter 'the true nature of the subject matter'.

[21] [1988] RPC 1, 6.

As a result, and applying the principles above, the system was held not to be an invention within the meaning of section 1(1).

The decision of the Examiner in *Merrill Lynch* was appealed in 1987 to the UK Patents Court in reliance on two arguments, each of which was supported by reference to the decision of the Technical Board of Appeal (TBA) in *VICOM/Computer-Related Invention*.[22] The first was that the Examiner had wrongly interpreted the 'as such' proviso as extending beyond claims restricted to excluded subject matter such as computer programs to include, for example, claims directed to an 'apparatus operating to rules contained in a computer program'.[23] And the second was that the Examiner had wrongly 'run together' the questions of industrial applicability and inherent patentability on one hand, and novelty and inventive step on the other.[24]

Falconer J rejected these arguments and approved the correctness of the Examiner's reasoning and conclusion. In so doing he affirmed the Examiner's interpretation of the phrase 'to the extent that' in the section 1(2) proviso as indicating an intention to exclude an invention compromising a section 1(2) item, regardless of its explicit restriction to the item per se.[25] He also confirmed as a corollary of this understanding the inability of a programmed computer ever to support a patent where it comprised 'a conventional computer operating in a conventional manner when carrying out the various steps of the program';[26] the only circumstances in which a programmed apparatus could support an invention being, in his view, where the apparatus itself produced a new and inventive technical effect.[27] On the basis of this view he supported the Examiner's two-stage test of inherent patentability that involved identifying the contribution made by the claimed subject matter, and determining whether the novelty and inventive step reflected in that contribution resided solely in an excluded item. In his concluding discussion Falconer J addressed the specific arguments of the appellant by referring to the decision in *Vicom*, which he reconciled with the Examiner's decision as follows:[28]

> [I]t is plain that in referring to a technical process [in its decision] the Board is referring to a process as considered in paragraph 5 [being a 'technical process' in contrast to a 'mathematical method'] providing a resulting change in the physical entity on which the process

[22] [1987] EPOR 74.

[23] See [1988] RPC 1, 11, Pat.

[24] ibid.

[25] ibid, 11–12.

[26] ibid, 12. Note in particular his Lordship's concern to avoid a result that would preclude protection for computer programs under section 1(2)(c) while permitting protection for the same programs when claimed as 'a computer programmed to carry out that program', on the ground that such result 'cannot have been intended by Parliament'. On what Parliament did intend, see Hansard HL (series 5) vol 379, cols 270, 301–302 (24 January 1977), 1408–1411 (15 February 1977); vol 381, cols 841–844 (29 March 1977); vol 383, cols 108–110 (9 May 1977).

[27] [1988] RPC 1, 12.

[28] ibid, 13–14.

is carried out . . . As I understand the reasoning of that decision of the Technical Board, their interpretation and application of Article 52(3), (corresponding to the qualification in section 1(2) of the Act) is not inconsistent with what I have held to be the proper construction to be given to the qualification and the approach required in considering its applicability, but, indeed, is in accord therewith.

By this passage Falconer J denied any substantive difference between his and the Examiner's decision and the decision of the TBA, implying a view of the latter as having turned on the *Vicom* system's production of a novel and inventive technical effect in the form of a technical process for producing physical entities. How this could be reconciled with his general conception of section 1(2) was not explained, but clearly depended on a view of section 1(1) and (2) as permitting patents for any method of producing digital or non-digital images, so that the novelty and inventiveness of the *Vicom* methods could be said to have derived from something other than a computer program or other section 1(2) item. Even if this view is accepted, however, and the methods of *Vicom* shown in fact to satisfy Falconer J's test of inherent patentability, there remains the difficulty of reconciling the cases at the level of principle; the TBA in *Vicom* having construed Article 52(2) as satisfied by the use of technical means to perform an industrially applicable process of manipulating and producing a physical entity, in contrast to the Examiner and Falconer J in *Merrill Lynch*, whose construction set a much higher standard requiring a novel and inventive technical effect not attributable to a section 1(2) item. The lacuna between these understandings is underscored by the different use made of the 'as such' proviso in formulating them; the TBA's decision in *Vicom* reflecting precisely the expansive view of the proviso that was rejected by the Examiner and Falconer J in *Merrill Lynch*.[29]

This analysis of Falconer J's treatment of *Vicom* begs the question why his Lordship followed the Examiner's reasoning on section 1(1) and (2) as closely as he did, rather than reformulate that reasoning to reflect more closely the then recently published language of the TBA. One possible answer is that he shared the Examiner's commitment to a literal interpretation of those provisions, and to deriving from them a coherent and exhaustive understanding of inherent patentability that was supported but not dictated by non-literal considerations of technical character. This is in contrast to the approach of the TBA in *Vicom*, whose non-exhaustive definition of the 'invention' was predicated on a purposive interpretation of Article 52(2) as precluding the protection of non-technical subject matter, albeit supported, paradoxically, by reference to a restrictive definition of 'mathematical method' under Article 52(2)(a).[30] A related possibility is that his Lordship shared the Examiner's commitment to giving full effect to the section 1(2) exclusions, and to supporting thereby a more restrictive understanding of inherent patentability

[29] This is supported by the TBA's use of the 'as such' proviso to support its distinction in *Vicom* between technical processes and mathematical methods (see [1987] EPOR 74, [5]–[6], discussed at pages 199–200).

[30] [1988] RPC 1, 13–14.

itself—whether for the specific purpose of limiting the availability of protection for business methods, or for the more general purpose of preventing section 1(2) from reducing to a requirement as vague as the 'manner of manufacture' requirement of pre-1977 law was thought to be. Of interest regarding the first of these suggestions is the decision in the EPC counterpart of *Merrill Lynch*, *CHRISTIAN FRANCERIES/Traffic Regulation*,[31] where the TBA adopted a similarly expansive view of Article 52(2) to refuse a patent for a method of controlling traffic using technical means, also in reliance on an expansive interpretation of the business methods exclusion. That this interpretation was subsequently relaxed in *Vicom* by a differently constituted Board of Appeal can be explained with reference to the nature of the subject matter there at issue and the terms in which the question of its inherent patentability was framed. Indeed, and as suggested in chapter 5, it was only in extrapolating its finding that the methods were not 'mathematical algorithms' to a finding that they were (therefore) 'inventions' that the TBA in *Vicom* articulated its expansive interpretation of inherent patentability. It can therefore be speculated whether the same expansive interpretation would have ensued had the subject matter not been a process for producing images which, considered as a non-automated method, is not a class of subject matter that is obviously excluded by section 1(2).[32]

The suggestion that the Examiner's and Falconer J's interpretation of section 1(2) in *Merrill Lynch* was fuelled by a concern to give full effect to the business methods exclusion of section 1(2)(c) in order to reject the application before them is consistent with the fate of that interpretation, and its almost immediate rejection by the Court of Appeal in separate proceedings involving, in place of an automated method for trading securities, a recombinant protein.

2. Inherent Patentability in Light of the Discoveries Exclusion

Genentech Inc's Patent[33] involved a patent application directed to a naturally-occurring human protein (t-PA) that had been produced by the applicants using known recombinant procedures following their discovery of the protein's structure and encoding DNA sequence.[34] In the Patents Court Whitford J rejected an argument that the protein was an inherently unpatentable discovery within the meaning of section 1(2)(a). According to his Lordship, while the protein was based on the novel discovery of its characterizing DNA and amino acid sequences, it was not that discovery to which the patent would relate, but rather the applicant's direction as to how the discovery might usefully be employed. The fact that such

[31] [1988] EPOR 65.

[32] On the consistency of *Vicom* and *Merrill Lynch* see further RJ Hart, 'Application of Patents to Computer Technology—UK and the EPO Harmonisation?' [1989] EIPR 42–46.

[33] [1987] RPC 553, Pat, aff'd [1989] RPC 147, CA.

[34] See [1987] RPC 553, 574–576.

employment was 'obvious enough' from the nature of the discovery was, he held, irrelevant:[35] the applicant's entitlement to the patent derived from the novelty of the discovery and the utility of its application.

The decision of Whitford J recalls the nineteenth century conception of the invention as a tangible manifestation of a novel principle.[36] In addition to contradicting the principles of biotech patenting applied by the UK courts in the (more restrictive) era of the twentieth century,[37] that conception conflicts with the decision in *Merrill Lynch*, by conceding that a subject matter is no less an invention if its novelty and inventiveness derive entirely from an item excluded by section 1(2).

Before the Court of Appeal this conflict was acknowledged and resolved in Whitford J's favour.[38] In addition to emphasizing the inconsistency of the *Merrill Lynch* approach with pre-1977[39] and TBA[40] jurisprudence, the Court rejected the 'broad expressions of policy' implicit in the interpretation of the 'as such' proviso on which the former was based,[41] and confirmed the ability of any novel and non-obvious discovery within the meaning of section 1(2)(a) to '[form] the substratum of invention so that if it is applied in a technique or process or incorporated in a product then it would be patentable'.[42]

Three points about this decision deserve emphasis. The first is that it proceeds from a view of chemical and biotechnological inventions as repositories of information of novel and practical significance. Hence the Courts' finding that it was the applicant's discovery of the protein's characterizing DNA sequence and the significance of that discovery for the protein's utility that justified treating the protein (thus characterized) as an invention.[43] Interestingly, it was precisely this structural view of chemical inventions as new compounds that led the Examiner in *Merrill Lynch* to venture that a computer system implementing an excluded method might nonetheless support a patent if it incorporated novel structural features. This raises the second point about *Genentech*, which concerns the reasons for the Courts' rejection of the *Merrill Lynch* approach to section 1(2) overall in preference for a more expansive understanding of inherent patentability. As suggested above, those reasons derived largely from the nature of the subject matter at issue and the Courts' view of biotech patenting generally. This is particularly evident in the judgments of the Court of Appeal, which emphasize (among other

[35] ibid, 566.

[36] See, eg *Hickton's Patent Syndicate v Patents & Machine Improvements Co Ltd* (1909) 26 RPC 359, CA.

[37] See, eg *American Cyanamid Company (Dann's) Patent* [1971] RPC 425, HL; *American Cyanamid Company v Berk Pharmaceuticals Limited* [1976] RPC 231, CA.

[38] See [1989] RPC 147, 207–208 per Purchas LJ, 239–240 per Dillon LJ. cf [1989] RPC 147, 270 per Mustill LJ.

[39] ibid, 207–214 per Purchas LJ, 239–240 per Dillon LJ, 264–269 per Mustill LJ.

[40] ibid, 207ff per Purchas LJ; see also ibid, 269–270 per Mustill LJ.

[41] ibid, 207 per Purchas LJ.

[42] ibid, 208 per Purchas LJ.

[43] See ibid, 203A, 213 per Purchas LJ, 240 per Dillon LJ, 269 per Mustill LJ.

things) the implications of *Merrill Lynch* for the availability of patents in the medical and biotechnological fields, and the unique challenge posed by biotechnology for patent law.[44] Of particular interest in this regard is Purchas LJ's opinion on the second of those issues, which he expressed as follows:[45]

> The difficulties in this case arise from the fact that in one sense 'nothing is new', to state a questionably dangerous aphorism. Recombinant DNA technology was recognized as possible and had been the subject of published reports before the priority date. The existence of a substance in nature described as t-PA obtainable from Bowes melanoma cells with its special therapeutic qualities was known and others besides [the applicant] had appreciated that recombinant DNA technology was a likely solution to the difficulty in obtaining commercially viable supplies of this important protein. In another sense it can equally be said that [the applicant] by the exercise of considerable skill, application and ingenuity discovered information about t-PA which had previously eluded the efforts of some of the finest scientific teams in the world and that this discovery enabled the development of a pharmaceutical process of immense value to the community at large and to [the applicant] in particular.

This recalls the decision of Lord Diplock in *American Cyanamid Co (Dann's) Patent*,[46] nearly twenty years earlier, in its expression of the following conundrum of biotech patenting: in the biotechnological field, the requirement for an invention potentially precludes the availability of patent protection for skilfully derived subject matter of use and value to the community, and to that extent runs counter to the (purported) objective of patent law in encouraging socially useful innovation. Read in the context of that conundrum, the Patents and Appeal Courts' rejection of the *Merrill Lynch* approach to section 1(2) implies the same preparedness as Lord Diplock to relax the standards of inherent patentability in the pursuit of that objective by focusing not on what subject matter takes from an excluded item, but rather on what it gives to the public.

This raises the third point about *Genentech*, which concerns the implications of that decision for subject matter other than biotechnology, including computer programs. On its face, the Court of Appeal's definition of inherent patentability as embracing any new discovery incorporated in a product or process of (new or improved) utility is sufficiently broad to cover most computer-related subject matter, including the automated trading system at issue in *Merrill Lynch*, particularly if 'discovery' is read interchangeably with the other section 1(2) items as the reasoning in *Genentech* suggests it ought to be. However, this is contradicted by the express statements of the Court of Appeal in *Genentech*, which affirmed the correctness of the outcome of *Merrill Lynch* notwithstanding the incorrectness of its legal foundation. For Purchas LJ the result in *Merrill Lynch* was 'probably correct' given the absence in that case of evidence of any inventive step, and the fact that

[44] See [1989] RPC 147, 203A per Purchas LJ, 239–240 per Dillon LJ. cf ibid, 269 per Mustill LJ.

[45] ibid, 203A per Purchas LJ.

[46] See [1971] RPC 425, 451–452; pages 96–97.

the case concerned—according to his Lordship—a 'mathematical formula' within the meaning of section 1(2)(a) and not a 'discovery' per se.[47] Dillon LJ agreed, albeit referring not to inventiveness specifically but rather to the principle 'that a patent for a computer when programmed or for the disc containing the program is no more than a patent for the program as such';[48] the point being that in *Merrill Lynch* there was nothing more than a subject matter excluded by section 1(2), unlike in *Genentech*, where the discovery of the DNA sequence was said to have been used to create a new object of utility, albeit an object characterized by the discovery itself.

Implicit in this reasoning is an understanding of the section 1(2) exclusions as distinct, so that the specific legal principles applicable in respect of one class of excluded subject matter are not necessarily applicable in respect of another.[49] Thus, while a 'discovery' presented as a novel product supporting a new or improved utility is an invention on account of its value to the public, a 'method for doing business' and/or 'computer program' presented as the same is not, on account of its technical conventionality and consequential identity with the underlying method/program itself. The judgment of Purchas LJ went even further by offering a rationale for this difference in the form of the public policy considerations discussed above, despite his express rejection, for all categories of subject matter, of the unnamed policy objectives attributed to the *Merrill Lynch* interpretation of the 'as such' proviso on which the restrictive definition of inherent patentability supported in that case was based.

3. Reconciling *Genentech Inc's Patent* with Section 1(2) of the Patents Act

A differently constituted Court of Appeal subsequently agreed with Dillon LJ when it convened to hear the final appeal in *Merrill Lynch* against the decision of the Patents Court.[50] In that appeal, Fox LJ, with whom the other members of the Court of Appeal agreed, summarized the position relating to inherent patentability after *Genentech* as follows:[51]

> *Genentech* decides that the reasoning of Falconer J. is wrong. On the other hand, it seems to me to be clear, for the reasons there indicated by Dillon L.J., that it cannot be permissible to patent an item excluded by section 1(2) under the guise of an article which contains that item – that is to say, in the case of a computer program, the patenting of a conventional computer containing that program. Something further is necessary. The nature of that addition is, I think, to be found in the *Vicom* case where it is stated: 'Decisive is what technical contribution the invention makes to the known art'. There must, I think, be some

[47] [1989] RPC 147, 206.
[48] ibid, 240.
[49] See also ibid, 268–269 per Mustill LJ.
[50] See [1989] RPC 561, CA.
[51] ibid, 569.

technical advance on the prior art in the form of a new result (*e.g.*, a substantial increase in processing speed as in *Vicom*).

Now let it be supposed that claim 1 can be regarded as producing a new result in the form of a technical contribution to the prior art. That result, whatever the technical advance may be, is simply the production of a trading system. It is a data-processing system for doing a specific business, that is to say, making a trading market in securities. The end result, therefore, is simply 'a method . . . of doing business' and is excluded by section 1(2)(c). The fact that the method of doing business may be an improvement on previous methods of doing business does not seem to me to be material. The prohibition in section 1(2)(c) is generic; qualitative considerations do not enter into the matter. The section draws no distinction between the method by which the mode for doing business is achieved. If what is produced in the end is itself an item excluded from patentability by section 1(2), the matter can go no further.

Two overriding principles emerge from this. The first, formulated with reference to *Vicom*, is that a computer or other product that incorporates an item excluded by section 1(2) can only be treated as an invention if it makes a technical contribution to the prior art in the form of a new result; a requirement suggested by Fox LJ to have been satisfied in *Vicom* by the effect of the program in increasing the computer's processing speed. And the second, formulated with reference to the terms of section 1(2) themselves, is that any subject matter that produces an excluded item as its result is not an invention regardless of the nature or extent of such contribution. Thus, after confirming the requirement for technical character in respect of physical products incorporating a section 1(2) item, his Lordship stressed that the converse did not apply, so that a product which satisfied that requirement might still be a non-invention if its result was in any case excluded by section 1(2). In this way he subjugated the requirement for a technical contribution to the individual exclusions of section 1(2), and pre-empted his agreement with the Principal Examiner and Falconer J that the subject matter before him was inherently unpatentable, irrespective of its technicality, on the ground that its end result was a method for doing business within the meaning of section 1(2)(c).

The question immediately raised by Fox LJ's reasoning concerns its consistency with the decisions in *Genentech* and *Vicom* with reference to which it was supported. In relation to *Genentech* his Lordship's reasoning can be regarded as consistent, but only if the first of his principles (requiring a 'technical advance on the prior art in the form of a new result') is understood to be satisfied for products having a new physical structure supporting a new or improved utility, and as not depending on additional evidence of technical inventiveness. While this definition is on its face satisfied by the system of *Merrill Lynch* (suggesting inherent patentability), there remained the caveat of his Lordship's second principle, that any subject matter the end result of which is a section 1(2) item is not an invention regardless of its technical character. This principle can also be reconciled with *Genentech*, but only if the rationale of that case is accepted, and the discoveries exclusion of section 1(2)(a) conceded to require a different approach from the business methods and computer programs exclusions of section 1(2)(c), so as to permit a lower threshold of inherent patentability for biotechnological than computer products on the basis, implicitly, of their relative value to society.

In contrast is the situation concerning *Vicom*, which is a more difficult case to reconcile with *Merrill Lynch* for two reasons. The first is Fox LJ's reliance on two statements from *Vicom* that do not obviously support the proposition for which they were advanced. Thus, the technical character of the computer program in *Vicom* was not treated (as Fox LJ suggested) as deriving from the program's effect in increasing substantially the apparatus's processing speed, but rather from the program's production and manipulation of digital images by means of a technical (in the sense of a mechanical) and industrially applicable process.[52] Indeed, the reference in *Vicom* to the impact of the process on the computer's processing speed alluded to by Fox LJ is contained in the reported summary of the parties' arguments, and does not form part of the TBA's reasons for decision.[53] Similarly, read in its entirety the TBA's reasoning does not support the proposition that a subject matter's inherent patentability depends on its technical advance but rather—and assuming the standards differ—that a subject matter must be technical overall to be an invention. Beyond this more expansive proposition, the only principles that can be derived with certainty from *Vicom* are those discussed in chapter 5, namely, that a mathematical algorithm is inherently patentable if it operates on and produces a physical entity using technical and industrially applicable means, and that the inventions represented by such algorithm will include, in addition to the algorithmic method itself, its material (software and hardware) embodiments. When read in the context in which it appears, the statement from *Vicom* quoted by Fox LJ—'[d]ecisive is what technical contribution the invention makes to the known art'[54]—was clearly intended to support these principles by entrenching the distinction between the technical substance of such inventions on one hand, and their material forms of embodiment on the other.

This raises the second reason for the difficulty in reconciling *Merrill Lynch* with *Vicom*, which derives from the second of Fox LJ's principles, that a subject matter producing as its end result an item excluded by section 1(2) is not an invention regardless of its technical contribution. At a factual level this principle can be reconciled with *Vicom* by accepting the correctness of the TBA's view in that case, that technical and industrially applicable methods for producing images are not 'excluded items'. At a doctrinal level, however, the situation is more complex. The reason is that while Fox LJ's principle receives theoretical support in *Vicom*,[55] in practice it runs counter to that decision for two reasons. In the first place, while the

[52] See pages 200–202.

[53] See [1987] EPOR 74, [IV].

[54] ibid, [15]–[16].

[55] See especially ibid, [3] where the TBA made it clear that giving a s 1(2) item technical form does not make it patentable.

TBA explicitly held that the presentation of an excluded item in the form of computer software or hardware did not make that item an invention, it nonetheless accepted the fact of the *Vicom* methods' implementation by computer means as evidence of their technical character and consequential avoidance of section 1(2). Similarly, while the TBA supported a substantive approach to inherent patentability based on a preliminary identification of a subject matter's true nature,[56] in practice it accepted a narrow formulation of inherent patentability premised on the *Vicom* methods' categorization either as mathematical algorithms or as inventions.[57] In so doing it discounted the other exclusions of section 1(2) in characterizing the methods specifically and defining inherent patentability generally, and ensured a very wide definition of the 'invention' to fill the ground left by its narrow definition of 'mathematical algorithm'. For these reasons it is difficult to view the TBA in *Vicom* as sharing the Court of Appeal's concern with upholding the methods exclusions of section 1(2) as much as maximizing the availability of patents for technical subject matter, including computer programs.

Overall, the decision in *Merrill Lynch* has less in common with *Vicom* than with the TBA's earlier decision in *Christian Franceries*, in which is reflected a similar emphasis on the importance of a subject matter's purpose and end result when determining its inherent patentability, and a concern generally to prevent considerations of technicality from diminishing the literal terms of section 1(2) by enabling the use of computer or other technical means to circumvent that section's exclusions.[58] For similar reasons it is also in keeping with the principle applied by the Examiner and Falconer J in *Merrill Lynch* itself, that subject matter the contribution of which to the prior art derives from an excluded item is inherently unpatentable, notwithstanding the explicit rejection of that principle by the Court of Appeal in *Genentech*, and the acceptance of that rejection by the Court of Appeal in *Merrill Lynch*.

4. Conclusion

By the time of the decisions in *Genentech* and *Merrill Lynch*, the courts' consideration of section 1(2) had produced three different interpretations of inherent patentability. The first, articulated by Falconer J (and the Examiner) in *Merrill Lynch*, reflects a restrictive understanding of inventions as precluding protection for subject matter the novelty or inventiveness of which can be traced to a section 1(2) item. The second, articulated by Whitford J and the Court of Appeal in *Genentech*, reflects an expansive understanding of inventions as including (allegedly) new and useful isolated products, whatever their relation to a section 1(2) item. And the

[56] See [1987] EPOR 74, [16].

[57] See ibid, [5].

[58] The Court of Appeal did not refer to the decision in *Christian Franceries* at all, presumably because it was not raised by the parties.

third, articulated by the Court of Appeal in *Merrill Lynch* in an attempt to reconcile the reasoning of *Genentech* with the conclusion of Falconer J, accepts the prima facie applicability of the contribution theory of inherent patentability, with an overriding exception for subject matter the end result of which is an item excluded by section 1(2).

That the decisions in *Merrill Lynch* and *Genentech* produced such divergent understandings of inherent patentability despite the evident belief of the Examiners and Judges in the fate of the subject matter at issue reflects two things. The first is the difficulty of reconciling section 1(2) with the EPO requirement for technical character in terms of a coherent and uniform definition of inherent patentability that explains, simultaneously, the patentability of a recombinant protein and the unpatentability of a programmed computer. And the second, implicit in the first, is the extent to which the courts' attempts at such reconciliation in *Merrill Lynch* and *Genentech* were driven precisely by those results, and more specifically, the extent to which the decisions in those cases reflect the *ex post facto* rationalization of foregone conclusions, namely, the allowance of a patent for a newly characterized protein (in *Genentech*), and the refusal of a patent for a computer programmed to perform a method of trade (in *Merrill Lynch*). These points are underscored by Purchas LJ's suggestion in *Genentech* that the provisions of section 1(2) might involve different policy considerations and therefore require different approaches. They are also underscored by Fox LJ's decision in *Merrill Lynch* that satisfying the definition of technical character derived by the EPO from the terms of section 1(2) was of itself not determinative of inherent patentability: an applicant still needed to show that the result produced by its subject matter was not an item excluded by those terms themselves.

C. Inherent Patentability as a Fact-Specific Enquiry: Computer-Related Subject Matter after *Merrill Lynch Ltd's Application*

Between the Court of Appeal's decision in *Merrill Lynch* and 2005, inherent patentability was the subject of judicial analysis in five main cases. Those cases are *Gale's Application*,[59] *Wang Laboratories Inc's Application*,[60] *Lux Traffic Controls Ltd v Pike Signals Ltd*,[61] *Raytheon Co's Application*,[62] and *Fujitsu Ltd's Application*,[63] and include two decisions of the Court of Appeal as well as four decisions of Aldous J (or Aldous LJ, as he became); the latter of whom played a significant role in shaping

[59] [1991] RPC 305, Pat, aff'd [1991] RPC 317, CA.
[60] [1991] RPC 463, Pat.
[61] [1993] RPC 107, Pat.
[62] [1993] RPC 427, Pat.
[63] [1996] RPC 511, Pat, aff'd [1997] RPC 608, CA. See also *Hitachi Ltd's Application* [1991] RPC 415, PO.

contemporary approaches to inherent patentability in the UK. The first of Aldous J's decisions was delivered from the Bench of the Patents Court in *Gale*, in a case that reflects the tensions inherent in the Court of Appeal's different standards of protection for methods and discoveries in *Merrill Lynch* and *Genentech* respectively, and the crystallization of that tension in the context of the computer programs exclusion of section 1(2)(c).

1. Conflicting Approaches to Computer Products: *Gale's Application* in the Court of Appeal

Gale's Application involved a read only memory device (ROM) that incorporated a computer program for calculating the square root of numbers.[64] The argument put in support of the application was that the physical nature of the ROM distinguished it from the other abstract or cognitive subject matter of section 1(2), thereby precluding its exclusion for inherent unpatentability.[65] This argument was rejected before the Principal Examiner, who found instead that the applicant's subject matter was an abstract entity in the form of a computer program—albeit presented in the physical guise of a ROM—that was inherently unpatentable under section 1(2)(c).[66] In so deciding he applied the principle of *Genentech* and *Merrill Lynch* that a computer program cannot escape exclusion under section 1(2) merely by being presented in the claims as a disk, suggesting that while there are many practical differences between a disk and a ROM, none affects the objects' character *qua* invention.[67]

On appeal to the Patents Court the decision of the Examiner was overturned.[68] While agreeing in principle with much of the Examiner's reasoning, Aldous J disagreed with his conclusion and the method by which it was reached. The first indication of that disagreement was his Lordship's understanding of how the question of inherent patentability should be approached, which he explained as follows:[69]

> From [the decisions in *Genentech* and *Merrill Lynch*] I conclude that the first task of the court is to construe the claim, as that is where the invention is defined. If the claim properly construed is drafted so as to relate to any of the matters disqualified by section 1(2) then the invention is not patentable. If however the claim is drafted to a process or technique or product and the basis of such process or technique or product is a disqualified matter, the court should go on to consider whether the claimed invention is in fact no more than a claim to an invention for a disqualified matter. It is a question of fact to be decided in each

[64] See [1991] RPC 305, 308.
[65] ibid, 308–309.
[66] ibid, 310.
[67] ibid.
[68] See ibid, 311ff.
[69] ibid, 315–316.

case, but if the claimed invention is more than a claim to an invention for a disqualified matter then it qualifies as a patentable invention.

By this passage, Aldous J reversed the first and second limbs of the Court of Appeal's test in *Merrill Lynch*, and recast the second as requiring, as a matter of form, that an applicant's claim not relate expressly to an excluded item, and the first as requiring, as a matter of substance, that the subject matter to which the claim did relate be 'more than' an excluded item.

Applying this approach, his Lordship accepted the applicant's characterization of his alleged invention as a physical object the technical properties of which were functionally defined by its underlying algorithm, and proceeded to consider whether the object thus characterized was 'no more than' an item excluded by section 1(2).[70] His Lordship decided it was more. Specifically, it was a 'new technical product'[71] as distinct from 'an ordinary product programmed in a different way',[72] being an article 'whose structure ha[d] been altered during manufacture so as to perform the function of [the] method or program defined by the claim',[73] making it an invention separate from that method or program itself.

Reflected in his Lordship's reasoning is the first of the *Merrill Lynch* principles, that a product incorporating an excluded item can nonetheless support a patent if it presents a technical advance in the form of a new result, where by a 'new result' he understood a new machine. To this extent his Lordship's decision gave effect to the principles of the Examiner in *Merrill Lynch* (that a computer system performing an unpatentable method might nonetheless be an invention in the form of a new structure) and of the Court of Appeal in *Genentech* (that an obvious use of an otherwise unpatentable subject matter will support a patent if it results in the same); the only difference between *Gale* and the latter case being that in *Gale* the excluded item was a computer product and not a discovery, and the product a ROM and not a recombinant protein. In so doing, however, his Lordship's decision went beyond *Genentech*, by applying the principle of that case to a computer product and thereby challenging the second of the *Merrill Lynch* principles, that no article can support a patent if its end result is (revealed in the specification to be) a program or other method excluded by section 1(2); the point being that the purpose and effect of the ROM in *Gale* was to provide a material record for a computer program so as to enable the execution of that program's instructions for performing a mathematical method. It also went beyond the Examiner's decision in *Merrill Lynch*, by contradicting the central proposition of that decision that a subject matter the novelty or inventiveness of which derives from an excluded item is, for that reason, inherently unpatentable.

[70] ibid, 316.
[71] ibid, 317.
[72] ibid, 316.
[73] ibid.

This raises an apparent ambiguity in Fox LJ's judgment in *Merrill Lynch*, the nature of which confirms also the basis for Aldous J's disagreement with the Examiner in *Gale*. The ambiguity is whether it is the 'new' or 'end' result of the subject matter that must not be a section 1(2) item in order for the subject matter to avoid exclusion, and the basis for the disagreement is how the subject matter in *Gale* was properly to be characterized, and the implications of its characterization for the specific terms in which the question of its inherent patentability was formulated and resolved. For the Examiner the test of inherent patentability proceeded not from the applicant's characterization of the ROM in the patent claims as a physical object with a new technical structure (as for Aldous J), but rather from the Examiner's own characterization of ROMs in general as functional objects the purpose of which is to record a computer program in a form in which it can subsequently be used. Similarly, the test applied by the Examiner in determining the ROM's inherent patentability was not whether the ROM was a new technical product (as for Aldous J), but rather whether it reflected more than the fulfilment of its inherent functional characteristics.[74] In doctrinal terms the effects of framing the test in the second of these ways were: (a) to pre-empt a finding of inherent unpatentability on account of the ROM's technical conventionality, (b) to accept as the subject matter's new and end result the computer program, and (c) to confirm generally that the inherent patentability of a product incorporating a computer program or other excluded item is not established by evidence of its structural novelty alone. Hence the significance of Aldous J's finding to the contrary, which lies precisely in its contradiction of each of these principles.

On appeal to the Court of Appeal,[75] Nicholls LJ, delivering the principal judgment for the Court, resolved the divergent approaches thus described in favour of the Principal Examiner. In commencing his reasoning his Lordship emphasized the declaratory nature of section 1(2), and its effect in affirming the pre-1977 axiom of inherent patentability that patents are not granted for ideas and discoveries, but rather for the industrially applicable products and processes to which they give rise.[76] In so doing he also gave explicit approval to the nineteenth century principle of UK law, approved in *Genentech*, that[77]

> the practical application of an idea or discovery . . . leads to patentability even if, as frequently happens, the practical application of the discovery is inherent in the discovery itself or is obvious once the discovery has been made and stated.

Unlike in *Genentech*, however, his Lordship's approval of this principle was not advanced to support a wide reading of inherent patentability (to cover any practical application of an idea or discovery) in justification of a conclusion of inherent

[74] [1991] RPC 305, 310.
[75] See ibid, 317, CA.
[76] ibid, 323.
[77] ibid, 323.

patentability, but rather, it seems, to support an understanding of computer programs as falling on the idea/discovery side of the pure/applied divide as a result of, or at least consistent with, their exclusion by section 1(2). Thus he held that a computer program within the meaning of section 1(2)(c) is a series of abstract instructions equivalent to a musical work that depends for its utility on, but that is not altered in substance by, its incorporation in a ROM or other computer-readable product.[78] Implicit in this finding is the same functional view of ROMs taken by the Examiner, which accounts for the following criticism by his Lordship of Aldous J's decision in *Gale*:[79]

> [Aldous J] drew a distinction between a claim relating to a disc containing a program and a ROM with particular circuitry. I feel compelled to part company with him. The ROM becomes a dedicated piece of apparatus, but the position remains that the only respects in which the structure of the ROM has been altered are those necessary to embody, in conventional form, a particular set of instructions. . . .
>
> I approach the substantial issue in this case, therefore, on the footing that it is convenient and right to strip away, as a confusing irrelevance, the fact that the claim is for 'hardware'. The claim in the specification is, in substance, a claim to a series of instructions which incorporate Gale's improved method of calculating square roots. It is a claim to electronic circuitry in the form of a ROM which is only distinguishable from other electronic circuitry in the form of a ROM by the sequence of instructions it contains. As such those instructions are not patentable, because they constitute a computer program.

Having rejected Aldous J's formalistic approach to the subject matter's characterization, his Lordship determined that the alleged invention was not the ROM at all but rather the instructions for performing the method of calculation recorded thereon, and proceeded to consider whether those instructions avoided prima facie exclusion 'as instructions' by possessing the characteristics of an invention.[80] Treating *Vicom* and two of the TBA's 'contribution' cases of the early 1990s as his guiding authorities,[81] he decided they did not.[82] The reason, however, was not that the instructions were a mathematical algorithm within the meaning of section 1(2)(a)—which he found they were not—but rather that they lacked the technical contribution required to make them 'something more than' a computer program.[83] The justification for the first of these findings was that the instructions were more

[78] See ibid, 325. cf the definition proposed for inclusion in the PA, reproduced in ch 4, n 216.

[79] ibid, 325–327.

[80] ibid, 325.

[81] T_22/85 (*IBM/Document Abstracting and Retrieving*) [1990] EPOR 98; T_65/86 (*IBM/Text Processing*) [1990] EPOR 181.

[82] See [1991] RPC 305, 326–327, citing also T_115/85 (*IBM/Computer-Related Invention*) [1990] EPOR 107; T_6/83 (*IBM/Data Processor Network*) [1990] EPOR 91 to support his finding that the ROM's instructions did 'not solve a "technical" problem lying within the computer as happened [in those cases]', while stating: 'I confess to having difficulty in identifying clearly the boundary line between what is and what is not a technical problem for this purpose. That, at least to some extent, may well be no more than a reflection of my lack of expertise in this technical field' ([1991] RPC 305, 328). See also [1991] RPC 305, 331 per Parker LJ.

[83] See [1991] RPC 305, 327.

than an algorithm, being the application of an algorithm 'for the purpose of writing computer instructions'.[84] The justification for the second was that they neither managed a technical process external to the computer nor solved a technical problem within it, denoting at best 'an improvement in programming' which, in his Lordship's view, was not an 'invention' within the meaning of section 1(1).[85]

The decision of Nicholls LJ affirmed the correctness of the Examiner's reasoning by rejecting the finding, implicit in Aldous J's decision, that the conventional adaptation of a known product to perform a new function in accordance with a computer program could of itself give rise to an invention in the form of a new machine regardless of that function.[86] In addition, while advancing as the reasons for that rejection the derivation of the machine's 'newness' from a computer program excluded by section 1(2)(c) and the non-technical nature of the program itself,[87] his Lordship was also clearly influenced by the lack of technical inventiveness involved in adapting the product, and the conventionality of the machine that such adaptation produced.[88] The rigorous view of product inventions implicit in this approach could be regarded as something of a paradox given his Lordship's restrictive understanding of the mathematical methods exclusion of section 1(2)(a),[89] which resulted directly from his distinction between inherently unpatentable subject matter on one hand, and the same subject matter when applied to an end of industrial applicability on the other.[90] In contrast, however, was his Lordship's understanding of the computer programs exclusion, which he treated as falling on the 'pure' side of the pure/applied divide and as therefore covering all methods for programming a computer that neither alter the conventional functionality of their implementing apparatus nor manage a technical process outside that apparatus, where by 'technical process' he clearly understood something other than a process (otherwise) excluded by section 1(2). Hence his finding, contrary to *Vicom* but consistent with *IBM/Document Abstracting and Retrieving* and *IBM/Text Processing*, that the *Gale* method's avoidance of the mathematical methods exclusion of section 1(2)(a) was not determinative of its inherent patentability per se.

Overall, Nicholls LJ's decision in *Gale* reflects the same restrictive understanding of the *Merrill Lynch* requirement for 'technical advance' as was supported by the Examiner, namely, as requiring that an article incorporating a section 1(2) item reflect a new result that is technically non-conventional. In so doing it confirmed the continuing relevance of the analogous use and collocation principles of nineteenth century UK law in the context of computer products,[91] albeit with

[84] See [1991] RPC 305, 327.
[85] See ibid, 327–328.
[86] See ibid, 326–327.
[87] See ibid.
[88] See ibid, 327–328.
[89] See ibid, 327.
[90] See ibid, 323.
[91] See at pages 45–47, 102–103.

reference to the exclusion of computer programs by section 1(2)(c) specifically and the requirement for an invention generally, and the effect of each in precluding protection for computer programs when conceived *qua* instructions 'as such'.[92] His Lordship's decision also confirms the non-applicability of the *Genentech* view of inventions (as repositories of useful information) and of inherent patentability generally (as including newly characterized objects supporting a new or improved utility) in the context of computer-related subject matter, thereby entrenching the higher standard of inherent patentability suggested in that case for computer than for biotechnological products. And finally, it contradicts the logic implicit in *Vicom*, that a computer program the instructions of which avoid the mathematical methods exclusion of Article 52(2)(a) also avoids the computer programs exclusion and is for that reason an invention. In this way his Lordship ascribed to the computer programs exclusion a role that was different from merely supporting the exclusion of other section 1(2) methods when presented in automated form. Pursuant to that role, the exclusion also confirms what his Lordship suggested to be implicit in the need for an invention per se, namely, that patents cannot be granted for subject matter the alleged advance of which lies in the field of computer programming, on the ground that such subject matter are manifestly technically conventional, and therefore incapable of offering the consideration required for a patent, ie incapable of contributing sufficiently to the technical arts to justify the grant of a monopoly right.

2. Fact-Specificity as a Mask for Legal Uncertainty

As has been seen, in *Merrill Lynch* the Court of Appeal reconciled the outcome of Falconer J with the reasoning of *Genentech* by interpreting inherent patentability as requiring a technical advance on the prior art, with an overriding exception for a subject matter producing as its result an item excluded by section 1(2). In *Gale*, however, the second limb of this test was lost in the decision-maker's reasoning, and the fate of the ROM ultimately determined with reference not to its production of a method for performing a mathematical calculation, but rather to its technical conventionality and consequential failure to advance the arts of a technical field. The result was a fourth test of inherent patentability, predicated on a view of section 1(2) as resolving to a uniform requirement for a technical contribution to the prior art or, and in pre-1977 terms, a new technical principle directed to the alteration or production of a physical product. In the majority of cases from *Gale* until 2005 it is this view, and the restrictive understanding of the methods exclusions of section 1(2) from which it proceeds, that is reflected, confirming the prima

[92] See [1991] RPC 305, 325, where Nicholls LJ made clear his view, albeit without deciding the issue, that non-technical programs within the meaning of his judgment were excluded from patentability by the requirement for an 'invention' independent of the computer programs exclusion of s 1(2)(c) itself.

facie commitment of the UK courts during this period to a restrictive interpretation of section 1(2) in the context of computer-related subject matter. The first indication of that commitment is contained, paradoxically, in a decision of Aldous J in *Wang Laboratories*.

(a) Inherent Patentability as a Fact-Specific Enquiry

As has been seen, in *Gale* Aldous J accepted the applicant's characterization of his subject matter as a physical object, and applied principles consistent with *Genentech* to support its inherent patentability. However, it is clear from his Lordship's subsequent decision that the motivation for that approach was an expansive attitude not in relation to inherent patentability generally, nor to the inherent patentability of new technical structures specifically, but rather to the inherent patentability of dedicated products, and that the principles driving that attitude were the same principles of traditional jurisprudence reflected in Nicholls LJ's judgment on appeal—albeit construed more liberally by his Lordship to permit the patenting of a structurally novel but technically conventional apparatus. The decision in question is in *Wang Laboratories*, and was decided after his Lordship's judgment in *Gale* but before the judgment of the Court of Appeal.

In *Wang Laboratories*, Aldous J, sitting again in the Patents Court, refused a patent for a conventional computer that was programmed, in the first claim, to store and dispense expert knowledge (the so-called 'expert system'), and in the second claim, to enable users to create their own expert system by entering their own body of expert knowledge onto the computer in a systemized manner following the prompts of the program (the so-called 'expert system shell').[93] According to the applicant both claims described inventions: the first in the form of a technical process, and the second in the form of a new machine that was capable of being used in a new way.[94]

It was Aldous J's view of the first of these arguments that was determinative of the case overall. While accepting that the process was 'technical' in the sense of comprising steps conceived specifically with automation in mind, his Lordship nonetheless held it to offend the mental methods exclusion of section 1(2)(c) by reason of its capacity to be performed mentally.[95] In reaching this conclusion he rejected the restrictive, intentional interpretation of the mental methods exclusion argued for by the applicant, pursuant to which any claim that has as its basis steps not intended for human performance will escape exclusion under section 1(2).[96] He also affirmed the correctness of his reasoning in *Gale*[97]—which he presented as

[93] See [1991] RPC 463, 465–467.

[94] See ibid, 473.

[95] ibid, 474.

[96] ibid, 472.

[97] See ibid, 468 (noting that his decision in *Gale* was under appeal, and declaring his intention to apply the reasoning from that decision regardless).

consistent with the then contemporary decisions of the TBA,[98] despite conceding both the inconsistency of those decisions[99] and his difficulty generally in extracting from them a clear indication of the Board's understanding of the word 'technical'[100]—and cast the question of inherent patentability as 'a question of fact to be decided in each case whether the claimed invention is more than a claim to an invention for a disqualified matter'.[101]

In discussing the kinds of facts that might support a disqualification under section 1(2), his Lordship gave the example of a computer product adapted to implement a new section 1(2) method in a conventional manner,[102] consistent with his interpretation of the mental methods exclusion itself, as covering new methods conceived specifically for technical implementation that are nonetheless capable of mental performance. While in combination these principles appeared to contradict his decision in *Gale* by precluding protection for novel technical products, on closer analysis they clarified the basis for that decision. That basis was not that the ROM implemented a method not excluded by section 1(2), nor that it incorporated a novel technical structure per se, but rather that its characterizing method was physically embedded at the point of the ROM's manufacture, so that the structure of the ROM supported a new and dedicated functionality, albeit characterized by the excluded method itself. The result is a very different proposition that recognizes a different standard of inherent patentability for dedicated and non-dedicated products respectively. Specifically, while the inherent patentability of a dedicated product depended on the combined characteristics of novelty and utility expressed as a positive requirement for an invention (consistent with *Genentech*), the inherent patentability of a non-dedicated product depended on the single characteristic of technical inventiveness expressed as a negative requirement for subject matter not derived from an item excluded by section 1(2) (consistent with *Merrill Lynch* and *Gale*). In addition, it seems clear from *Wang Laboratories* and *Gale* that the basis for that difference derived from his Lordship's understanding of conventional non-dedicated products as collocations of inherently unpatentable software and hardware unsuitable for protection on traditional UK principles, in contrast to dedicated products, which he presumed to be inherently patentable, regardless of their conventionality, as 'combinations' of the same.[103] This was supported by his Lordship's reasons for rejecting the second class of claims in *Wang Laboratories*, which, he suggested, were directed to precisely a collocation in the

[98] ibid, 471–472 (drawing support from the TBA's conclusions in *IBM/Document Abstracting and Retrieving* [1990] EPOR 98 and *IBM/Computer-Related Invention* [1990] EPOR 107, while also stating that he found 'it difficult to understand what the Board meant by the word "technical"').

[99] See [1991] RPC 463, 471 (discussing *IBM/Data Processor Network* [1990] EPOR 91 and *IBM/Document Abstracting and Retrieving* [1990] EPOR 98).

[100] See [1991] RPC 463, 472 (discussing *IBM/Computer-Related Invention* [1990] EPOR 107).

[101] ibid, 473.

[102] ibid.

[103] cf the language used by his Lordship in *Gale's Application* [1991] RPC 305, 316–317.

sense of this distinction. Specifically, the expert system shell was a conventional apparatus that remained 'the same' even when programmed, so that its conjunction of hardware and software could not be described as a 'combination making a different whole' as required to avoid the computer programs exclusion.[104]

The approach of Aldous J in *Gale* and *Wang Laboratories* reflects the tension created by the simultaneous recognition of expansive and restrictive standards of protection for products and methods, and the crystallization of that tension at the site of the computer programs exclusion. It also reflects, however, the extent to which that tension is obscured by emphases on the factual nature of the inherent patentability enquiry, as a result of the effect of such emphases in localizing that enquiry to the realm of individual subject matter and thereby obscuring the policies and principles that drive its construction and resolution. Indeed, this is underscored by the decision of the Patent Office immediately after *Wang Laboratories*, in *Hitachi Ltd's Application*,[105] where the Officer applied a mix of the TBA's reasoning in *Vicom* and Aldous J's reasoning in *Gale* and *Wang Laboratories* to refuse a patent for automated methods for handling and executing computer-programming languages.[106] According to the Officer, the failure of the methods to operate on any physical entity analogous to the digital images of *Vicom* distinguished them from the methods of that case, and required their exclusion under section 1(2)(a) and (c) as an inherently unpatentable algorithm and computer program that failed to provide the technical result of an invention.[107] However, and lest his decision be attacked on appeal, the Officer also referred to Aldous J's description of the word 'technical' in *Wang Laboratories* as a 'word of art', and to his Lordship's emphasis generally on the factual nature of the inherent patentability enquiry, which he construed to have the following import for substantive readings of section 1(2): what is and is not technical—and by implication what is and is not an invention—is a question of fact that depends ultimately for its resolution on the expert intuition of individual decision-makers,[108] and that cannot be determined analogously from antecedent authorities however closely related are their circumstances.[109] Thus, according to the Officer, even if the methods in question were accepted as being 'very close' to the methods of *Vicom*, on the principles of Aldous J it would not follow that they were required to be treated in the same way.

The significance of *Hitachi* lies in its demonstration of the effect of Aldous J's understanding of inherent patentability as a factual enquiry to be undertaken entirely within the four corners of the patent claims. That effect was to cast inherent

[104] See [1991] RPC 463, 473.
[105] [1991] RPC 415, PO.
[106] ibid, 416–418.
[107] ibid, 420.
[108] ibid, 422 ('By [Aldous J's statement] I take him to mean that the term ['technical'] does not admit of precise definition but that given a particular invention an experienced patent practitioner will be able to come to a decision upon whether an invention does provide a technical result').
[109] See ibid, 420–421.

patentability as a discretionary question, and thereby insulate decision-makers from the criticisms for deviating from the approach of the EPO and misapplying UK statutory law that were canvassed in the appeals from the decisions of Falconer J and the Examiner in *Merrill Lynch*. In addition, however, the case also underlines the scope for such criticisms as a result of the divergent approaches to inherent patentability since *Merrill Lynch*, and the effect of decision-makers' confinement to fact-based reasoning in perpetuating the same. Hence, for example, the Officer's contradiction of the findings of the Courts of Appeal in *Gale* and *Merrill Lynch*: (a) that the mathematical methods exclusion of section 1(2)(a) did not apply to computer programs and (b) that the TBA's decision in *Vicom* turned on the methods' improvement of their machine's technical functionality (by increasing its processing speed) and not their operation on or production of a physical object per se. Hence also his support for a wider reading of the section 1(2) exclusions than had theretofore been supported, as excluding computer systems that obviated the manual performance of an 'intellectual task',[110] where by 'intellectual task' he appears to have understood any non-manufacturing method capable of performance by, and not merely in, the human mind.[111]

(b) Fact-Specificity as a Mask for Legal Uncertainty

The decision in *Hitachi* exemplifies the complex relationship between fact and law based reasoning in UK decision-makers' treatment of section 1(2), and the double-edged sword of the former in both obscuring and trivializing inconsistencies in the latter and thereby perpetuating uncertainties in relation to the meaning and scope of inherent patentability.

Similar themes pervade the remaining pre-2005 cases in which the inherent patentability of computer-related subject matter was considered in detail. Those cases are *Lux*, *Raytheon*, and *Fujitsu*, and reflect the uneasy relationship between UK and EPO law on account of the UK courts' disinclination to concede any disagreement with TBA authority, notwithstanding its inconsistency, and the difficulty thereby created for their development of a coherent jurisprudence under section 1(2).

(i) ***Lux Traffic Controls Ltd v Pike Signals Ltd***

Lux Traffic Controls Ltd v Pike Signals Ltd involved an appeal to the Patents Court against the refusal of a patent for two traffic light control systems that were similar in form if not in substance to the inherently unpatentable systems of *Christian Franceries*.[112] In characterizing the systems Aldous J relied on two central princi-

[110] See ibid, 422.

[111] See ibid, 422–423.

[112] See [1993] RPC 107, 111–122; cf *CHRISTIAN FRANCERIES/Traffic Regulation* [1988] EPOR 65, [VI].

ples, which he formulated with explicit reference to the judgments of Nicholls LJ in *Gale* and Fox LJ in *Merrill Lynch*, and the TBA in *Christian Franceries* and *IBM/Text Processing*.[113] The first of those principles, consistent with *Christian Franceries*, was that section 1(2) excludes from patentability 'mere ideas not normally thought to be the proper subject for patents which are concerned with manufacturing',[114] including 'methods of controlling traffic as such' whether or not within the category of methods for doing business.[115] And the second of those principles, consistent with *IBM/Text Processing* and the Court of Appeal's decision in *Gale*, was that inherent patentability could not be determined exclusively with reference to the description of the subject matter in the patent claims, but rather depended on whether the subject matter reflected in substance a contribution to the prior art in a technical field so as to take it outside the scope of section 1(2).[116] Applying these principles his Lordship held the systems before him to be inherently patentable. In particular, looking beyond the systems' description in the patent claims revealed an underlying discovery that distinguished them from the subject matter of *Christian Franceries*[117] and demonstrated their nature generally as machines manufactured to a new technical principle,[118] which in his Lordship's judgment made them inherently suited to patent protection however characterized and whatever the method they performed.

The result, despite his purported application of the principles of *Merrill Lynch* and *Christian Franceries*, was a decision that conflicted with those principles by treating section 1(2) as having no application in respect of technical subject matter, including products characterized by their performance of a business method. In reaching this result Aldous LJ demonstrated the interpretive freedom conferred by the factual nature of the inherent patentability enquiry in enabling the simultaneous distinction and contradiction of past authorities, and the application of different standards of patentability depending on the terms of a subject matter's description in the claims. Thus, having refused a patent in *Wang Laboratories* for functionally characterized products the purpose and effect of which were to perform (and enable others to create a product to perform) a mental method, Aldous J allowed a patent in *Lux* for a functionally characterized product the purpose and effect of which were to perform a method for doing business. Despite this, and unlike in *Gale*, it seems clear from *Lux* that his Lordship based his finding of technical character on more than the dedicated functionality of the product itself. Hence his emphasis on the system's novel technical principle and consequential contribution to the technical arts, which emphasis supported the restrictive definition of technical character advanced by the Court of Appeal in *Gale* (and, in the context of

113 See [1993] RPC 107, 137–139.
114 ibid, 138.
115 ibid.
116 ibid, 139.
117 ibid.
118 ibid, 139–140.

a general-purpose computer, by the TBA in *IBM/Text Processing*), rather than his own expansive definition of the same in *Gale*. The result—consistent also with his Lordship's emphasis on 'manufacturing' and acceptance generally of the expansive definition of 'business methods' supported in *Christian Franceries*—was a decision that reflected the tenor of the Court of Appeal's judgment in *Gale*, and its commitment to applying the standards of early twentieth century UK law in the context of non-manufacturing methods and computer-related subject matter.

Like the decisions before it, the judgment in *Lux* exploited the factual nature of the inherent patentability enquiry to simplify that enquiry, and render nugatory inconsistencies in its construction and determination by successive past decision-makers. However, it also revealed the difference in the results produced by this approach from the results produced by the TBA's own fact-based approach to inherent patentability. That difference lay in the greater transparency of the UK decisions as a result of being reasoned in a more coherent, detailed, and self-referential manner. Rather than resolving the uncertainties underlying the TBA's treatment of inherent patentability, however, the greater transparency of UK decisions exacerbated that uncertainty by exposing its depth. This is apparent from the five decisions generated by the last two cases for discussion in this section, which are *Raytheon* and *Fujitsu*. These cases are of particular interest for having involved methods of digital processing similar to the methods of *Vicom*, the inherent patentability of which wa s determined by a single Examiner, two Judges of the Patents Court, and, in *Fujitsu*, Aldous LJ, who had by then moved to the Court of Appeal.

(ii) *Raytheon Co's Application*

Raytheon Co's Application involved a patent application directed to a multi-device computer system for classifying and identifying ships by a process of image comparison.[119] At the first level of appeal the Principal Examiner rejected the applicant's submission that the system comprised a combination of hardware and software components, construing it instead as a conventional computer running a program in a conventional manner, the inherent patentability of which depended exclusively on the program itself.[120] Addressing that question the Examiner commenced by asking whether the program involved a technical advance sufficient to make it more than a computer program within the meaning of section 1(2)(c). The applicant argued that it did; the process represented a 'technical process in the industrial field of pattern recognition', similar to the methods of *Vicom*.[121]

In considering this argument and the implications of *Vicom* generally for the application before him, the Examiner hesitated. Confronted with the difficult task set by *Merrill Lynch* of extracting from *Vicom* some guide as to where the 'technical advance' in the system was to be found, he declared the question to be an

[119] See [1993] RPC 427, 429–430.
[120] See ibid, 432–433.
[121] ibid, 433.

open one, citing the different conceptions that the applicant's description of the system supported.[122] Thus, he held, while the system could be viewed as a means 'of image formation, sensing, measuring and comparison' that produced a technical effect and advance in 'the provision of an accurate comparison',[123] it could also be viewed as a new means of recognizing a shape, the advance of which he was not prepared to accept as technical.[124] In the result, he decided that the question was in any case academic by reason of the second of the *Merrill Lynch* principles, that a subject matter producing as its end result an item excluded by section 1(2) is in all cases inherently unpatentable.[125] According to the Examiner, the scheme was precisely such a subject matter; its end result being a mental method within the meaning of section 1(2)(c) so as to require the scheme's exclusion under that paragraph.[126] In reaching this conclusion he dismissed as irrelevant the fact that the steps of the system were all steps of an industrial process that differed from the steps involved in pattern recognition by the human mind,[127] citing in doing so the expansive interpretation of the mental methods exclusion adopted by Aldous J in *Wang Laboratories*.[128]

The Examiner's decision in *Raytheon* was the first reported decision in which full effect was given to the *Merrill Lynch* view of inherent patentability, including specifically its principle that the section 1(2) exclusions are not reducible to a requirement for technical character. However, the decision also reflects the difficulty of deciding when one (or more) hardware product controlled by a computer program presents a technical advance so as to avoid initial exclusion as a computer program on the first of the *Merrill Lynch* principles, due in part to the possibility of different conceptions of an individual subject matter *qua* invention. Having accepted the hardware's technical conventionality, the Examiner in *Raytheon*—consistent with the Court of Appeal's decision in *Gale*—looked to the technical character of the process managed by the hardware, relying heavily in doing so on the TBA's reasoning in *Vicom*. The problem he thereby encountered, however, was that unlike in *Vicom*, in *Raytheon* the methods operated on a so-called physical entity in the form of a digital image, but did not produce any such entity as their end result. Hence the significance of his resort to the second of the *Merrill Lynch* principles, which obviated the need to decide directly whether the methods could nonetheless be accepted to advance the technical arts, and whether the answer to that question required his acceptance of the TBA's view in *Vicom* that the digital representation of an image or other physically-realized object was a physical entity. Indeed, even his consideration of the second of the *Merrill Lynch* principles did not

122 [1993] RPC 427, 435 (accepting the argument of the applicant).
123 ibid.
124 ibid.
125 ibid.
126 ibid, 436.
127 ibid, 435.
128 ibid, 436–437.

require resolution of that issue, on account of Aldous J's expansive interpretation of the mental methods exclusion articulated in *Wang Laboratories*, and its inclusion of methods performed on physical objects to produce an abstract result.

Before the Patents Court, the Examiner's decision in *Raytheon* was upheld,[129] albeit without endorsement of his reasons or resolution of the issues they raised. Indeed, while accepting the mental methods exclusion as precluding the system's patentability, Mr Jeffs QC reasoned that conclusion in a way that supported the *Merrill Lynch* understanding of *Vicom* as having turned on the existence of a technical advance,[130] and the *Lux* understanding of section 1(2) generally as having no application in respect of subject matter presenting such an advance.[131] Unlike in *Lux*, however, in *Raytheon* the Patents Court approached the question from within section 1(2) itself.[132] Thus, in searching for the required technical character, Mr Jeffs QC's focus was on the meaning of the mental methods exclusion itself, which, like the Examiner before him (and following Aldous J in *Wang Laboratories*), he construed independently of considerations of technical form explicitly to embrace methods conceived with automation in mind.[133]

This raises the most interesting aspect of Mr Jeffs QC's decision, which lies in his discussion of the nature of a method for performing a mental act within the meaning of section 1(2)(c). According to his analysis, mental methods occupy a spectrum that is bounded at one end by sensory or perceptual acts such as smelling an orange, and at the other by analytical acts such as performing a mathematical calculation, with the difference between them being their relative capacity for objective explication and automation.[134] In presenting this spectrum Mr Jeffs QC made in substance the following point: the mere fact that a person operating a computer achieves the same result as a person performing a mental act does not make the computer method itself a mental method; the question whether a human-implemented or computer-implemented process being rather such a method is an objective one that depends not only on the purpose and end result of the method, but also on the steps adopted to achieve them. This supported the conclusion that while identifying an object by smell is a mental method, identifying the same object by a process of chemical synthesis is not.[135]

Two points follow from this view and its context. The first is that the mental methods exclusion is restricted to methods performed in (and not merely by) the human mind, and thus does not cover methods that operate on (as distinct from in

[129] See ibid, 438ff.

[130] See ibid, 449.

[131] See ibid, 447.

[132] Mr Jeffs QC also held the system in *Raytheon* to be excluded as a computer program under s 1(2)(c), on the basis that it did 'no more than provide a briefing for a computer programmer to write a program that would enable conventional pieces of apparatus to perform a particular mental act' (ibid, 451).

[133] See ibid, 446–447.

[134] See ibid, 442.

[135] ibid.

and with the aid of) the physical world—a point that contradicts the Officer's suggestion to the contrary in *Hitachi*, while emphasizing again the significance of *Wang Laboratories* in having entrenched an interpretation of section 1(2)(c) that excludes from protection all methods that are in principle capable of mental performance, regardless of whether they are actually or even easily so performed. And the second point is that his decision suggests as one reason for the mental methods exclusion the need for a subject matter to be inherently capable of full explication (and therefore sufficient description), so as to be independently repeatable by a third party. In this way Mr Jeffs QC offered an explicit justification for the mental methods exclusion, beyond the technical character axiom itself, premised on the need for inventions to be prima facie capable of repetition, and thus of utility and potential industrial applicability.

(iii) ***Fujitsu Ltd's Application***

Like *Raytheon*, *Fujitsu Ltd's Application* involved a method for processing digital images, albeit this time in the form of crystal structures, and this time producing as their end result a new digital image.[136] Also like *Raytheon*, much of the argument in *Fujitsu* proceeded from the applicant's submission that the methods were sufficiently similar to the methods of *Vicom* to warrant the same finding of inherent patentability.[137] Interestingly, however, in *Fujitsu* the Principal Examiner's consideration of that submission displayed none of the uncertainty of his consideration of the equivalent submission in *Raytheon*, largely, it seems, on account of his acceptance since that case of the justification for *Vicom* offered by the TBA in *IBM/Homophone Checker*.[138] According to the TBA, the inherent patentability of the methods in *Vicom* turned not on their use of industrially applicable technical means to produce a physical entity, but rather on their technical contribution of 'a more efficient restoration or enhancement of the technical quality of an image'.[139] Accepting this, the Examiner held the methods of *Fujitsu* and *Vicom* to be easily distinguished with reference to the different roles performed by their underlying algorithms.[140] Thus whereas the significance of the *Fujitsu* algorithms lay exclusively in their determination of the intellectual content of the digital images, in *Vicom* the algorithms were additionally responsible for both the technical means by which the images were displayed, and the technical quality of the images once displayed.[141] It followed in the Examiner's view that any finding that the *Fujitsu* methods possessed the 'technical character' or 'practical and technical impact' required to avoid the computer programs exclusion would necessarily derive from

[136] [1996] RPC 511, 513–515.
[137] ibid, 516.
[138] T_65/86 (*IBM/Homophone Checker*) [1990] EPOR 181; see ch 5, n 29.
[139] [1990] EPOR 181, [21]. See [1996] RPC 511, 517.
[140] See [1996] RPC 511, 517.
[141] ibid.

'the fact that the representation is of a technical artefact, ie, a crystal structure'.[142] According to the Examiner, that fact was not sufficient to create an invention, as the methods were still 'no more than a computer programmed to display particular information in a technically conventional way', and thus still a computer program excluded by section 1(2)(c).[143] This conclusion was supported also by his view of the methods' utility, including particularly their concern with product design rather than product construction, which was regarded by the Examiner as confirming their 'essentially intellectual character'[144] and lack of any necessary connection with—and consequential failure to make any necessary contribution to—the art of crystal manufacture.[145] As a result of that failure, he held, the methods were excluded not only as computer programs, but also as methods for performing a mental act, being methods that obviated by technically conventional means a known manual process for performing an intellectual task that was capable of being performed 'by the human mind'.[146]

Three points can be derived from this reasoning. The first is that section 1(2) excludes from patentability any technically conventional method the steps of which are directed and controlled by the human mind, albeit performed and realized on physical objects, consistent with *Hitachi* but inconsistent with the Examiner's own prior decision in *Raytheon*. The second is that digital, and possibly also non-digital, visual representations of such objects are 'information' and not 'physical entities' as was held in *Vicom*, so that a computer program that mimics a process of physical construction is—on the second of the *Merrill Lynch* principles at least—inherently unpatentable. And the third point is that for a computer program or other method to support a patent it needs to contribute to the technical arts, where by 'technical' the Examiner understood 'manufacturing', consistent with Aldous J's interpretation of section 1(2) in *Lux* as covering 'mere ideas not normally thought to be the proper subject for patents which are concerned with manufacturing',[147] and confirming generally the adherence of contemporary UK decision-makers to the pre-*NRDC* conception of inherent patentability predicated on the need for subject matter of exclusively mechanical or chemical utility directed to advancing the manufacturing arts.

In addition to these points, the Examiner's decision in *Fujitsu* is of interest as one of few decisions to its date in which the divergence of UK and EPO approaches to inherent patentability was acknowledged; the Examiner having conceded the inconsistency of his decision with the TBA's practice of accepting the fact of a method's automation as precluding its exclusion from patentability as a

[142] ibid.
[143] ibid.
[144] ibid, 518–519.
[145] ibid, 519.
[146] ibid, 520–521.
[147] See n 114.

mental method.[148] As noted, the reluctance of UK decision-makers to concede such divergence pervades the cases prior to *Fujitsu*, and as Falconer J's decision in *Merrill Lynch* first reflects, was a source of substantial uncertainty. Indeed, the appeal from the Examiner's decision in *Fujitsu* provides an even starker example of this.

In that appeal, Laddie J for the Patents Court upheld the Examiner's decision by reasoning that reflects a very different approach to the *Fujitsu* methods and the question of their inherent patentability from the approach of the Examiner.[149] Dealing again with submissions that drew heavily on *Vicom*,[150] his Lordship gave detailed consideration to the 'most important decisions in this area' raised by the parties' submissions, which he identified as *Vicom*, *IBM/Homophone Checker*, *Merrill Lynch*, and *Gale*. In so doing he presented those cases as evidence of the consistency of UK and EPC law, reflected, for example, in his following commentary on the outcome of *Gale*:[151]

> [I]t was inevitable . . . that the court came to the decision it did. Just as it would not be possible to obtain patent protection for a literary work by recording it on paper and then putting in a claim for the article consisting of the piece of paper bearing the work, so it is not possible to secure patent protection for a program by claiming a standard recording medium on which the program is recorded. In coming to its decision, the court cited without criticism both [*VICOM/Computer-Related Invention* and *IBM/Homophone Checker*]. As such the case does not seem to me to take the jurisprudence on this issue any further. It is merely a case in which the *VICOM* and *IBM* principles are applied.

However, in presenting *Gale* and *Vicom* as applications of the same, general principle—that reducing a computer program to a material form does not per se create an invention—his Lordship ignored the important caveat to that principle recognized in those cases—unless the program itself is inherently patentable—and obscured the divergence of opinion among EPO and UK decision-makers in relation to that caveat, as well as the difficulty of the question, at the heart of the divergence, regarding what constitutes the reduction of a program to material form. More importantly, however, he attributed a consistency to UK and EPO law that masked his own departure from EPO understandings of inherent patentability. The source of that departure was his Lordship's understanding of the 'cumulative effect' of the UK and TBA cases regarding inherent patentability for section 1(2) as a whole. According to Laddie J, what those cases reveal is that section 1(2) contains individual exclusions each of which is premised on its own public policy rationale that does not reduce to a generic requirement for technicality.[152] By this

[148] See [1996] RPC 511, 519. The decision referred to by the Examiner was T_59/93 (*IBM/Rotating Displayed Objects*) (TBA, 20 April 1994).

[149] See [1996] RPC 511, 521ff.

[150] ibid, 523.

[151] ibid, 530.

[152] ibid ('The types of subject matter referred to in section 1(2) are excluded from patentability as a matter of policy. This is so whether the matter is technical or not').

finding his Lordship explicitly rejected the EPO practice of reducing the section 1(2) exclusions to such a requirement, and accepted instead the primacy of those exclusions under the PA as independent statements of policy the satisfaction of each of which is necessary to show the existence of an invention. The implications of this approach for the 'technical contribution' test attributed to *Vicom* by the Court of Appeal in *Merrill Lynch* were described by his Lordship as follows:[153]

Is it enough for [the applicant] to demonstrate that [his] advance is of a technical character and relates to a technical field? The answer depends on what is meant by the word 'technical'. Mr. Young appeared to me to use it as if it covered all areas of practical development which were not associated with the liberal arts. That is not what I understand to have been the way in which it was used in the decided cases. In particular in *VICOM* and *IBM* it is only processes or methods which do not fall foul of the express exclusions from patentability under section 1(2) which are treated as suitably 'technical'. Therefore use of the word 'technical', the meaning of which takes colour from its context, is apt to confuse. What counts in this area is whether the method or process controlled by the program or the computer running it is one which itself is excluded from patentability.

Thus, according to Laddie J, a subject matter does not avoid exclusion under section 1(2) by being shown to be technical, but rather is shown to be technical if it properly avoids exclusion under section 1(2). By this reasoning his Lordship vindicated the reasoning of Purchas LJ in *Genentech* by supporting an approach that differed according to the category of subject matter in question and the policy reasons for its exclusion.

This raises the additional point of interest about Laddie J's decision in *Fujitsu*, which concerns his own application of these principles to the subject matter before him.[154] According to his Lordship, the question for consideration in *Fujitsu* was whether the system was a method for performing a mental act excluded by section 1(2)(c). Seeking again to emphasize the consistency of his judgment with *Vicom*, he treated that question as embracing largely if not entirely the question whether the system was a mathematical method within the meaning of section 1(2)(a),[155] and as determinative also (in the case before him) of the applicability of the computer programs exclusion of section 1(2)(c).[156] From this position he imported the TBA's reasoning in relation to mathematical methods in *Vicom* to support his own interpretation of the mental methods exclusion, as follows:[157]

In *VICOM*, the Board explained that a mathematical method could be distinguished from a patentable process based on it in that the former involved an abstract concept in which

153 ibid, 531.
154 See ibid, 532–533.
155 ibid, 532.
156 ibid.
157 ibid.

numbers were worked on to produce new numbers whereas in a patentable process a physical entity was worked on and a new physical entity was produced. Very similar concepts apply to the distinction between methods for performing mental acts and processes, methods or apparatus based upon such acts. Excluded mental acts must include those mental activities which involve a significant level of abstraction and intellectual generality. Rules as to the planting of potatoes in which the operator is instructed to measure and evaluate matters such as the type of soil, location, weather and availability of irrigation is a method for performing a mental act. Directions to plant one seed potato every metre is not. It is a precise process.

Thus, according to Laddie J's decision, the *Vicom* distinction between methods performed on abstract objects (such as numbers) and methods performed on physical objects (such as digital images) supported a view of the mental methods exclusion specifically, and inherent patentability generally, as importing a requirement for a particular level of concreteness and specificity, so that methods dependent on the intellectual input of their implementer were incapable of being an invention. Applied to the subject matter of *Fujitsu*, the result was a finding of inherent unpatentability. Decisive for the Judge was the fact that certain integral steps of the system—including particularly the selection of data to be worked on, the means of working on the data, and the means of assessing the results and utility of that work[158]—were left to the discretion of the individual operator, making the process 'abstract' and its result 'undefined'.[159] As he stated, '[w]hat is produced is not an inevitable result of taking a number of defined steps but is determined by the personal skill and assessment of the operator.'[160] His conclusion, consistent with the conclusion of Aldous J in relation to the expert system shells of *Wang Laboratories*,[161] was that it thus consisted in substance of a scheme or method for performing a mental act or computer program, and was inherently unpatentable on that basis. In reaching his decision his Lordship expressly rejected the suggestion of the Examiner before him that the method's exclusion from patentability turned on its effect in obviating a manual process; a suggestion he described as reflecting a confusion of the requirement for an invention on one hand, and the requirement for novelty on the other, and noting that his conclusion in *Fujitsu* would have been the same regardless of whether the process had previously been performed manually.[162]

The ultimate significance of Laddie J's reasoning on this point is two-fold. First, and consistent with Mr Jeffs QC's treatment of sensory and perceptual acts in *Raytheon*, it suggests as a rationale for the mental methods exclusion the need for all patentable subject matter to be of a type that is capable of complete explication

[158] [1996] RPC 511, 532.

[159] ibid.

[160] ibid.

[161] The expert system shells in *Wang Laboratories* were, like the methods of *Fujitsu*, methods for enabling a third party to produce an object.

[162] [1996] RPC 511, 533.

and thus of repetition, reflecting a view of inherent patentability generally as interdependent with the requirement not only of utility, but also, and in his Lordship's judgment more so, of sufficient description. And second, and despite his suggestion to the contrary, it further widened the gap between UK and EPO law by contradicting the principle of the latter that subject matter the utility of which depends on the making of certain technical decisions at the point of replication is, precisely by reason of such dependence, 'technical', and thus (on the TBA's understanding) inherently patentable.[163]

This analysis of Laddie J's reasoning in *Fujitsu* brings into relief the significance also of the decision of the Court of Appeal in the final appeal in that case.[164] Delivering the principal judgment of the Court, Aldous LJ rejected the Judge's view of 'technical' as connotative rather than denotative, and explicitly reaffirmed his decision in *Lux* that inherent patentability under section 1(2) is solely a question of technical character. In so doing Aldous LJ relied principally on the decision of Fox LJ in *Merrill Lynch*, and its requirement for a technical contribution, which he presented as a restriction on the literal terms of section 1(2) justified by the historical principle of UK law that mere discoveries or ideas are not patentable, in contrast to discoveries and ideas 'which have a technical aspect or make a technical advance'.[165] In addition to confirming (by omission) the redundancy of the second limb of the *Merrill Lynch* test of inherent patentability, this description challenged the first limb of that test by suggesting the sufficiency, for inherent patentability purposes, of showing subject matter to have a 'technical aspect'.[166] The result, on its face, was a more dilute standard of patentability reminiscent of his Lordship's prior decision in *Gale*.

It is perhaps ironic given Aldous LJ's interpretation of section 1(2) as denoting a generic requirement for a technical aspect or advance that he nonetheless expressed a concern that the mental methods and computer programs exclusions be considered separately when assessing the *Fujitsu* system's eligibility for a patent. Indeed, in addition to noting the failure of Laddie J explicitly to have decided whether the system was a computer program within the meaning of section 1(2)(c),[167] Aldous LJ emphasized that his own affirmative resolution of that question said nothing on the separate issue of whether the system was also a method for performing a mental act,[168] thereby undermining Laddie J's view that the choice between those exclusions was, in the context of the *Fujitsu* application, a choice of semantics. Nonetheless, Aldous LJ also made it clear that his reservation of judgment on that issue did not indicate a retreat from his expansive understanding of the mental methods exclusion in *Wang Laboratories*. On the contrary, he explicitly rejected the

[163] See T_769/92 (*SOHEI/General-Purpose Management System*) [1996] EPO 253.
[164] [1997] RPC 609, CA.
[165] ibid, 614.
[166] cf ibid.
[167] ibid, 617.
[168] ibid, 619.

applicant's invitation to overrule that understanding in favour of a narrow construction, proposed in reliance on the TBA's decision in *PETTERSSON/Queuing System*,[169] that would have covered only such methods as the human mind actually performs, confirming instead the correctness of *Wang Laboratories* from the Court of Appeal.[170] In so doing his Lordship cited three justifications for the expansive construction. First, it obviates the need to resort to evidence of how the human mind actually works.[171] Second, it does not require considerations of novelty covered by paragraph 1(1)(a).[172] And third, it reflects the ordinary language meaning of 'mental act', as covering any mental act regardless of its actual prior performance.[173] On the basis of these justifications, and despite agreeing 'as a matter of semantics' that a claim to a method for using a computer cannot also be a claim to a method for performing a mental act, Aldous LJ confirmed in substance that a claim to the latter is no more patentable when performed on a computer than when performed on a piece of paper.[174]

Beyond this reaffirmation of *Wang Laboratories*, however, Aldous LJ was not prepared to venture, thereby leaving unresolved the questions of how his apparently expansive definition of inherent patentability was to be reconciled with his expansive definition of the mental methods exclusion, and which if any aspects of the analyses of the Examiner and Laddie J on the latter he supported. Indeed, and emphasizing again the 'difficulty in identifying clearly the boundary line between what is and is not a technical contribution', and the need generally to decide each case 'upon its own facts',[175] in reasoning his decision Aldous LJ adhered closely to the applicant's two central submissions, offering little additional insight to inherent patentability beyond the principles already discussed. The first of those submissions was that the case stood on all fours with *Vicom*, which, it was said, was decided on the basis of the technical contribution manifested by the method's processing of real images or structures.[176] And the second submission was that the subject matter in any case 'provided a new tool for modelling crystal structure combinations' that 'relieved the chemist of the laborious task of building a model', and that 'provided a [technical] solution to the problem of providing a quick and error free way of creating real images of two crystal structures joined together'.[177] On the first of these arguments his Lordship disagreed, construing the technical contribution of *Vicom* as having resided not in the fact that an image was being

[169] T_1002/92 (*PETTERSSON/Queuing System*) [1996] EPOR 1, discussed at pages 218–220.
[170] [1997] RPC 609, 620.
[171] ibid.
[172] ibid.
[173] ibid.
[174] ibid, 620–621.
[175] ibid, 616.
[176] ibid, 615, 618.
[177] ibid, 618.

produced but rather in the enhancement of the images produced,[178] consistent with the finding of the Examiner, but inconsistent with previous understandings of that case. On the second he also disagreed, treating the advantages described by the applicant as non-technical on account of having been derived from the program's inherent functional features as a computer program.[179] How this second finding could be reconciled with his definition of inherent patentability above, however, is unclear, and reflects the wider tension in his Lordship's judgment between the two limbs of that definition: the first (technical aspect) being consistent with his suggestion that inherent patentability should be kept separate from the considerations of novelty that pervade the second, and the second (technical contribution) reflecting the higher standard required to reconcile his Lordship's positive conception of inherent patentability with his interpretation of the mental methods exclusion as including methods conceived specifically with automation in mind.[180]

(iv) Conclusion

The decisions in *Lux*, *Raytheon*, and *Fujitsu* are unified in conclusion but divergent in reasoning, and reflect in their extremes the struggle of the courts to derive from section 1(2) a unified and coherent understanding of inherent patentability that precludes protection for technically conventional subject matter while embracing the theory and practice of EPO jurisprudence. At one extreme lies the view of the Patents Court in *Fujitsu*, reminiscent of Purchas LJ's decision in *Genentech* and Fox LJ's decision in *Merrill Lynch*, that section 1(2) comprises a series of individual exclusions grounded in separate policy justifications that do not resolve to a positive requirement for technical character. And at the other extreme lies the contrasting and prevailing view of Aldous LJ in the same case, reminiscent of his Lordship's previous decisions in *Gale's Application*, *Wang Laboratories*, and *Lux*, that section 1(2) denotes a positive requirement for a subject matter having a technical aspect or contribution, the fulfilment of which depends entirely on the facts of individual cases and is sufficient to establish that an invention exists.

The decision of Aldous LJ in *Fujitsu* was delivered in 1997, and was the last authoritative word on the inherent patentability of computer-related subject matter in the UK before the Court of Appeal's 2005 decision in *Aerotel Ltd v Telco Holdings Ltd*.[181] In the period between those two cases, however, significant developments took place with respect to inherent patentability as a result of a series of decisions involving biotechnology.

[178] ibid, 616. cf his Lordship's description of *Vicom* in the same section of his judgment as having turned on 'the way the enhanced image was produced'.

[179] ibid, 618.

[180] See ibid, 619–620, discussed above, and his Lordship's conclusion of his analysis in relation to the mental methods exclusion with the statement: 'Methods of performing mental acts, which means methods of the type performed mentally, are unpatentable, unless some concept of technical contribution is present.'

[181] [2006] EWCA Civ 1371.

Diagram A Main cases regarding inherent patentability prior to 2005

Case	Alleged invention (main claims only)	Subsection 1(2) items	Decision	Result	Précis of Judicial Reasoning
Merrill Lynch Inc's Application [1988] RPC 1, Pat; [1989] RPC 561, CA (cf T_16/83 (CHRISTIAN FRANCERIES/Traffic Regulation) [1988] EPOR 65)	System for trading in securities	Computer program Business method	The end result is a business method. cf Falconer J: The system's novelty and inventiveness derive from a business method.	Unpatentable (business method)	PE and Falconer J: A system that produces a technical effect— ie an effect the contribution (novelty and inventiveness) of which to the prior art is not attributable to the system's implementation or incorporation of an excluded item—is an invention. [PE: A subject matter that has as its end result the implementation of an excluded item, or that fails to produce a technical advance, is not an invention. Falconer J: A subject matter the novelty or inventiveness of which derives from an excluded item is not an invention.] CA: A subject matter that makes a technical contribution to the art in the form of a new result, which result is not an excluded item, is an invention.
Genentech Inc's Patent [1987] RPC 553, Pat; [1989] RPC 147, CA (cf **Howard Florey**)	Human protein	Discovery	The protein is a novel structural product of improved utility.	Patentable	Whitford J and CA: A product based on a novel discovery regarding the structure of a natural substance is an invention; Falconer J in *Merrill Lynch* was wrong.

Gale's Application [1991] RPC 305, Pat; [1991] RPC 317, CA (cf T_208/84 (VICOM/ Computer-Related Invention) [1987] EPOR 74)	ROM that incorporates a program for calculating square roots	Computer program Mathematical method Mental method	The ROM is indistinguishable from the program it incorporates. *Contra* Aldous J: The ROM is a novel structure.	Unpatentable (computer program) *Contra* Aldous J: patentable (new machine)	Aldous J: A system that performs a mathematical calculation is a new technical product and thus an invention. CA: A system programmed to perform a mathematical calculation that neither manages a technical process external to the computer nor solves a technical problem internal to the computer makes no technical contribution to the art and is therefore not an invention. Aldous J was wrong.
Wang Laboratories Inc's Application [1991] RPC 463, Pat (cf VICOM/Computer-Related Invention)	System for providing and storing expert knowledge	Computer program Mental method	The system is conventional but for the method it performs, which is a mental method.	Unpatentable (mental method)	Aldous J: A system programmed to perform a method that can be performed mentally is not an invention.
Hitachi Ltd's Application [1991] RPC 415, PO (cf VICOM/Computer-Related Invention)	System for translating computer languages	Computer program Mathematical method	The system is conventional but for the method it performs, which does not operate on a physical entity.	Unpatentable (algorithm; computer program)	PO: A method that fails to produce a technical result is not an invention.
Lux Traffic Controls Ltd v Pike Signals Ltd [1993] RPC 107, Pat (cf CHRISTIAN FRANCERIES/Traffic Regulation)	Systems for controlling traffic	Business method	The systems involve new technical machines, and are thus not methods for controlling traffic.	Patentable (new machines)	Aldous J: An invention is a subject matter that makes a technical contribution to the art, such as a system that incorporates a new technical principle.

(*Cont'd*)

Diagram A Main cases regarding inherent patentability prior to 2005 (*Cont'd*)

Case	Alleged invention (main claims only)	Subsection 1(2) items	Decision	Result	Précis of Judicial Reasoning
Raytheon Co's Application [1993] RPC 427, Pat (cf VICOM/Computer-Related Invention)	System for identifying ships by a process of image comparison	Computer program Mental method	The system has as its end result a mental method, and is thus not an invention.	Unpatentable (mental method)	PE and Patents Court (Mr Jeffs QC): A method that has as its end result an excluded item is not an invention.
Fujitsu Ltd's Application [1996] RPC 511, Pat; [1997] RPC 608, CA (cf VICOM/Computer-Related Invention)	System for processing digital images of crystals	Computer program Mental method	The system does not contribute to the technical arts.	Unpatentable (computer program; mental method)	PE: The system was merely a computer programmed to display information in a technically conventional way. Laddie J: An invention is a subject matter not excluded by s1(2), cf a subject matter having 'technical' character, which a method dependent on the skill and assessment of a person for its execution is not. CA (Aldous LJ): An invention is a subject matter having technical character, being discoveries or ideas that have a technical aspect or make a technical advance; a mental method is no more patentable when performed on a computer than on a piece of paper; Laddie J was wrong.

Chiron v Murex Diagnostics Ltd [1996] RPC 535, CA	Immunoassay for detecting antibody hepatitis C	Discovery	The isolated polypeptides and immunoassay have technical aspects and are not discoveries.	Patentable	Aldous J and CA: As subject matter having technical aspects cannot be a discovery.
Biogen Inc v Medeva plc [1995] RPC 25, CA; [1997] RPC 1, HL	Recombinant DNA molecule	Inherent patentability not argued	The subject matter is a decision to try an experiment which others thought would not succeed, and a commercial decision is not an invention.	Implied: unpatentable	Aldous J: The invention (which needed to be sufficiently supported under ss 14(5)(c) and 72(1)(c)) was the production of the relevant recombinant DNA molecule. CA: According to Aldous J, the inventive concept was the idea and a decision to use expression to obtain polypeptides. But this makes the invention the decision to try an experiment which others thought would not succeed, and a mere commercial decision is not an invention. The claimed molecule is simply a molecule consisting of an available vector, designed by others, with an undetermined fragment of natural DNA inserted into it. CA and HL (Lord Hoffmann): A claim for an invention must be sufficiently supported by the disclosure in the specification or it will be revocable under s 72(1)(c).

(Cont'd)

Diagram A Main cases regarding inherent patentability prior to 2005 (*Cont'd*)

Case	Alleged invention (main claims only)	Subsection 1(2) items	Decision	Result	Précis of Judicial Reasoning
					Pre-1977 principles of fair basis and speculative claims apply under the PA, consistent with the principle of T_409/91 (*EXXON/Fuel Oils*) [1994] EPOR 149 that the extent of the monopoly, as defined by the claims, must correspond to the technical contribution to the art. Lord Hoffmann: The requirement for an invention will invariably be academic; if the other patentability requirements are satisfied, there will be an invention. However, identification of the invention is critical in determining sufficient support. Lord Mustill: In some cases—eg cases involving new technologies—the invention might require close conceptual analysis.

Kirin-Amgen Inc v Hoechst Marion Roussel Ltd [2002] EWCA Civ 1096; [2004] UKHL 46	DNA sequence encoding for the human protein erythropoietin	Discovery	The subject matter makes a technical contribution.	Patentable	Neuberger J: The discovery of the DNA sequence made a technical contribution by enabling the production of a recombinant protein that supported various new and specified industrial applications. (Triggered a change in PO practice; technical character now sufficient for inherent patentability, with the *Merrill Lynch* support for the exclusion of certain subject matter regardless of its technical contribution no longer applied.) CA: The invention was not the discovery, but rather the method of producing EPO. HL (Lord Hoffmann): The CA was right, based on claim construction principles, the nature of 'inventions', and the 'social contract' which the patent represents. Identification of the *invention* is critical in determining support and the other patentability requirements.

D. Inherent Patentability as a Requirement for Subject Matter Sufficiently Supported *Qua* Invention: Biotechnology after *Genentech Inc's Patent*

Since *Genentech*, the inherent patentability of biotechnological products has been considered in three main cases. The first and second of those cases involved an essentially identical decision of Aldous J that was appealed successfully in the first case, and unsuccessfully in the second, to differently constituted Courts of Appeal. Those cases are *Biogen Inc v Medeva plc*[182] and *Chiron Corporation v Murex Diagnostics Ltd*[183] respectively, the first of which—*Biogen*—was subsequently appealed to the House of Lords. Similarly for the third case, which is *Kirin-Amgen Inc v Hoechst Marion Roussel Ltd*,[184] and which represents with *Biogen* a shift in judicial understandings of inherent patentability.

1. Conflicting Approaches to Biotechnology: *Biogen* and *Chiron* in the Court of Appeal

Biogen involved an action for infringement of a European patent that had been granted to Biogen in 1990, and a counterclaim for revocation of the patent by Medeva. The subject matter of the patent was a recombinant DNA molecule characterized by a DNA sequence that encoded for a protein displaying HBV antigen specificity.

At first instance, counsel for Medeva raised inherent patentability as a ground for revocation, but subsequently declined to press the issue on accepting, in Aldous J's words, 'that I would come to the same conclusion as I did in the *Chiron* case for basically the same reasons'.[185] In *Chiron*, Aldous J had rejected an argument that claims for certain isolated polypeptides encoding for the hepatitis C virus (HCV) were invalid for comprising a 'discovery as such' within the meaning of section 1(2)(a) of the PA. In the defendant's argument, the presentation of the products in the patent claims with reference to their use rather than their chemical or physical structure meant that any patent granted for them would monopolize the discovery on which they were based, making the patent for a 'discovery as such'.[186] Aldous J rejected that argument, describing it as 'untenable' in light of the subject matters' technical aspects.[187] Further, in the Judge's view the defendant's complaint was less inherent patentability than sufficient support, and the fact that the protection

182 [1995] RPC 25, CA, aff'd [1997] RPC 1, HL.
183 [1996] RPC 535, CA.
184 [2002] EWCA Civ 1096, aff'd [2004] UKHL 46.
185 [1995] RPC 25, 40.
186 See [1996] RPC 535, 574.
187 See ibid.

conferred by the patent covered more than was enabled by the specification. This raised the pre-1977 principles of fair basis and speculative claims, and whether they were grounds for revocation of a patent under section 72(1)(c) of the Act. According to that section, a patent may be revoked where its specification 'does not disclose the invention clearly enough and completely enough for it to be performed by a person skilled in the art'. This supports the section 14(3) PA requirement that '[t]he specification of an application shall disclose the invention in a manner which is clear enough and complete enough for the invention to be performed by a person skilled in the art'. By implication of *Asahi Kasei Kogyo KK's Application*,[188] however, section 72(1)(c) also allows revocation of a patent for breach of the section 14(5)(c) requirement that the claims 'shall . . . be supported by the description' contained in the patent specification.[189] Hence the question for the *Chiron* Court, which was whether section 14(5)(c) required compliance with the pre-1977 principles above. The implication of the Judge was that it did not,[190] with the result that lack of fair basis and/or equiparation in the claim could not be raised as a ground for revoking the patent.

Similarly in *Biogen*, a central issue before the Court was sufficient support.[191] Relying on section 125 of the PA and its definition of the protection conferred by a patent with reference to the invention 'specified in a claim',[192] Aldous J interpreted his task as being to identify the subject matter described in the claim, and ensure that it was sufficiently supported by the specification.[193] In the result, he accepted the patentee's argument that its claim was for only one subject matter—'the production of recombinant DNA molecules having the claimed characteristics using a suitable host'—which was therefore the only 'invention' for which a supporting disclosure was required.[194] That finding reduced the burden of establishing validity, which was consequently determined in Biogen's favour.[195]

On appeal, however, this decision was overturned, following the Court's rejection of the Judge's interpretation of the invention, and dismissive treatment of Medeva's section 1(2) objection. Having been discouraged from arguing inherent patentability at trial, Medeva came to the Court of Appeal conceding that the DNA molecule was an invention. The Court made it clear, however, that it regretted that concession.[196] While acknowledging its inability given that concession to consider revocation of the patent for lack of an 'invention' within the

188 [1991] RPC 485, HL.

189 See s 14(2)(b) (requiring that a patent application contain a specification which describes the invention); ibid, 536 (discussed in *Biogen Inc v Medeva Plc* [1997] RPC 1, 47).

190 See [1996] RPC 535, 578.

191 [1995] RPC 25, 41.

192 PA, s 125(1) is reproduced at page 192.

193 [1995] RPC 25, 41.

194 ibid; see also ibid, 55, 56, 58.

195 ibid, 50.

196 [1995] RPC 25, 83.

meaning of section 1(2), it noted the need nonetheless to characterize the *Biogen* invention in order to determine the validity of its patent. In undertaking that characterization, the Court doubted that there was an 'invention' at all. In its view, Aldous J's finding of inventive step depended on a conception of the invention as 'the idea *and decision* of using expression to obtain polypeptides', which was not a subject matter for which a patent could validly be granted.[197] As the Court said, the decision 'to try an experiment which others thought would not succeed'—which was what the Trial Judge had described as meriting the patent—was '[a] mere commercial decision', and 'not an invention'.[198] By implication, the claim did not 'as such disclose any invention' within the meaning of section 1(1) of the PA.[199]

Central to this decision was the Court's rejection of an approach to conceiving subject matter with reference exclusively to the claims. In its judgment, the question of patentability needed also to take account of 'the description in the specification and any other relevant evidence'.[200] However, in finding this the Court was also implying something else, namely, that in determining claim validity, responsibility for conceiving the invention lies with the courts and not with the claimants. Put differently, the 'invention' is not simply the subject matter described in the claims, but rather the subject matter described in the claims when properly conceived *qua* 'invention'. It followed that any subject matter capable of being conceived as an invention (eg an isolated DNA sequence) which depended for its entitlement to a patent on being conceived as something other than an invention (eg a decision to use a particular method to isolate a DNA sequence) would not be capable of supporting a patent, as it would not merit protection *qua* invention.[201]

This reading of the Court of Appeal's reasoning is consistent with its findings. While agreeing that the description of the invention in the plaintiff's specification did not need to support a claim across the breadth of its scope in order for the claim to be valid under section 14(5)(c) of the PA, the Court did not agree that the disclosure of one enabling method for realizing the invention was necessarily sufficient. Consistent with this, it rejected Aldous J's implication regarding the validity of pre-1977 principles of fair basis and speculative claims, and affirmed the statement in the headnote of *Eastman Kodak Co's Application*[202] that 'an application could be rejected on the basis of insufficiency where claims were broad and indeterminate and of a speculative character'.[203] In addition, it noted the

[197] ibid, 90, emphasis in original.
[198] ibid, 90–91.
[199] ibid, 91.
[200] ibid, 87.
[201] ibid, 87, 88–89. The result was a judgment that recalled Nicholls LJ's criticisms of Aldous J's approach in *Gale* (see page 269).
[202] [1970] RPC 548, PAT.
[203] [1995] RPC 25, 96.

consistency of this reasoning with contemporary EPO jurisprudence, including the decision in *EXXON/Fuel Oils*, as follows:[204]

The underlying purpose of the requirement of support by the description, insofar as its substantive aspect is concerned, and of the requirement of sufficient disclosure in the same, namely to ensure that the patent monopoly should be justified by the actual technical contribution to the art. . . . [T]he disclosure of the claimed invention is only sufficient if it enables the skilled person to obtain substantially all embodiments falling within the ambit of the claims.

The Court of Appeal agreed with this passage, noting that '[w]hat will suffice to satisfy this criterion will vary depending upon the nature of the claim that has been made'.[205] As a result, it held, it is 'essential to apply the test having regard to the extent of the claim',[206] and by implication from its analysis of claim construction above, having regard to the subject matter claimed when conceived *qua* invention. In *Biogen* the claim failed this test, and was accordingly revoked for insufficient support.

Soon after the Court of Appeal's decision in *Biogen*, a differently constituted Court heard an appeal from Aldous J's judgment in *Chiron*, and reached a different decision from that reached in *Biogen*. With respect to inherent patentability, the defendant in *Chiron* argued that the subject matter claimed related to a 'discovery as such' within the meaning of section 1(2)(a) of the Patents Act. The basis for that argument was that the claims failed to disclose the products' physical or chemical structure, and thereby 'foreclosed [their] use . . . in making other inventions'.[207] In rejecting this argument the Court of Appeal referred to *Genentech* and *Gale's Application*, which it described as containing 'nothing . . . to suggest that the physical structure or chemical formula of the product resulting from the discovery must be disclosed'.[208] Indeed, according to the Court in *Chiron*, '[i]t would be inconsistent with both of [those cases] that that should be so if the useful application does not itself have to be inventive or novel'.[209] Given this, the Court decided that the appellants' objection was not that there was no invention, but rather that the invention was not sufficiently supported by the specification. Further, in resolving the issue of sufficient support, pre-1977 authorities would need, he suggested, to be 'treated with caution':[210]

A speculative claim, a claim to an obvious desideratum and a result claim are all alternative ways of describing a claim framed by reference to the result to be achieved rather than prescribing the means of achieving that result. The vice may lie in the fact that the specification

[204] T_409/91 (*EXXON/Fuel Oils*) [1994] EPOR 149, 155–156; see [1995] RPC 25, 97–98.
[205] [1995] RPC 25, 98.
[206] ibid.
[207] [1996] RPC 535, 605.
[208] ibid.
[209] ibid.
[210] ibid, 621.

and the claims give no instruction on how to achieve the result or in the fact that they teach one way of achieving the result but claim a monopoly to all ways of doing so. All the authorities to which we were referred in this connection were decided under the Patents Act 1949 or its predecessors. Thus they must be treated with caution in view of the fact that fair basis and best method were then grounds for revocation but are not now.

According to the Court, none of the 'epithets' suggested by the pre-1977 authorities in question were applicable in the *Chiron* case, leading to a finding that the claims were valid.[211]

2. Restricting Protection to Subject Matter Sufficiently Supported *Qua* Invention: *Biogen v Medeva* and *Kirin-Amgen v Hoechst Marion* in the House of Lords

The treatment of the issues of inherent patentability in the lower court decisions in *Biogen* and *Chiron* as issues of sufficient description instead was prescient of the House of Lords decision in *Biogen*. The principal judgment in that case was written by Lord Hoffmann, and included the following analysis of the requirement for an invention:[212]

What is an invention?
[Section 1(1)] might suggest that logically one should first decide whether the claimed invention can properly be described as an invention at all. Only if this question receives an affirmative answer would it be necessary to go on to consider whether the invention satisfies the prescribed conditions for being 'patentable'. In practice, however, I have no doubt that in most cases this would be a mistake and cause unnecessary difficulty.

The Act does not define the concept of an invention. Section 1(1) was intended to reflect, 'as nearly as practicable', art 52 of the [EPC]: see s 130(7) of the 1977 Act. Article 52 also has no definition of an invention. It seems the parties to the EPC were unable to agree upon one . . . But the reason the parties were content to do without a definition was that they recognised that the question would almost invariably be academic. The four conditions in s 1(1) do a great deal more than restrict the class of 'inventions' which may be patented. They probably also contain every element of the concept of an invention in ordinary speech. I say probably, because in the absence of a definition one cannot say with certainty that one might not come across something which satisfied all the conditions but could not be described as an invention. But the draftsmen of the Convention and the Act, as well as counsel at the bar, were unable to think of any examples. Just in case one should appear, s 1(5) gives the Secretary of State power to vary the list of matters excluded by para (d) to 'for the purpose of maintaining them in conformity with developments in science and technology'.

As the four conditions are relatively familiar ground, elucidated by definitions in the Act and the jurisprudence of the courts and the EPO, it will normally be more convenient to start by deciding whether they are satisfied. In virtually every case this will be the end of the

[211] [1996] RPC 535, 621.
[212] [1997] RPC 1, 41–42.

inquiry. There may one day be a case in which it is necessary to decide whether something which satisfies the conditions can be called an invention, but that question can wait until it arises.

One can of course imagine cases in which the alleged subject matter is so obviously not an invention that it is tempting to take an axe to the problem by dismissing the claim without inquiring too closely into which of the conditions has not been satisfied. So in [*Genentech*] Mustill LJ said, by reference to the ordinary speech meaning of 'invention': 'You cannot invent water, although certainly you can invent ways in which it may be distilled or synthesised.'

This is obviously right and in such a case it may seem pedantic to say that water fails the condition in para (a) of s 1(1) because it is not new. Unfortunately, most cases which come before the courts are more difficult. Judges would therefore be well advised to put on one side their intuitive sense of what constitutes an invention until they have considered the questions of novelty, inventiveness and so forth. In the present case, I think that Medeva's counsel was right to resist the invitation of the Court of Appeal to make submissions on whether the claims constituted an invention.

Two central points emerge from this reasoning. The first is that the requirement for an invention will rarely if ever add anything to the other requirements of patentability, so that subject matter that satisfies the latter should be treated as an 'invention' within the meaning of section 1(1). Indeed, according to Lord Hoffmann, it is precisely the 'academic' nature of the inherent patentability enquiry that accounts for the historical failure of law-makers to have agreed upon an exhaustive definition of the 'invention'. And the second point is that judges should approach the question of patentability without regard to their 'intuitive sense of what constitutes an invention', at least until they have determined whether the subject matter is otherwise entitled to a patent; the implication being that 'intuition'—the very faculty charged in *Hitachi* with responsibility for determining inherent patentability—is, like inherent patentability itself, an unhelpful distraction. It is the first of these points that caused Lord Mustill to refrain in *Biogen*. Thus, in a tentatively framed decision, his Lordship rejected the view of inherent patentability as 'academic', and defended the appropriateness of a 'close conceptual analysis' of that requirement in certain cases, including cases involving 'new technologies, beyond contemplation when the foundations of [the] law were established'.[213]

Having all but rejected the need to show an invention in establishing a valid claim to a patent, Lord Hoffmann went on in *Biogen* to consider the question of sufficient support; and on this there was no disagreement among the Lords. After hearing submissions from counsel on pre-1977 authorities regarding '[w]ide or speculative claims,'[214] including the decisions in *Mullard Radio Valve Co Ltd v Philco Radio & Television Corp*[215] and *Pottier's Application*,[216] Lord Hoffmann

213 ibid, 31–32.

214 See ibid, 22.

215 [1936] 2 All ER 920, HL, discussed at pages 86–87; see [1997] RPC 1, 30.

216 [1967] RPC 170, Pat, discussed at page 97; see [1997] RPC 1, 16.

affirmed the Court of Appeal's decision. Specifically, he found that section 72(1)(c) enabled the revocation of a patent in any case where 'the extent of the monopoly claimed exceeds the technical contribution to the art made by the invention as described in the specification'.[217] Aside from being consistent with pre-1977 law, and EPO authorities such as *EXXON/Fuel Oils*,[218] this result was consistent with the policy objective of ensuring that 'research and healthy competition' are not stifled by allowing the first who achieves an obviously desirable goal to monopolize every way of achieving that goal. As Lord Hoffmann said:[219]

> It is inevitable in a young science, like electricity in the early nineteenth century or flying at the turn of the last century or recombinant DNA technology in the 1970s, that dramatically new things will be done for the first time. The technical contribution made in such cases deserves to be recognised. But care is needed not to stifle further research and healthy competition by allowing the first person who has found a way of achieving an obviously desirable goal to monopolise every other way of doing so.

In 1999, this reasoning was embraced by Pumfrey J in the Patents Court in a case involving a claim for chemical products. That case was *Monsanto Co v Merck & Co Inc*,[220] where Pumfrey J construed Lord Hoffmann's judgment in *Biogen* as affirming the continued validity of *Olin Mathieson Chemical Corp v Biorex Laboratories Ltd*,[221] based on the principle of *Mullard Radio Valve Ltd v Philco Radio & Television Corp*. This enabled the Judge to support the *Olin Mathieson* test of claim validity discussed above, with its emphasis on restricting patent protection having regard to the likely impact of the patent on the art.[222] By supporting the validity of this test under post-1977 law, Pumfrey J underlined the expansive potential of Lord Hoffmann's decision in *Biogen*, and of the post-1977 requirement for sufficient support as a means of enabling UK courts to mediate the bargain struck by a patent. However, and as the *Chiron* litigation demonstrated, Lord Hoffmann's reasoning did not have the support of all UK judges; a fact further underlined by the decision of Aldous LJ in the appeal in *Monsanto v Merck* itself:[223]

> I do not believe it is helpful to assume that the test of sufficiency under section 72(1)(c) of the Act is that applied in *Olin Mathieson* which was decided under the 1949 Act. The issue in that case was whether the claim was fairly based (see section 32(1)(i) of the Patents Act 1949).... The true test is that laid down in the section 72(1)(c) of the 1977 Act. Further, I cannot endorse, as a general proposition of law, the judge's conclusion that if a claim covers compounds which do not satisfy the representations made about them in the specification it is likely to be invalid. That was the law under the 1949 Act in that a patent could be revoked if it was obtained under a false representation. The test of sufficiency under the

[217] See [1997] RPC 1, 54.
[218] ibid.
[219] ibid, 52.
[220] [2000] RPC 709, Pat, aff'd [2001] EWCA Civ 1610.
[221] [1970] RPC 157, Ch, discussed at pages 97–98.
[222] See ibid, 192–193.
[223] [2001] EWCA Civ 1610, [54].

1977 Act is concerned with enablement. The specification must contain an enabling disclosure of the invention. If the invention requires satisfaction of representations, then the specification will not be sufficient unless those representations are satisfied.

With respect, it is difficult to accept the reasoning in this passage. For a start, the principle applied in *Olin Mathieson* was not the statutory principle of fair basis per se, but rather the common law requirement for an equiparating claim.[224] And second, the passage ignores Lord Hoffmann's understanding of 'enabling disclosure' as requiring (consistent with pre-1977 law) that 'the claims . . . be supported by the description', including in the sense of *EXXON/Fuel Oils* above, namely, in the sense of corresponding to the relevant contribution to the art. As Pumfrey J's decision in *Monsanto* made clear, this latter principle of *EXXON/Fuel Oils* was simply a restatement of the *Mullard Radio* rule.

The decision of the House of Lords in *Biogen* was delivered in October 1996, five months before the Court of Appeal's judgment in *Fujitsu*. Despite this, it was not referred to by Aldous J in *Fujitsu*, and leaves no evident mark on his judgment in that case. It was also not referred to by Neuberger J in the only other case in this period involving inherent patentability. That case was *Kirin-Amgen Inc v Hoechst Marion Roussel Ltd*,[225] which concerned allegations of infringement and validity in respect of a patent claiming for the DNA sequence encoding for the human protein erythropoietin. Rejecting the defendant's submission that the sequence was an inherently unpatentable discovery within the meaning of section 1(2)(a), Neuberger J relied primarily on the decisions of Whitford J in *Genentech* and Aldous LJ in *Fujitsu* to support the conclusion that while the patent's essential feature was a discovery in the form of a DNA sequence, such discovery nonetheless made a technical contribution by enabling the production of a recombinant protein that supported various new and specified industrial applications.[226] In this way, his Lordship applied the principles of Aldous LJ in *Fujitsu* to reach the same conclusion as was reached in *Genentech*. In so doing, he took express comfort from the long-standing practice of the EPO and the United States Patent Office of granting patents for DNA sequences, and from the failure of any other European country to have impugned the *Kirin-Amgen* patent on inherent patentability grounds in their own consideration of its validity.[227] He also eschewed the need to reconcile his decision with the statement of Mustill LJ in *Genentech* that 'a particular DNA sequence [is] an "existing fact of nature, newly discovered"' on the ground, among others, that his task was merely to apply the 'tolerably clear' principles of inherent patentability under UK law to the facts before him, and not to

[224] See the discussion at pages 86–88, 97–98, 105–106.
[225] [2001] EWHC 433, Pat.
[226] ibid, [532]–[541].
[227] ibid, [541].

'[delve] into detailed judgments in other cases merely to see how the principles were applied to the facts of those other cases'.[228]

It is paradoxical given Neuberger J's suggestion in *Kirin-Amgen* that his decision involved no more than the application of established principles that the decision caused the UK Patent Office formally to change its approach to determining inherent patentability the following year. That change was recorded in its notice of 24 April 2002, in which the Comptroller-General of Patents, Designs and Trade Marks advised that the Office would no longer construe the Court of Appeal's decision in *Fujitsu* as being subject to the earlier comments of the Court of Appeal in *Merrill Lynch* so as to exclude certain subject matter from patentability regardless of its technical contribution.[229] According to the Comptroller, the failure of Neuberger J to have recognized any equivalent qualification made it appropriate for the Office to accept technical character as the overriding criterion of inherent patentability in contemporary law, so that, in the Comptroller's own words, 'inventions which involve a technical contribution will not be refused a patent merely because they relate to business methods or mental acts'.

In 2003 the decision of Neuberger J was appealed to the Court of Appeal, and from the Court of Appeal to the House of Lords. As in *Biogen*, the focus of the appellate Courts in *Kirin-Amgen* was on the proper interpretation of the patent claims, and the nature of the subject matter which they described when properly conceived *qua* invention.

As noted, at first instance in *Kirin-Amgen* Neuberger J had treated the invention as the patentee's discovery of the erythropoietin sequence and its associated information. By contrast, the Court of Appeal treated it as the patentee's particular method of making erythropoietin, ie its expression of an exogenous DNA sequence encoding for erythropoietin.[230] The House of Lords agreed with that treatment. According to Lord Hoffmann, who wrote the judgment, the Court of Appeal's decision was correct for three central reasons.[231] First, it was based on a construction of the claims and not the specification. Second, it was consistent with section 1(2) of the PA, and the nature of an invention within the meaning of that section as 'a practical product or process, not information about the natural world'.[232] And third, it 'accord[ed] with the social contract between the state and the inventor which underlies patent law':[233]

> The state gives the inventor a monopoly in return for an immediate disclosure of all the information necessary to enable performance of the invention. That disclosure is not only to enable other people to perform the invention after the patent has expired. If that were all, the inventor might as well be allowed to keep it secret during the life of the patent. It is also

[228] [2001] EWHC 433, [539].
[229] United Kingdom Patent Office, *Patents Act 1977: Interpreting section 1(2)* (24 April 2002).
[230] [2002] EWCA Civ 1096, [23], [32].
[231] [2004] UKHL 46, [77].
[232] ibid.
[233] ibid.

to enable anyone to make immediate use of the information for any purpose which does not infringe the claims.

In light of this understanding of the function of patents, and the importance of the specification in ensuring it is met through the immediate disclosure of information, it 'could not be right', Lord Hoffmann suggested, to give [a patentee] a monopoly of the use of [that] information as such'.[234] Applying these principles, the invention—and therefore (by section 125) the patent—was not sufficiently wide to include the defendant's own (invented) method of creating erythropoietin.

In combination, Lord Hoffmann's decisions in *Kirin-Amgen* and *Biogen* can be read as having signalled a shift in the courts' understanding of the role played by the requirement for an invention. Before the decision in *Biogen*, inherent patentability was understood to denote a threshold requirement that the claims of a patent disclose an 'invention'. After *Biogen*, it was understood to denote a (further) requirement that the subject matter claimed for a patent depend for its entitlement to protection on that which constituted it *qua* 'invention'. Among other things, this gave meaning to the *Mullard Radio* rule by making it clear that that which a patent protects—and therefore, that which a patent must secure for the public—is not the subject matter claimed and accepted to be an invention, but rather the subject matter entitled to protection, and actually protected, when properly conceived *qua* invention. Hence the importance of the question—paradoxically underlined by Lord Hoffmann himself[235]—'what constitutes a subject matter *qua* invention?' In *Generics Ltd v Lundbeck A/S*,[236] his Lordship considered that question in the context of newly isolated chemical products.

3. Chemical Products *Qua* Invention: *Generics v Lundbeck* in the House of Lords

Generics v Lundbeck was delivered in February 2009, less than six weeks after the case was heard. It involved an appeal from a decision of the Court of Appeal,[237] delivered by Lord Hoffmann and Jacob LJ, that a patent for an isolated chemical product comprising a (+) enantiomer was not revocable under section 72 of the PA. Before the priority date of the patent, the enantiomer (known as escitalopram) was only available in racemic form as part of the compound citalopram. After discovering a way to resolve that compound and thereby produce escitalopram,

[234] ibid.

[235] In addition to the cases above, see *Merrell Dow Pharmaceuticals Inc v H N Norton & Co Ltd* [1996] RPC 76, HL, 88 per Lord Hoffmann (stating that legal definitions of novelty depend on the law's answer to the question, '*what does it mean to know something, so that it can be part of the state of the art?*').

[236] [2009] UKHL 12, affirming *Lundbeck A/S v Generics Ltd* [2008] EWCA Civ 311, overruling [2007] EWHC 1040, Pat.

[237] See [2008] EWCA Civ 311.

Lundbeck obtained a European patent, which, being a patent for the enantiomer as such, covered every method of making the product.

Before the courts, Generics applied for revocation of the patent under section 72(1) of the PA, on the grounds that escitalopram was not 'a patentable invention', and that the patent claim in respect of the drug was not sufficiently supported by the specification. In particular, it argued that at the priority date of the *Lundbeck* patent, escitalopram formed part of the state of the art and was also lacking an inventive step. Further, even if escitalopram were a patentable product, it exceeded the technical contribution to the art disclosed in the patent specification, and was thus not sufficiently supported in law within the meaning of sections 14(5)(c) and 72(1)(c) of the PA.

At first instance, Kitchin J rejected the patentability point.[238] In his judgment, the claim was for the (+) enantiomer as such, which was novel on the principles of contemporary law,[239] namely, that the prior art neither disclosed the product, nor enabled an ordinary skilled person to produce the product. Further, while producing the enantiomer was an obviously desirable goal, Lundbeck's method was not obvious to try, which made the enantiomer an inventive product. This left the issue of sufficient support, and the proportionality of the rights conferred by the patent in light of the technical contribution to the art.

As in the biotech cases above, the question of sufficiency was central to the case. Important was the principle of *EXXON/Fuel Oils*, that 'the extent of the monopoly, as defined by the claims, should correspond to the technical contribution to the art in order for it to be supported or justified'.[240] As has been noted, that principle was accepted in *Biogen*, and described by Lord Hoffmann in that case as requiring revocation of a patent for a class of recombinant antigens on the ground of insufficient support. Considering the implications of his reasoning in that case, Kitchin J noted the emphasis which Lord Hoffmann had placed on restricting the availability of patent protection to ensure that 'research and healthy competition' are not stifled by allowing the first who achieves an obviously desirable goal to monopolize every way of achieving that goal.[241] The Judge reasonably concluded from this dictum, read in the context of *Biogen* as a whole, that being the first to isolate a chemical product, the properties of which were previously known, might entitle a person to a patent for the method but not for the isolated product as such. The reason was the inventive step, which was restricted to the method of isolating the product. That such step might be sufficient to establish the product's inventive step did not, he suggested, detract from this, for a patent for a novel and inventive product might still be invalid for excessive breadth. The implication for the (+)

[238] [2007] EWHC 1040, Pat.

[239] See *Synthon BV v SmithKline Beecham Plc* [2005] UKHL 59.

[240] T_409/91 (*EXXON/Fuel Oils*) [1994] OJ EPO 653, [3.3].

[241] [2007] EWHC 1040, [262].

enantiomer was clear: it was patentable under section 1(1) of the Act, but insufficiently supported in the *Biogen* sense.

On appeal this decision was overturned.[242] According to Lord Hoffmann in the Court of Appeal, with whom Smith and Jacob LJJ agreed, Kitchin J erred in tying the requirement for sufficient support to the inventive step disclosed in the patent, rather than to the actual 'invention' itself.[243] In his Lordship's judgment:[244]

> In order to decide whether the specification is sufficient, it is . . . first necessary to decide what the invention is. That must be found by reading and construing the claims, in which the inventor identifies what he claims to be his invention. As the Board of Appeal of the European Patent Office said in *Exxon/Fuel Oils* . . . 'It is the definition of the invention in the claims that needs support'.

In *Biogen* the invention was a class of recombinant products, while in *Lundbeck* it was a chemical product as such independent of any method of making the product. As a result, the requirements of sufficiency differed between the cases:[245]

> Section 60(1) of the Act makes it clear that a claim may be either to a product or a process. In the case of a product claim, performing the invention for the purposes of section 72(1)(c) means making or otherwise obtaining the product. In the case of a process claim, it means working the process. A product claim is therefore sufficiently enabled if the specification discloses how to make it. There is nothing to say that it must disclose more than one way.

While 'sympathetic to the feeling' of Kitchin J below that the distinction between the *Biogen* and *Lundbeck* patents owed little to their respective contributions to the art,[246] the statutory allowance of patents for products determined the validity of the Lundbeck grant, and made it (according to Lord Hoffmann in his judgment) 'too late to have regrets about the breadth' of the claim:[247]

> Parliament has chosen to allow product claims and the jurisprudence of the EPO, which we have always regarded as carrying great weight, shows that such claims can be made in the latter case as well. It is too late to have regrets about the breadth of the monopoly which such claims confer.

The House of Lords affirmed this decision, seeming to support the Court of Appeal's view that in a case involving a chemical product, the technical contribution made to the art is by definition the (patentable) product itself. The exception to this was Lord Walker, for whom it was not enough that the product contribute to the art: it needed to do so sufficiently to justify a patent. In *Lundbeck*,

[242] [2008] EWCA Civ 311.
[243] ibid, [41].
[244] ibid, [29].
[245] ibid, [30].
[246] ibid, [42].
[247] ibid, [46] per Lord Hoffmann; see [54]–[57] per Jacob LJ.

he implied, the contribution was sufficient, as escitalopram was a thing of 'lasting importance'.[248]

The question at the heart of the *Lundbeck* case was how does one determine the contribution to the art in the case of a method of first making a product, and how does one decide whether that contribution justifies a patent for the product as such? According to Kitchin J, the contribution to the art is the inventive step, which must therefore lie in the product itself in order for that product to support a patent. According to the higher courts, however, the contribution to the art is the invention claimed, which must therefore be a product as such in order to support a patent for the same. While this seems to subvert the *Exxon* principle, it proceeds from the existence of a 'patentable *invention*', which requires a contribution to a technical art.[249] Hence, even on the view of the House of Lords, a product ought only to support a patent if, when conceived *qua* invention, it makes a contribution of a technical nature.

The central point to emerge from *Lundbeck* is the need to consider sufficient support with regard to the category of invention claimed. Consider for example this statement by Lord Walker:[250]

> Judges have often observed that the wide abstract terms in which patent law is expressed must always be related to the facts of the particular case. That is especially true in relation to the sufficiency of a product claim, since the term 'product' covers such an extremely wide variety. A product may be as simple as a baby's disposable diaper (see *Mölnlycke AB v Procter & Gamble Ltd* [1992] FSR 549) or a corkscrew (see *Hallen Co v Brabantia (UK) Ltd* [1991] RPC 195) or as complex as an 'heavier-than-air flying machine' referred to by Lord Hoffmann in *Biogen*, or a class of microscopic organisms, produced by recombinant DNA technology, such as was considered by this House in *Biogen* and *Kirin-Amgen*.

The same is true, it is submitted, beyond sufficient support, with respect to the section 1(1) requirements. Further, in underlining the need to conceive a subject matter *qua* invention, these points reveal the error that was made by the Courts when conceiving escitalopram *qua* invention. According to each of the *Lundbeck* Courts, for patentability purposes escitalopram was not a chemical product, but rather an *isolated* chemical product. Indeed, this distinction was central to their Lordships' finding that the invention was not anticipated by the racemate, notwithstanding the principle of *Merrell Dow Pharmaceuticals Inc v Norton & Co Ltd*[251]—consistent with the principle of *Exxon* above—that 'the novelty of [an] invention must be co-extensive with the monopoly',[252] and its express implication for products as follows: 'If there is any method of manufacture or form of the product

[248] [2008] EWCA Civ 311, [33].

[249] *Aerotel Ltd v Telco Holdings Ltd* [2006] EWCA Civ 1371, [40]; *Symbian v Comptroller General of Patents* [2008] EWCA Civ 1066, [7].

[250] [2009] UKHL 12, [22].

[251] [1996] RPC 76, HL.

[252] ibid, 82.

which is part of the state of the art, then to that extent the invention is not new'.[253] In *Lundbeck*, a racemic form of escitalopram was part of the state of the art, but this did not deprive escitalopram of novelty, for when conceived correctly *qua* invention, that product was confined to its isolated form.[254]

Given this judicial conception of escitalopram, it is arguable that *Biogen* ought not to have been distinguished. The reason is that for patentability purposes, an isolated product is analogous to a recombinant product in its importation of a process element. The result is that the principles of *Biogen* ought to have applied, and Kitchin J's decision of insufficiency upheld. Implicit is a further criticism of the Courts for having inconsistently conceived escitalopram *qua* invention, namely, as an *isolated product* for novelty purposes, and as a *product as such* for sufficiency purposes.

Putting aside the nature of the invention claimed, and whether it was a product or an isolated product, there remains the wider implications of the judgments in the case for the UK law of inherent patentability. Those implications are ambivalent. On its face, the case resolves a longstanding tension between the granting of patents for chemical products, and the requirement that the scope of monopoly rights not exceed the disclosure in the specification. On the other hand, it proceeds from a conception of the product *qua* invention as existing independently of the method by which it was made. However, that conception can be questioned on philosophical and policy (as well as doctrinal) grounds. Further, even if it is accepted as correct, by affirming the need to conceive the subject matter of a claim *qua* invention within the relevant meaning of the PA,[255] the decision leaves scope for adopting a different approach in respect of different categories of patent claims, including product per se claims. In so doing, it leaves scope for paying more attention to the policy considerations in *Biogen*, by adopting an approach to claim validity that looks further beyond section 1(1) of the PA to the impact of the patent on research and competition in the field. Aside from its consistency with *Biogen*, and the decision of Pumfrey J in *Monsanto v Merck*, this approach has support in contemporary EPC law, pre-1977 patent jurisprudence, and the intent of the European drafters themselves, as shown by the analysis in chapters 3 and 4 above.

[253] ibid, 82–83.

[254] See [2007] EWHC 1040, [62]–[64]; [2008] EWCA Civ 311, [9]–[13]; [2009] UKHL 12, [6].

[255] Aside from the Courts' distinguishing of *Biogen* on account of its different type of claim, see [2009] UKHL 12, [22]–[25] per Lord Walker (emphasizing the need to approach product claims on a case-by-case basis given the 'extremely wide variety' of claims which they cover); [95] per Lord Neuberger (emphasizing that his comments applied only to 'a simple product claim such as the present' and 'in the absence of special factors').

4. Conclusion

Reading *Lundbeck*, and the other cases involving inherent patentability above, one is struck by the role played by a small number of Judges in developing contemporary patent law. For example, it was Lord Hoffmann who authored the *Biogen* judgment, and entrenched the principle from *EXXON/Fuel Oils* that in order for a claim to be a valid one it must correspond to the technical contribution to the art. It was also his Lordship who came down from the House of Lords in *Lundbeck* to clarify that principle in deference to the Act, and to the inherent patentability of chemical products. And finally, it was his Lordship who, in *Biogen* again, described inherent patentability as an 'academic' concept.[256] This last is challenged by the *Lundbeck* case, which underlines the centrality of the requirement of an invention to the law, and vindicates the view of Lord Mustill in *Biogen* that inherent patentability is not an 'academic' issue, but one sometimes requiring 'close conceptual analysis', particularly in the context of new technologies 'far distant from the mechanical and chemical inventions to which so much of traditional patent law relates'.[257] Influence of a similar order has been extended by other Judges, particularly in this context Aldous LJ, and after 2005 the late Pumfrey LJ.

E. Revisiting the Computer Programs Exclusion in Light of Harmonization: 2005 to Present

Whatever the case in the context of biotechnological products, Lord Hoffmann's description of the threshold requirement for an invention as 'invariably academic' is difficult to accept in the context of computer-related subject matter, if only on account of the amount of litigation involving section 1(2)(c) of the PA.[258] On the other hand, his Lordship's suggestion that the protection conferred by a patent is limited to the subject matter of a claim when conceived *qua* invention has support in the post-2005 computer-program cases. Of central importance among those cases are *Halliburton Energy Services Inc v Smith International Inc*[259] and *CFPH's Application*,[260] which involved separate judgments delivered on the same day in July 2005 by Pumfrey J and Mr Prescott QC respectively.

256 [1997] RPC 1, 41.

257 ibid, 31–32.

258 See *Shopalotto.com Ltd's Application* [2005] EWHC 2416, Pat, [6]–[7] per Pumfrey J; *Research in Motion UK Ltd v Inpro Licensing SARL* [2006] EWHC 70, Pat, [185] per Pumfrey J; cf *CFPH's Application* [2005] EWHC 1623, Pat, [9]–[10] per Mr Prescott QC.

259 [2005] EWHC 1623, Pat. The patent at issue related to drill bits for drilling in rock that were claimed (among other things) to reduce vibration.

260 [2005] EWHC 1589, Pat. The patent at issue related to a networked interactive wagering system that was claimed to reduce delays in the processing of information.

Diagram B Main cases regarding inherent patentability since 2005

Case	Alleged invention (main claims only)	Subsection 1(2) items	Decision	Result	Précis of Judicial Reasoning
CFPH's Application [2005] EWHC 1589, Pat See also **Halliburton Energy Services Inc v Smith International Inc** [2005] EWHC 1623, Pat	Networked interactive wagering system alleged (among other things) to reduce vibration	Computer program Business method	The advance of the art claimed by the patent is properly classified 'under the description' of a business method and not 'under the description' of technology.	Unpatentable	Mr Prescott QC: The source of UK law is Article 52(2) and (3) not section 1(2), and those paragraphs must be interpreted purposively, as discrete policy-based exclusions falling into two broad categories: 'soft' exclusions restricting protection, and 'hard' exclusions precluding protection. An invention is a subject matter the (novel and inventive) contribution of which can be classified 'under the description' of technology. The EPO and UK approaches are not consistent.
Shopalotto.com Ltd's Application [2005] EWHC 2416, Pat	Automated, Internet-based lottery game	Game	The subject matter makes no technical contribution to the art.	Unpatentable	Pumfrey J: Restating the principles from *Halliburton v Smith*, broadly consistent with the principles in *CFPH's Application*. An invention is not a subject matter having a technical effect, but rather a subject matter the contribution of which consists in substance of a technical contribution or effect.

(Cont'd)

Diagram B Main cases regarding inherent patentability since 2005 (*Cont'd*)

Case	Alleged invention (main claims only)	Subsection 1(2) items	Decision	Result	Précis of Judicial Reasoning
Aerotel Ltd v Telco Holdings [2006] EWCA Civ 1371	1. Telephone exchange system 2. Automated method of acquiring documents	1. Computer program Business method 2. Mental method Business method Computer program	1. The system comprises a new physical combination of hardware and is therefore more than just a method of doing business as such, and is technical in nature. 2. The contribution lies in the automation of an existing method, which is a contribution relating to a business method and which is not technical in nature.	1. Patentable 2. Unpatentable	CA (Jacob LJ): Adopting reasoning similar to that in *CFPH's Application* and *Halliburton v Smith* to support a 4-step ('structured') approach to identifying an invention, involving a) construing the claim, b) identifying the actual contribution made by the claim, c) asking whether the contribution falls solely within the excluded subject matter, and d) checking whether the actual or alleged contribution is technical in nature. Criticizing the EPO approaches as 'mutually inconsistent' and in one case intellectually dishonest, and requesting a referral of the issue to the EBA.
Symbian Ltd v Comptroller-General of Patents [2008] EWCA Civ 1066	Method of accessing data in a dynamic link library in a computer device	Computer program	The subject matter makes a technical contribution; it does not embody any of the excluded Article 52(2) items, improves the efficiency of a computer, and solves a technical problem within a computer.	Patentable	CA (Lord Neuberger): The EPO and UK cases can be reconciled as supporting a requirement for a subject matter having technical character. The *Aerotel* test is not to be followed blindly in every case. Harmonization requires consistent decision-making, achieved via dialogue and mutual compromise.

Generics Ltd v Lundbeck A/S [2009] EWHL 12	Isolated chemical compound	Inherently patentability not argued	The invention is the isolated product as such, capable of supporting a product patent.	Assumed: patentable as a product per se	CA (Lord Hoffmann, Jacob LJ): Determinations of sufficient support depend on the category of invention. Where the invention is a product, the claims will be sufficiently supported by one enabling disclosure, even though the resulting patent will protect more than that disclosure, because the contribution to the art represented by a product is the product as such. *Biogen* distinguished. HL: Agree with the CA.

1. Inherent Patentability as a Requirement for a (Potentially) Patentable *Invention*: *CFPH* and *Halliburton v Smith*

The reasoning in *CFPH* and *Halliburton* proceeded from the same critical starting position, supported with reference to section 130(7) of the PA, and the difference between the texts of section 1(2) of that Act and Article 52(2) and (3) of the EPC. That position was that the applicable source of UK law in respect of inherent patentability is Article 52(2) and (3), and not section 1(2).[261] Further, the significance of this was recognized as more than textual, due to the nature of the EPC as an international treaty created by a committee of people from different linguistic and legislative backgrounds, rather than of professional parliamentary draftsmen. According to Mr Prescott QC, in light of these considerations particularly, it is appropriate to adopt a teleological interpretation of the Article, rather than one based on the 'well-understood semantic conventions' of British legislation.[262]

The question of the *telos* of Article 52(1) and (2) thereby arose. According to both Judges, consistent with the approach of Laddie J in *Fujitsu*,[263] there is no single *telos* as such, as the exclusions are not 'members of the same logical class',[264] but rather discrete policy-based exclusions supporting potentially different degrees of application.[265] Further, in the Judges' analysis it was the TBA's failure to recognize this that accounted for its inconsistent decision-making in this area in the past.[266] Consistent with that analysis, in considering the meaning of Article 52(2) and (3), the Judges distinguished between their absolute and limited threshold exclusions. On their conception, the so-called hard exclusions[267] preclude protection for the relevant subject matter however conceived, while the soft exclusions restrict protection for subject matter conceived *qua* invention.[268] On the classification of the individual exclusions themselves, the Judges seem to have differed. According to Mr Prescott QC,[269] the discoveries and games exclusions were soft exclusions which restricted protection to that which could 'be made and supplied commercially'.[270] By contrast, the aesthetic creations, computer programs, presentations of information, and business methods exclusions were hard exclusions, that precluded protection for the subject matter however conceived, on the basis of

[261] See [2005] EWHC 1589, [22]–[23] (citing *Markem Corporation v Zipher Ltd* [2005] EWCA Civ 267, [94]); also [2005] EWHC 1623, [210].

[262] [2005] EWHC 1589, [26]–[27]; see also [2005] EWHC 1623, [212].

[263] [2005] EWHC 1589, [21].

[264] ibid, [28]; see also [2005] EWHC 1623, [212].

[265] See [2005] EWHC 1589, [30]–[31]; also [2005] EWHC 1623, [212].

[266] See [2005] EWHC 1623, [212]; also [2005] EWHC 1589, [43]–[46].

[267] The language of 'hard' and 'soft' exclusions is from Mr Prescott QC's judgment.

[268] See [2005] EWHC 1589, [33]; see also [2005] EWHC 1623, [212].

[269] See [2005] EWHC 1589, [30]–[42].

[270] ibid, [38].

their absolute unsuitability for a patent. According to Pumfrey J, on the other hand, the aesthetic creations, discoveries, mathematical methods, and scientific theories exclusions were hard exclusions that precluded protection either because the subject matter had alternative forms of protection, or because they had 'never been accepted as suitable subjects of monopolies on obvious, but different, policy grounds'.[271] By contrast, Article 52(2)(c) and (d) contained soft exclusions that restricted protection for no 'obvious underlying policy'.[272] When it came to applying those latter sub-paragraphs in the case before him, however, Pumfrey J found it unnecessary to attempt to derive such a policy in order to determine their proper scope of application, on the basis that that scope was in any case clear: 'the contribution the inventor makes must lie in a technical effect, and not merely in excluded subject matter'.[273] Drawing on the approach of the TBA in *IBM/Method for physical VLSI-chip design*,[274] he then identified the distinction required by those sub-paragraphs as being between claims which are, and are not, 'tethered' to a specific industrial activity.[275] In his judgment, if a claim is so tethered it will support a patent, but the scope of the protection conferred by the patent 'should be restricted to [the subject matter's] technical field'.[276] Implicit is a conception of the invention as a subject matter of industrial applicability directed to advancing the technical arts. Also implicit is a conception of the requirement for an invention as a requirement having two primary functions: to exclude certain categories of subject matter from the patent system, and in respect of all subject matter not so excluded, to restrict patentability and the scope of resulting protection to the subject matter when properly conceived *qua* invention. Mr Prescott QC clearly agreed, and expressed the latter implication as follows:[277]

> The legislation requires that, in order for something to be patentable, it must *be* an invention . . . By '*be* an invention' I mean not be excluded by Article 52(2). If something is an invention in that sense we can call it 'technology' for short. . . . It must also be the law that, for something to count as new and non-obvious, it must do so *under the description* 'invention', in that sense. It is not enough that something is new and not obvious if the sole novelty or inventive step consists of something that is excluded from basic patentability by Article 52 . . . Tomorrow's newspaper will be a manufactured article; it will certainly be a new article; and at least some of its contents will not be obvious. But the newspaper will not be patentable on mere account of the novel and non-obvious information it presents. For the news itself will not satisfy those conditions under the description 'invention', but only under the description 'excluded matter'. There must be unity under the four conditions.

271 [2005] EWHC 1623, [212].
272 ibid.
273 ibid, [215].
274 T_453/91 (*IBM/Method for physical VLSI-chip design*) (TBA, 31 May 1994).
275 [2005] EWHC 1623, [215]–[218].
276 ibid, [216].
277 [2005] EWHC 1589, [91]–[93].

In a further decision of 2005, *Shopalotto.com Ltd's Application*,[278] Pumfrey J repeated the central points from *Halliburton*. Specifically, he confirmed that the relevant source of law for the UK courts is Article 52(2) and (3) of the EPC,[279] that the EPO cases on those paragraphs are inconsistent,[280] and that it is 'not possible to provide an exhaustive definition of "invention"'[281] due partly to the absence of commonality among the Article 52 exclusions. Further, he emphasised that inherent patentability does not depend on a subject matter's 'technical effect', as recent EPO cases suggest, but rather on the *nature* and *extent* of that effect, and whether (in the context of a computer program) it sufficiently advances the technical arts.[282] Hence his support in that case for the following approach to identifying an invention:[283] 'First, determine what the inventor has contributed to the art . . . and, second, determine whether this contribution lies in excluded matter or, on the contrary, whether it consists [as a matter of substance] in a technical contribution or effect'. According to his judgment, the first of these steps does not require an assessment of the prior art, or a decision of novelty and inventive step, for the threshold question is not whether the patent *actually* contributes to the technical arts, but whether it *claims* to contribute to the technical arts—or in the language used in the discussion above, whether it is *directed* to advancing the technical arts—where such claim, by necessary implication, is not revealed on the face of the specification to be entirely without legal or factual foundation. This view was upheld in two subsequent High Court cases,[284] before being considered by the Court of Appeal in *Aerotel*.[285]

2. Requesting an Enlarged Board Decision on the EPC Requirement for an Invention: *Aerotel v Telco* in the Court of Appeal

Aerotel involved appeals in respect of two different patents for a telephone exchange system and an automated method of acquiring certain documents respectively. In a judgment written by Jacob LJ, the Court supported a new approach to determining inherent patentability based on an analysis that was broadly consistent with that of Pumfrey J and Mr Prescott QC above. Specifically, his Lordship confirmed that Article 52(2) and (3) ought to apply in place of section 1(2),[286] and be interpreted in accordance with the teleological principles contained in

[278] [2005] EWHC 2416, Pat.
[279] ibid, [5].
[280] ibid, [8].
[281] ibid, [5].
[282] ibid, [8]; see also [9]–[10].
[283] ibid, [9]–[10].
[284] See *Crawford's Application* [2005] EWHC 2417, Pat; *Research in Motion UK Ltd v Inpro Licensing SARL* [2006] EWHC 70, Pat.
[285] [2006] EWCA Civ 1371.
[286] ibid, [6].

Articles 31 to 33 of the Vienna Convention.[287] Further, he agreed that the difficulties raised by Article 52(2) and (3) derive from their absence of any 'evident underlying purpose', and the impossibility of 'form[ing] an overall approach to [their] categories' or linking them by means of a uniform concept, including a uniform policy basis.[288] And finally, he described the approaches of the EPO in respect of those paragraphs as 'mutually contradictory',[289] and one approach as being 'not intellectually honest'.[290]

With respect to the method for identifying an invention, Jacob LJ accepted the four-stage test proposed to the Court by the Comptroller-General of Patents.[291] According to that test, to establish an invention the Court must construe the claim, identify the actual contribution made by the claim, ask whether the contribution falls solely within the excluded subject matter, and, in a step described as mandated by *Merrill Lynch*, check whether the actual or alleged contribution is technical in nature. In his Lordship's judgment, this approach was a 're-formulation in a different order of the *Merrill Lynch* test'.[292]

Of immediate note in respect of this 're-formulation' is the exclusive focus of its second step on the 'actual [as distinct from the alleged] contribution'. On its face, this distinguishes the *Aerotel* approach from that of *CFPH* and *Halliburton*, by suggesting a need to examine the prior art in order to establish whether an invention exists; a view supported by Jacob LJ's following remarks:[293]

> Mr Birss [the Comptroller] added the words 'or alleged contribution' in his formulation of the second step. That will do at the application stage – where the Office must generally perforce accept what the inventor says is his contribution. It cannot actually be conclusive, however. If an inventor claims a computer when programmed with his new program, it will not assist him if he alleges wrongly that he has invented the computer itself, even if he specifies all the detailed elements of a computer in his claim. In the end the test must be what contribution has actually been made, not what the inventor says he has made.

It is not clear, however, why his Lordship took this approach, rather than adding the rider suggested above, namely, that an 'alleged contribution' under step two might be challenged by an opponent as falsely made, ie as revealed on the face of the specification—or indeed by a prior art search itself—to be without legal or factual foundation. While such a rider would raise evidentiary issues, such issues ought not to preclude recognition of the principle that all that is required to establish an invention is to show that the subject matter defined in the claims depends for its entitlement to patent protection on its proper conception *qua* invention.

[287] Vienna Convention on the Law of Treaties (Vienna, 22 May 1969, 8 ILM 679), discussed at pages 110–111; [2006] EWCA Civ 1371, [10].
[288] [2006] EWCA Civ 1371, [9].
[289] ibid, [25].
[290] ibid, [27].
[291] ibid, [40].
[292] ibid, [41].
[293] ibid, [44].

This view is consistent with that later taken by Mr Prescott QC in *IGT/Acres Gaming Inc's Application*, as follows:[294]

> After hearing argument in this case I wondered what is meant by the second paragraph I have quoted, namely paragraph 44 [reproduced at page 315]. Does it mean that the Patent Office is bound to accept the applicant's assertion (save in blatant cases)? Or can the Patent Office do a prior art search to find out what has the inventor really added to human knowledge? I therefore invited further submissions in writing.
>
> Although there was some disagreement, both parties accepted that the Patent Office is entitled to do a prior art search and that if it turns out that the alleged contribution was already known, or was obvious, there can hardly be a contribution to human knowledge. In my judgment that is correct. And there will be no patentable contribution to human knowledge if what is new and not obvious relates solely to a business method as such.

In effect, the principle supported by the Judge in this passage, and in *CFPH's Application* itself, is that expressed in the pre-1977 cases,[295] and supported by modern statutory definitions of the 'invention' itself.[296] A central implication of the principle, as Mr Prescott QC made clear, is the 'unity under the four [patentability] conditions',[297] and the *telos* of the European Patent Convention in its restriction of protection to *patentable inventions*, as distinct from subject matter that are inventions and (independently) novel etc.

An important aspect of the *Aerotel* judgment is the criticism which Jacob LJ made of the EPO jurisprudence regarding Article 52(2) and (3) of the EPC. According to his Lordship, that jurisprudence supports six mutually inconsistent approaches to the exclusions contained in the Article.[298] The first is a 'contribution approach,' focused on whether the inventive step resides only in the contribution of excluded matter. The second is a 'technical effect approach,' focused on whether the invention as defined in the claim makes a technical contribution to the known art. The third is an 'any hardware approach,' focused on whether the claim involves the use of, or is, a piece of physical hardware, however mundane, such as a computer or pencil and paper. And the fourth, fifth, and sixth are variants of this latter approach: the *PBS PARTNERSHIP/Controlling Pension Benefits System*[299] approach, whereby a method consisting of an excluded category is not an invention, but an apparatus implementing the method is; the *HITACHI/Auction Method*[300] approach, whereby a method implemented by an apparatus is an invention, even if it consists of an excluded category; and the *MICROSOFT/Data*

[294] [2008] EWHC 568, Pat, [23]–[24].

[295] See, eg *Cooper's Application* (1901) 19 RPC 53, AG, discussed at pages 73–75. The subject matter at issue in this case, a newspaper, was similar to the hypothetical relied on by Mr Prescott QC to explain his reasoning (see [2005] EWHC 1589, [23]).

[296] See Patents, Designs, and Trade Marks Act 1883, s 46; page 71.

[297] See [2005] EWHC 1589, [91]–[93].

[298] [2006] EWCA Civ 1371, [26]. For Jacob LJ's critique of the approaches, see ibid, [27]–[37].

[299] *PBS PARTNERSHIP/Controlling Pension Benefits System* [2002] EPOR 52.

[300] T_258/03 (*HITACHI/Auction Method*) [2004] OJ EPO 12/575.

Transfer[301] approach, whereby anything concrete is an invention. It was the last of these which his Lordship singled out as being intellectually dishonest. Further, in formulating the approach required by UK law as 'the technical effect approach with the rider . . . that a contribution which consists solely of excluded matter will not count as a technical contribution',[302] his Lordship emphasized its difference from each of the EPO approaches above.

It was on the basis of these criticisms of EPO jurisprudence that Jacob LJ requested the President of the EPO to refer the meaning of Article 52(2) and (3) of the EPC to the Enlarged Board of Appeal (EBA) for clarification. In his judgment, he presented that request in the following way:[303]

> It is formally no business of ours to define questions to be asked of an Enlarged Board of Appeal. What we say now is only put forward in case the President of the EPO thinks it helpful. If he thinks it pointless or arrogant of us to go this far, he is of course entirely free to ignore all we say. Nonetheless in the hope that there is a spirit of cooperation between national courts and the EPO we ventured to ask the parties what questions might be posed by the President of an Enlarged Board pursuant to Article 112. We have said the British Comptroller of Patents has encouraged us in this course. . . . The Comptroller and Mr Thorley provided a joint suggestion and Mr Macrossan helpfully his. Having considered the drafts, the questions which we think might be put are as follows: (1) What is the correct approach to adopt in determining whether an invention relates to subject matter that is excluded under Article 52? (2) How should those elements of a claim that relate to excluded subject matter be treated when assessing whether an invention is novel and inventive under Articles 54 and 56? (3) And specifically, (a) Is an operative computer program loaded onto a medium such as a chip or hard drive of a computer excluded by Article 52(2) unless it produces a technical effect, if so what is meant by 'technical effect'? (b) What are the key characteristics of the method of doing business exclusion?

A month after the judgment in *Aerotel* was given, the decision in *DUNS LICENSING ASSOCIATES/Estimating Sales Activity*[304] was published. In it, the opponent requested that the questions outlined above be referred by the Board to the EBA.[305] The TBA, however, refused the request. In its view, neither of the bases for referral under Article 112(1)(a) of the EPC was made out: a decision of the EBA

[301] T_424/03 (*MICROSOFT/Data transfer*) [2006] EPOR 39.

[302] [2006] EWCA Civ 1371, [38], [47]. The reason for his Lordship's support of this approach was that it was the same approach adopted by the Court of Appeal in *Merrill Lynch*, *Gale*, and *Fujitsu*, and therefore the approach required by precedent (see ibid, [38]).

[303] ibid, [75]–[76].

[304] T_154/04 (*DUNS LICENSING ASSOCIATES/Estimating Sales Activity*) [2007] EPOR 38, discussed at pages 238–246.

[305] The appellant also submitted the following three questions: (4) Is a system for estimating product distribution for non-reporting outlets based on weighing factors that are a function of the distance between the non-reporting outlets and sample reporting outlets and the characterization, eg size, of the non-reporting and reporting outlets of a technical nature? (5)(a) Are the exclusions of Art 52 to be treated differently from each other in the way that obviousness is assessed? (b) Are inventions alleged to be within Art 52 to have a different test for obviousness than other inventions not alleged to be within Art 52? See ibid, [V] (summary of Facts and Submissions).

was not required to ensure uniform application of the law, nor to resolve an important point of law arising in the case.[306] The mere fact of differing TBA decisions and/or EPO and national jurisprudence was not (it held) sufficient, notwithstanding considerations of harmonization, which did 'not exonerate a board of appeal from its duty as an independent judicial body to interpret and apply the European Patent Convention and to decide in last instance in patent granting matters'.[307]

It is clear from its judgment that the TBA in *Duns Licensing Associates* bristled at Jacob LJ's criticisms. Indeed, its response to those criticisms was to describe the jurisprudence of the Boards of Appeal as 'constant', and capable of succinct summary in the manner discussed in chapter 5 above,[308] namely, as requiring that a subject matter be shown to have technical features, whether or not they predominate over its non-technical features, and without regard to considerations of the prior art. Returning Jacob LJ's charge, it described the UK approach in *Aerotel* itself as *mala fides*: not 'consistent with a good-faith interpretation of the European Patent Convention'.[309]

Despite this, in October 2008 Jacob LJ was vindicated in his request when, nearly two years after *Duns Licensing Associates*, the President of the EPO herself responded, by referring the matter to the EBA for clarification.[310] At the time of writing, the decision of the Board has not been delivered.

3. Retreating from *Aerotel v Telco* (and *CFPH* and *Halliburton v Smith*) in the Interests of Harmonization: *Symbian v Comptroller-General* in the Court of Appeal

Soon after *Aerotel* was decided, Pumfrey J delivered a further judgment involving Article 52(2) and (3), in *Cappelini & Bloomberg's Applications*.[311] Applying the *Aerotel* structured approach, the Judge agreed with the submission of Counsel that determining whether the contribution made by a claim falls solely within an Article 52(2) exclusion (step three) will generally be easier than determining whether the actual or alleged contribution is technical in nature (step four) for the reasons he had explained in *Shopalotto.com*,[312] namely, 'that not all "technical effects" are relevant', and in particular, 'that where computer programs are under consideration, the fact that a computer with an *ex hypothesi* new program will be a

306 [2008] OJ EPO 2/46, [2].

307 ibid, [3].

308 See at page 238; ibid, [5].

309 [2007] EPOR 38, [70].

310 See G_3/08 (*PRESIDENT'S REFERENCE/Computer Program Exclusion*) [2009] EPOR 9. This case will be the first of an Enlarged Board on the meaning of Art 52(2) and (3).

311 [2007] EWHC 476, Pat.

312 [2005] EWHC 2416, Pat, discussed at page 314.

new machine and so a technical effect' is not sufficient.[313] According to Pumfrey J, his suggestion in *Shopalotto.com* that 'the technical effect to be identified had to be a technical effect over and above that to be expected from the mere loading of a program into a computer' survived the decision in *Aerotel*, and was thus still valid in UK law;[314] a view later affirmed in *Oneida Indian Nation's Application*.[315]

However, when the Court of Appeal revisited the issue in *Symbian v Comptroller-General*,[316] it suggested a different approach from that of either Pumfrey J or the Court of Appeal in *Aerotel*. In its view—which is plainly inconsistent with the judicial exchange above, and Jacob LJ's concern to differentiate the UK from the EPO approaches—the cases involving Article 52(2) and (3) can all be reconciled by conflating steps three and four of the *Aerotel* test, and treating them as synonymous with the *Duns Licensing Associates* test.[317] Pursuant to that approach, the question under Article 52(2) and (3) is whether the contribution made by a subject matter can be characterized as having a 'technical' nature. While conceding that this begs the meaning of 'technical', the Court of Appeal in *Symbian* regarded it as appropriate nonetheless for its reconciliation of the conflicting UK/EPO jurisprudence.[318]

Such reconciliation, however, involved a test expressed at such a high level of generality as to be unhelpful in resolving the parties' dispute. In the submissions of the Comptroller, Article 52(2)(c) had the effect of excluding from patentability any computer program not producing a novel effect outside the computer. In the submissions of the applicant, on the other hand, the provision had the more limited effect of excluding programs not providing a technical solution to a technical problem.[319] According to the Court of Appeal, either submission could be accepted on legislative, policy, and pragmatic grounds.[320] Given this, it decided to proceed in the following way. First, it considered its constitutional obligations, and the role of precedent in the context of the EPC.[321] In so doing, it accepted the view of Jacob LJ in *Actavis UK Ltd v Merck & Co Inc*[322] that the UK Court of Appeal is free to depart from a previous decision if satisfied that the TBA has formed a settled view on the relevant point. Noting the consistency of that view with the recognition by the House of Lords (in *Merrell Dow Pharmaceuticals Inc v Norton & Co Ltd*[323] and

313 [2007] EWHC 476, [5]; see [2005] EWHC 2416, [8].
314 [2007] EWHC 476, [5].
315 [2007] EWHC 954, Pat, [28].
316 [2008] EWCA Civ 1066.
317 ibid, [11]–[17]; cf ibid, [30].
318 ibid, [11]–[17].
319 ibid, [19].
320 ibid, [20].
321 ibid, [33]–[36].
322 [2008] EWCA Civ 444.
323 [1996] RPC 76, HL, 82.

Conor Medsystems Inc v Angiotech Pharmaceuticals Inc[324]) of the desirability of consistent UK/EPO decision-making, it concluded as follows:[325]

> Given that there are decisions of this court and of the Board which relate to the ambit of art 52, the right basis for assessing that ambit in this court should be as follows. If the judgments in the Court of Appeal cases give tolerably clear guidance which would resolve the issue on this appeal, then we should follow that guidance, unless it is inconsistent with clear guidance from the Board, in which case we should follow the latter guidance unless satisfied that it is wrong.

Second, the Court considered the implications of these obligations for Article 52(2) and (3).[326] Specifically, it noted the inconsistency of the past decisions in the area, the Court of Appeal's judgment in *Aerotel*, and the three decisions of the EPO subsequent to *Aerotel* approving the *Microsoft* approach, which the *Aerotel* Court had singled out for particular criticism. It then asked whether, on account of those three decisions, the Court ought to depart from *Aerotel* and adopt *Microsoft* in its place. The Court decided that it ought not to do so in light of the following five considerations: (a) the absence of an EBA decision in the area, and the unsettled state of EPO law in general; (b) the inconsistencies between the post-*Aerotel* EPO decisions; (c) the effect of at least one of those decisions in depriving Article 52(2) of any meaning; (d) the concerns regarding EPO jurisprudence expressed by other EPC Contracting States' courts; and (e) the risk of UK law being thrown into disarray if the Court of Appeal were 'seen to depart too readily from its previous, carefully considered, approach'.[327] Having thus decided against adopting the *Microsoft* approach, the Court of Appeal proceeded to consider whether the application in question revealed a technical contribution, identifying as its 'guidance' on that question a mix of EPO and UK authorities. In the result, it committed itself to the following approach:[328]

> [A]t least in this court at this stage, we should try to follow previous authority, we should seek to steer a relatively unadventurous and uncontroversial course, and we should be particularly concerned to minimise complexity and uncertainty. These aims are not necessarily mutually consistent, but, on this occasion, we believe they are achievable, namely by following the analysis adopted in *Vicom* and the two *IBM Corp.* cases, and of the Court of Appeal in *Merrill Lynch* and *Gale*.

The Court also eschewed (as 'dangerous') the suggestion that 'there is a clear rule available to determine whether or not a program is excluded by art 52(2)(c)',[329] stating that each case needs to be decided on its facts, and affirming the decision of Patten J in the decision on appeal that the claimed invention in *Symbian* made a

324 [2008] UKHL 49, [3].
325 [2008] EWCA Civ 1066, [36].
326 ibid, [37]–[47].
327 ibid, [46].
328 ibid, [51].
329 ibid, [52].

technical contribution to the art.[330] It then reconciled that finding with the cases above, the structured approach in *Aerotel*, and the decision by the EPO Examining Division to grant a patent for the same invention. Finally, the Court of Appeal made a plea for dialogue and compromise in the EPO and national courts' determination of substantive patent law issues:[331]

It is, of course, inevitable that there will be cases where the EPO will grant patents in this field when UKIPO should not, at least so long as the view in *Pension Benefit* and *Hitachi* is applied by the Board and is not applied here. The fact that the two offices and their supervisory courts have their own responsibilities means that discrepancies, even in approach or principle, are occasionally inevitable. However, the fact that such discrepancies have been characterised as 'absurd' by Nicholls LJ, and the reasoning in [*Angiotech*] emphasise the strong desirability of the approaches and principles in the two offices marching together as far as possible. This means that there is a need for a two-way dialogue between national tribunals and the EPO, coupled with a degree of mutual compromise. More directly relevant to the present appeal, it means that, where there may be a difference of approach or of principle, one must try to minimise the consequent differences in terms of the outcome in particular patent cases.

4. Conclusion

Since *Symbian*, litigation involving Article 52(2)(c) and (3) of the EPC has continued unabated, in the UK courts as well as in the EPO. That it has done so reflects the courts' and Boards' failure in the years since *CFPH* and *Halliburton* to have clarified the law, including particularly the meaning of Article 52(2) and (3), notwithstanding the steps taken by the Judges in those cases to do so by starting precisely the 'dialogue' advocated in the *Symbian* case. The same was alluded to and described as 'depressing' by Lewison J in *AT&T Knowledge Ventures LP*.[332] After noting the effect of the reasoning of Lord Neuberger in *Symbian* in casting doubt over the *Aerotel* structured approach, and the further uncertainty introduced by his Lordship's ambivalent treatment of the language of 'technical', the Judge in that case returned to Pumfrey J's test in *Shopalotto.com*, affirming what is in any case apparent, namely, that that test offers the clearest guidance from a UK court to date on the proper conception of an invention under the EPC. According to that conception, an invention for European patentability purposes is a subject matter of industrial applicability directed to advancing the technical arts. Of note is the broad similarity of this conception with that of pre-1977 law; the differences being the focus on subject matter of industrial applicability rather than mechanical or chemical utility, and concern with the technical rather than the manufacturing, industrial, or practical arts. What remains, however, are the questions of what

330 ibid, [53].

331 ibid, [61].

332 [2009] EWHC 343, Pat, [15].

constitutes a 'technical art', and what are the implications of conceiving the 'invention' with reference to such an art for the conception of individual subject matter *qua* invention, and the scope of protection which they consequently attract?

In conclusion, the decision in *Symbian* represents a retreat from the dialogue regarding the meaning and basis of the law of inherent patentability that was begun in *CFPH* and *Halliburton*, and thus a regression in the development of the jurisprudence in this area. Further, by making it clear that any future dialogue must be geared towards harmonization—conducted within a framework of 'mutual compromise' and recognition of the need to 'march[ ] together as far as possible' and, where that is not possible, 'to minimise . . . differences in . . . the outcome in particular patent cases'[333]—the decision constrains the scope for principled policy and doctrinal analysis of the kind undertaken in those cases, and clearly envisaged by the UK parliamentary drafters,[334] with disappointing implications for contemporary patent law, and understandings of European harmonization.

[333] [2008] EWCA Civ 1066, [61].

[334] See, eg Hansard HL (series 5) vol 379, cols 1410–1411 (15 February 1977).

Conclusion

A. The Requirement for an Invention in Historical Context

When Alfred Daniell attributed the long association of the term 'invention' with patent law to the breadth of its early conception, and its value in having 'shortened expression and imported no theory',[1] he alluded to its nature as a blank legal slate on which the policy of the patent system could be written. In reflecting on that slate and what *has* been written, it is instructive to return to the origins of the system.

The origins of UK statutory patent law lie in section 6 of the Statute of Monopolies 1623.[2] That section sought to codify Elizabethan and common law patentability principles, by authorizing the grant of a patent for 'any manner of new manufactures . . . not contrary to the law, nor mischievous to the state by raising prices of commodities at home, or hurt of trade, or generally inconvenient'.

The language and context of this section, as an exception to a prohibition against monopoly grants, raise an issue central to contemporary law. That issue is the relationship between the invention on one hand, and novelty, inventive step and utility/susceptibility of industrial application[3] on the other; and the constraints imposed on judicial conceptions of the former by the text and context of legislation.

Under the European Patent Convention (EPC),[4] the basic subject matter of patent protection is the 'patentable invention', which is defined in Article 52(1) as an invention that is new, inventive, and susceptible of industrial application. As a result, and as the courts have consistently affirmed, it is possible to have an invention that is not new, inventive, and/or susceptible of industrial application.

Historically, however, the position was different. Before 1883, the courts distinguished the requirement for an invention from the requirements for novelty and utility, even as the statute ran them together in its requirement for a 'manner of *new* manufacture . . . not . . . generally inconvenient', and treated mechanical or chemical ingenuity as variously implicit in the requirement for each. Further, even after 1883 when all four requirements were given separate expression in the

[1] A Daniell, 'Inventions and Invention' (1899) 11 Juridical Rev 151, 151.

[2] 21 Jac I c 3.

[3] On the history and relationship of the concepts of utility and susceptibility of industrial application, see C Wadlow, 'Utility and Industrial Applicability' in T Takenaka (ed), *Patent Law and Theory: A Handbook of Contemporary Research* (Cheltenham: Edward Elgar, 2009) 355–382.

[4] Convention on the Grant of European Patents (European Patent Convention, as amended) (Munich, 5 October 1973, 13 ILM 268).

legislation,[5] the statutory definition of the 'invention' remained essentially unchanged.[6]

The immediate importance of this is two-fold. First, it underlines the fluidity in the relationship between the concepts of the invention on one hand, and of novelty, inventiveness, and industrial applicability/utility on the other. And second, it alerts us to the conceptual confusion introduced by post-1883 legislation, due to its definition of the 'invention' as any subject matter entitled to a patent under the Statute of Monopolies, notwithstanding that such entitlement depended on requirements which the legislation treated as independent of the requirement for an invention itself.

Given this, it is necessary when considering the history of legal conceptions of the invention to impose some stability on the object of our enquiry, to ensure that we are in fact studying a single concept, and neither several concepts nor the word 'invention' itself. In the current book I have sought to achieve this by understanding the 'invention' as a subject matter the constitutive properties of which are regarded at law to make it suitable for patent protection, whatever additional restrictions on patentability the law might impose. Thus conceived, an invention is an inherently patentable subject matter: a subject matter regarded at law to fall within the sphere of subject matter appropriate for protection by the patent system.

Importantly, this definition preserves the distinctions of Article 52(1), as any contemporary (EPC-compliant) definition must. However, it does not follow from these distinctions that the requirement for a patentable invention is a requirement for a subject matter that is an invention and new and inventive and susceptible of industrial application, which it is not. Rather, it is a requirement for an invention that is new and inventive and susceptible of industrial application; the invention thus constituting the lynch-pin of the patent system, to which the other patentability concepts all relate, and on which those other concepts fundamentally depend, consistent with the policy orientation of the definition of the 'invention' above. Further, while Article 52(1) precludes novelty, inventiveness, and susceptibility of industrial application from being constitutive elements of the invention, it does not preclude newness, ingenuity, or utility from being such elements, provided they are conceived differently from their contemporary (Convention) counterparts.

These points receive support from the history of legal conceptions of the invention in UK law prior to the ratification of the EPC. That history reveals the courts' long-standing recognition, first, of the need to conceive a subject matter *qua* invention when determining its novelty, inventiveness, and utility, and second, of the requirement for some element of newness, ingenuity, and utility as constitutive of the invention properly conceived, and the threshold exclusion of subject matter lacking such elements from the regulatory sphere of the patent system.

[5] See Patents and Designs Act 1932, s 3(d)–(g).

[6] See Patents, Designs, and Trade Marks Act 1883 (46 & 47 Vict c 57) s 46; Patents and Designs Act 1907, s 93; Patents Act 1949, s 101.

B. Legal Conceptions of the Invention Prior to the EPC 1973

1. The Invention at the Time of the Statute of Monopolies 1623

The earliest patents for inventions were granted under the Elizabethan patent policy of 1561, for the purpose of encouraging people to 'study and seke for the knowledge of . . . good engines and devyses',[7] and other things capable of supporting a new industry. Initially, the majority of patents granted were for methods and devices imported from the continent, including such non-manufacturing methods as teaching and printing shorthand,[8] drainage,[9] mining,[10] and conserving fuel.[11]

By a series of sixteenth century decisions, improvements on established devices were declared incapable of supporting a patent, even if commercially significant, on the grounds that 'it is easier to add than to invent',[12] and that allowing patents for an improvement would prejudice established industrial practice.[13] Those decisions survived the first common law statement of patentability, as encompassing 'any new trade . . . or any engine tending to the furtherance of a trade that never was used before: and that for the good of the realm'.[14]

In 1623, these principles were purportedly codified in the Statute of Monopolies. On its face, however, the conception of the 'invention' in that Statute is narrower than that suggested by the relevant decisions, due to its restriction to methods of *manufacture*. Nonetheless, Hindmarch writing in the 1840s supported it as a true codification, describing an invention in 1623 as an art of manufacture, not extending to the product resulting from its practice, and not including the practice of a trade itself.[15] A more accurate codification, however, would have conceived the invention as an ingenious method of working materials to produce a practical result in the industrial arts, ie as an industrial art, with 'art' in this context requiring

[7] See EW Hulme, 'The History of the Patent System under the Prerogative and at Common Law' (1896) 12 LQR 141, 142, 145 (reproducing the text of an early Elizabethan patent for the importation of a dredging machine, dated 26 May 1562).

[8] See, eg Patent No XLIV (dated 26 July 1588) 'to teach, print, and publish works in shorthand, reproduced in EW Hulme, 'The History of the Patent System under the Prerogative and at Common Law: A Sequel' (1900) 16 LQR 44, 49.

[9] See, eg Patent No XXVIII (dated 8 July 1573) to 'put into practice an instrument for land drainage, and for the stopping of breaches in dams', reproduced ibid, 46.

[10] See, eg Patent Nos IX (dated 10 October 1564) 'for mining in eight English Counties' and XXI (dated 20 April 1569) 'to use the arte of myninge', reproduced in Hulme, 'The History of the Patent System under the Prerogative and at Common Law', 147, 149.

[11] See, eg Patent No XVIII (dated 26 August 1567) 'to cut iron, save fuel and extract oil', reproduced ibid, 149.

[12] *Bircot's Case* (1573) Co 3rd Inst 181 (Ex Ch).

[13] *Matthey's Patent* (c1580) Web Pat Cas 5, 6–7, PC.

[14] *Darcy v Allin* (*Case of Monopolies*) (1602) 77 ER 1260, KB, 1263 (footnotes omitted).

[15] WM Hindmarch, *A Treatise on the Law Relating to Patent Privileges for the Sole Use of Inventions; and the Practice of Obtaining Letters Patents for Inventions: With an Appendix of Statutes, Rules, Forms, &c. &c.* (London: Stevens and Norton, 1846) 56–57.

an ingenious purposive human action on the material world. This conception does not include methods of teaching shorthand, but does include methods of drainage, mining, and fuel conservation. It also includes improvements to an existing art, recalling that in the sixteenth century such improvements were excluded from patentability not only for lack of ingenuity, but also for lack of utility, and the prohibition against patents for any subject matter the establishment of which would prejudice existing industrial practice.

2. The Invention in the Pre-Modern Era of Chemical and Mechanical Manufacture

More than a century after the Statute of Monopolies, the validity of patents for additions was recognized on the pragmatic ground that to conclude otherwise would go to repeal 'almost every patent that ever was granted'.[16] Subsequently that decision was affirmed on condition that the addition be 'material and new', and reflect an 'improvement of the trade'.[17] Consistent with the analysis above, these decisions are best read as reflecting a change in the view of utility itself, rather than in conceptions of the (patentable) invention.

In the late eighteenth century, the courts conflicted over the proper sphere of the patent system, and the subject matter for which a patent might validly be granted. Importantly, they also conflicted over the conception of such subject matter *qua* invention. In dispute was whether a patent might validly be granted for something other than a 'manufacture'—ie a product—and in the case of such a manufacture, whether the invention was the product or the method of its production.[18] What the courts *were* clear on in the late eighteenth and early nineteenth centuries was the need for a patentee to give full consideration for a patent, by disclosing accurately and completely the subject matter claimed. Thus in *Brunton v Hawkes*,[19] the Court declared that 'a patent granted by the Crown cannot extend beyond the consideration of the patent,'[20] with the result that that which was claimed could not exceed 'that, which being both matter of actual discovery and of useful discovery, is the only proper subject for the protection of a patent'.[21]

After 1830, a change in attitudes to inventions occurred, consistent with a declining hostility to patents in general. Following this, the courts conceived inventions for patentability purposes more expansively than before, not as 'manufactures', but rather as principles 'turned to account' through their 'direction to the business

[16] *Morris v Bramsom* (1776) G 311, NP, 311 (set of needles to be applied to a stocking frame for making eyelet-holes).

[17] *R v Arkwright* (1785) 1 Web Pat Cas 64, NP, 71.

[18] See *Boulton v Bull* (1795) 126 ER 651, CP.

[19] (1821) 106 ER 1034, KB.

[20] ibid, 1038 per Abbott CJ.

[21] ibid, 1039 per Best J, quoting *Hill v Thompson* (1817) 171 ER 367, NP.

of human life'.[22] It was accordingly the invention thus conceived that was protected by a patent, and the invention thus conceived that needed to be new and of (industrial) utility. The result was the recognition in *Losh v Hague*[23] of a threshold exclusion in respect of subject matter insufficiently distinguished in their constitutive (mechanical or chemical) principle to contribute to the learning of a practical field. In effect, this was a restatement of the requirement for ingenuity expressed in the sixteenth and eighteenth centuries, and reflected a conception of the invention as a mechanical or chemical art.

As in the sixteenth century, the recognition in *Losh v Hague* of a requirement for ingenuity was expressed as a threshold exclusion in respect of a category of subject matter. In the sixteenth century the category was improvements/additions, and in the nineteenth century it was analogous uses. This encouraged a conception of the invention in terms of *categories* of subject matter, rather than individual subject matter themselves, and enabled the requirement for ingenuity to be distinguished in cases involving categories perceived by the courts to merit a patent. One important such category in the nineteenth century was that of the so-called combination.[24]

In 1852, new patent legislation was introduced, including the first statutory definition of the 'invention'.[25] According to that definition, an invention was 'any Manner of new Manufacture the Subject of Letters Patent and Grant of Privilege within the Meaning of the Act of the Twenty-first Year of the Reign of King James the First, Chapter Three'—ie any subject matter entitled to a patent under the Statute of Monopolies. The effect of defining the 'invention' expressly in the legislation was to draw attention to the word 'invention', and the language of the Statute of Monopolies itself. Given this, it is not surprising that after 1852 the distinction between uses and combinations was affirmed with reference to the requirement for a subject matter of mechanical or chemical utility directed to advancing the industrial arts.[26]

Also in this period, the House of Lords upheld a decision of the Court of Appeal that an improved method of purifying gas could not support a patent, despite its use and commercial value, because of the inconvenience its establishment would cause to English workers. In the view of the Court, and reflecting pre-1623 concerns with preventing patents from causing 'general inconvenience', including by superseding manual labour,[27] 'no one has a right to prevent a workman from using

[22] *Househill Iron Co v Neilson* (1843) 1 Web Pat Cas 673, HL, 683–684; see also *Jupe v Pratt* (1837) 1 Web Pat Cas 145, Ex, 146.

[23] (1838) Web Pat Cas 202, NP.

[24] See *Crane v Price* (1842) Web Pat Cas 393, CP.

[25] The Patent Law Amendment Act 1852 (15 & 16 Vict c 83), s LV.

[26] See, eg *Harwood v Great Northern Railway Co* (1860) 121 ER 1044, QB; cases cited at page 67, nn 108–110.

[27] See *Morgan v Seaward* (1837) 150 ER 874, Ex, 881, discussed at page 100, n 195; also Hulme, 'The History of the Patent System under the Prerogative and at Common Law', 152 (citing the

care to keep his tools in the most efficient state' or 'from cleansing his vessels and throwing away the useless contents whenever he likes'.[28]

The history of legal conceptions of the invention from 1623 to 1883 is largely a history of the courts' attempt to adapt sixteenth and seventeenth century principles of patentability to accommodate the patenting of derivative chemical and mechanical subject matter perceived to be of sufficient public utility to justify the restraint on public freedoms which a patent entailed.[29] The result, by the nineteenth century, was a conception of the invention as a mechanical or chemical art, and a practice of conceiving individual subject matter *qua* invention with reference to their constitutive (mechanical or chemical) principle. Subject matter the principle of which was by definition not novel (such as analogous uses) were consequently not 'inventions'. Implicit was the idea that it was not merely the subject matter described in the patent application which needed to be novel, but rather that subject matter conceived *qua* invention, just as it was that subject matter conceived *qua* invention that a patent protected. Motivating this idea may have been a judicial desire to impose stability on the object of patent protection, so as to ensure proportionality between that which a patent protected, and that which it secured for the benefit of the public. The requirement for such proportionality was underlined in *Crossley v Potter*,[30] where the Court described as void any patent the object of which was not to benefit the world by its communication, but rather to obstruct by the general character of its claims.

Still, the courts' enforcement of this requirement was substantially affected by the reality of the patents in existence. Dissatisfaction regarding that reality came to a head in 1865, when concern over the 'multiplicity of monopolies' for (among other things) 'alleged inventions of a trivial character' and their 'obstruction of the progress and improvement of arts and manufactures'[31] prompted the first extensive review of the UK patent system, and its modernization.

3. The Invention in the Modern Era of 'New' Technologies

The vehicle for that modernization was the Patents, Designs, and Trade Marks Act 1883, which extended the Statute of Monopolies-based definition of the 'invention', to include 'an alleged invention',[32] consistent with the earlier judicial requirement for a subject matter directed to advancing a relevant art. In 1901,

Elizabethan concern with protecting existing industry as a reason for its unwillingness to grant patents for 'first-finders' or 'inventors').

[28] *Patterson v The Gas Light and Coke Co* (1875) LR 2 Ch D 812, CA, 834, aff'd (1876) 3 App Cas 239, HL.

[29] Note in this context the emphasis on public utility in the prescribed form of individual grants; see, eg Patent Law Amendment Act 1852, schedule.

[30] (1853) Macr 239, NP.

[31] United Kingdom, *Report of the Commissioners appointed to inquire into the Working of the Law relating to Letters Patent for Inventions* (C (1st series) 5974, 1865) v.

[32] Patents, Designs, and Trade Marks Act 1883, s 46.

nineteenth century interpretations of that definition were tested in a case involving a method of typographically arranging a newspaper page to enable it to be folded without disrupting the line of its text.[33] According to the Court, the method involved 'invention with reference to a manufacture', and was therefore an 'invention' within the meaning of the Act, namely, and on the Court's understanding, a method of producing a material product of a substantial character the 'alleged utility' of which was mechanical.[34] In effect this was the *Losh v Hague* conception, restricted from a mechanical or chemical art to a manufacturing art. In the decades after it was restricted further by a requirement that the material product be vendible,[35] and then relaxed somewhat by a definition of 'vendible product' to mean any article of manufacture.[36] The result prior to 1958 was a denial of protection (for lack of an invention) to business methods and other schemes, arrangements of aesthetic and informational content, methods of coloration, navigation, transference, extinguishing fires, and methods of treating the human body—the latter seemingly on social morality grounds, notwithstanding explicit denials of that.[37]

It was also in the early years of the 1883 Act that the question of a product *qua* invention was revisited. That occurred in *Kopp v Rosenwald Bros*,[38] where it was held that the phrase 'manner of new manufacture' imported a process element, such that products per se could not constitute inventions. This decision was consistent with the concerns of the day regarding obstructive claims, particularly in the chemical field. Indeed, in 1919 those concerns prompted a statutory restriction on the scope of protection for newly manufactured chemical products, to the products when produced by a specific (disclosed) method.[39]

Soon after this, the courts developed their own mechanism for dealing with obstructive patents. That mechanism, formulated in *Mullard Radio Valve Co Ltd v Philco Radio & Television Corp*,[40] was a requirement that the inventive step disclosed in a patent correspond to the monopoly claimed for protection, such that the invention and claim would 'equiparate'. In 1949 this common law requirement was supplemented by a statutory requirement that claims be 'fairly based' on the specification,[41] in the same year as the statutory restriction of chemical product protection was abolished, and the nature of such products as 'inventions' apparently affirmed.[42]

[33] *Cooper's Application* (1901) 19 RPC 53, AG.

[34] ibid, 54.

[35] See *G E C's Application* (1942) 60 RPC 1, PAT.

[36] See *The Cementation Co's Applications* (1945) 62 RPC 151, PAT.

[37] See, eg Hansard HL (series 5) vol 379, cols 290–291 (24 January 1977) (commenting on the PO rule of not supporting contraception with the grant of a patent, discussed at pages 88–90).

[38] (1900) 19 RPC 205, Ch.

[39] Patents and Designs Act 1907, s 38A(1), inserted by the Patents and Designs Act 1919, s 11(1).

[40] [1936] 2 All ER 920, HL.

[41] Patents Act 1949, s 4(4).

[42] See C Robinson, 'Patent Protection for Chemical Products in Canada, Great Britain and the United States' (1972) 2 IIC 139, 141. See further Committee of Experts, Reply to the Questionnaire

In the meantime, in Australia, the High Court was considering the English conception of the invention in the context of a method for treating soil to control weeds.[43] The Court rejected that conception, defining an invention instead as any 'proper subject of letters patent according to the principles which have been developed for the application of section 6 of the Statute of Monopolies'.[44] This was an almost literal reading of the statutory definition, and one which supported (the High Court said) a conception of the invention as any human action on the physical world producing an artificial end of practical utility, which it treated as synonymous with an end of economic significance.[45] According to the Court, this conception did not cover essentially non-economic subject matter, including perhaps methods of medical treatment, nor subject matter that lacked artefactuality, or were characterized by their informational or aesthetic content alone.[46]

The Court's focus in *NRDC* was on subject matter eligible for patent protection, and on formulating an expansive, technology-neutral definition of the same. Consistent with this, it affirmed that horticultural and agricultural processes were inventions, and by implication biotechnological processes as well. The question remained as to the nature of the invention in the case of such a process, which question the House of Lords considered in 1971.

That consideration occurred in *American Cyanamid Co (Dann's) Patent*,[47] involving a method of producing an antibiotic from a naturally occurring strain of a known micro-organism. The issue was the invention for which a patent might validly be granted, and which therefore needed to be sufficiently disclosed in the patent. According to the majority, it was only the method of isolating the strain, and not the resulting antibiotic itself. In an important judgment, Lord Diplock dissented, finding it to be the antibiotic as well, and citing in support three specific factors. The first was the value of antibiotics to society.[48] The second was the cost of 'searching for and finding hitherto unidentified strains of microorganisms' from which antibiotics might be produced.[49] And the third was the original purpose of the patent system, 'to provide a material inducement to persons to undertake whatever effort and expense might be needed to introduce into the United Kingdom

drawn up by the Bureau of the Committee of Experts of the Council of Europe from the point of view of legislation in the United Kingdom, EXP/Brev (53) 7 (17 March 1953) 6; Committee of Experts, Observations of the United Kingdom Delegation: Industrial Character, EXP/Brev (56) 3 (3 May 1956) 1–2; United Kingdom, *United Kingdom Patent Law: The Effects of the Strasbourg Convention of 1963* (Cmnd 2835, 1965) [33] (proposing alternative statutory definitions of inherent patentability, both of which covered chemical products). The same view was taken of the impact of the repeal in 1967 of the German equivalent of section 38A(1); see K Nastelski, 'Product Protection for Chemical Inventions in Germany' (1972) 3 IIC 267, 270.

43 *National Research Development Corp's Application* (*NRDC*) (1959) 102 CLR 252, HCA.

44 ibid, 269.

45 ibid, 276–277.

46 ibid, 275.

47 [1971] RPC 425, HL.

48 ibid, 451.

49 ibid.

some useful new process or new product with a view to its becoming generally available for use or manufacture by the inhabitants of the realm'.[50] Of interest here is the implication that the conception of a subject matter *qua* invention depends in part on policy considerations, and the utility of the subject matter in question. In particular, one can read Lord Diplock as having used the requirement for an invention as a mechanism for orienting the system around its purpose, and that purpose itself as a mechanism for testing (and thereby ensuring) the technological neutrality of that system. On the other hand, whether it is ever scientifically or philosophically appropriate to conceive a method of isolating a strain of micro-organism *qua* product is another matter, as is the 'material inducement' which the 'social policy' of the patent system justifies conferring on inventors, and how that inducement differs (if at all) between practical arts.[51]

In the years after *Dann's Patent*, on the eve of the EPC's introduction, the courts moved towards the Australian view of the invention above, by revisiting the status of subject matter previously regarded as ineligible for patent protection. Their conclusions were: (a) that methods of medical treatment might be inventions, as methods of skill and effort capable of contributing to the learning of a practical art;[52] (b) that methods of altering the operability of a physical object might be inventions, when claimed in combination with the object itself;[53] and (c) that business methods were not inventions, unless automated and claimed as a programmed apparatus.[54]

Thus, by 1977, an invention in UK law was something close to a practical art. Important issues of uncertainty remained, however, including the inherent patentability of the subject matter in (a) to (c) above, and of chemical and other products as such. Of particular difficulty in the latter respect was the question first considered in *Boulton v Bull*: what was the invention in the case of such a product?

C. Concluding Reflections on the Requirement for an Invention in Patent Law

The question what is a product *qua* invention is an ontological one: what does it mean for a product to exist *qua* invention? Its answer depends on the nature of inventions, and the reason for requiring an invention at all.

A central theme in this book has been that, properly construed, the requirement for an invention serves two specific functions: first, it determines (and thereby

[50] ibid, 451–452.

[51] See also *Olin Mathieson Chemical Corp v Biorex Laboratories Ltd* [1970] RPC 157, Ch; *Kirin-Amgen Inc v Hoechst Marion Roussel Ltd* [2004] UKHL 46, [77].

[52] See *Schering AG's Application* [1971] RPC 337, PAT.

[53] See *Burroughs Corporation (Perkins') Application* [1974] RPC 147, PAT; *IBM's Application* [1980] FSR 564.

[54] See *IBM's Application* [1980] FSR 564.

limits) the subject matter to which the patent system is capable of applying; and second, it ensures the restriction of patent protection to individual subject matter conceived *qua* invention. The first of these functions is generally accepted, and is explicit in the threshold (Article 52(1)) requirement for an invention. The second, by contrast, has been largely ignored, and at times seemingly outright rejected. Nonetheless, it is central to the logic of the patent system.[55] Among other things, this is apparent from the following three specific factors.

The first of those factors is the statutory restriction of patent protection to 'patentable inventions', as distinct from subject matter that are inventions and patentable;[56] noting the effect of the latter conception in creating a gap between the existence of an invention and the determination of its patentability through which individual subject matter can readily fall.[57] By way of comparison, an equivalent gap was closed in the literary copyright context in *Exxon v Exxon Insurance Corp*,[58] where the Court held that a literary work is not a 'work' that is 'literary', but rather a 'literary work'.[59]

The second factor is the possibility of dual patent and copyright protection for certain subject matter, which possibility presumes different conceptions of such subject matter, reflecting the different purposes of patent and copyright law. Again, this was recognized in the copyright context in *Autodesk Inc v Dyason (No 2)*[60] and *Data Access v Powerflex Services*,[61] where the necessity to conceive computer programs *qua* expressive works (in contrast to technical artefacts) for copyright infringement purposes was emphasized, along with the effect of not doing so, namely, excessive copyright protection, by reason of any part of a computer program being a (technologically) 'substantial' part.

And the third factor is the two statutory uses and definitions of the 'invention' in sections 1(2) and 125(1) of the PA, following Articles 52 and 69 of the EPC; the former to denote inherently patentable subject matter, and the latter to denote '[the invention] specified in the claim'.[62] It is submitted that these uses can only be reconciled if the 'invention' defined in section 125(1) is the subject matter specified in the claim when properly conceived *qua* 'invention' as defined in section 1/ Article 52. That submission is consistent with the view of Lord Hoffmann in *Kirin-Amgen Inc v Hoechst Marion Roussel Ltd*,[63] that the conception of the subject matter *qua* invention in that case needed to take account of the Article 52(2)

[55] T_154/04 *DUNS LICENSING ASSOCIATES/Estimating Sales Activity* [2007] EPOR 38.

[56] See EPC, Art 52(1).

[57] An example is escitalopram in *Generics Ltd v Lundbeck A/S* [2009] UKHL 12, discussed on the following page.

[58] [1981] 3 WLR 541, CA.

[59] ibid, 548 per Stephenson LJ, others agreeing.

[60] (1993) 176 CLR 300, HCA, [11].

[61] See (1999) 202 CLR 1, HCA, [80]–[84].

[62] PA, s 125(1). cf EPC, Art 60 ('The extent of the protection conferred by a European patent or a European patent application shall be determined by the terms of the claims').

[63] [2004] UKHL 46.

definition of the 'invention',[64] in addition to the 'social contract between the state and the inventor which underlies patent law'.[65] And as his Lordship went on in *Kirin-Amgen* to make clear, identifying the notional skilled addressee of the claimed subject matter, and situating the subject matter in its appropriate scientific and creative tradition, are central to the conception of any subject matter *qua* invention:[66]

In the case of a patent specification, the notional addressee is the person skilled in the art. He reads the specification with common general knowledge of the art, and the assumption that its purpose is to describe and demarcate an invention – 'a practical idea which the patentee has had for a new product or process' – and not to be a textbook in mathematics or chemistry or a shopping list of chemicals or hardware. It is this insight which lies at the heart of 'purposive construction'.

One can also find EPO support for this view, including in *DUNS LICENSING ASSOCIATES/Estimating Sales Activity*[67] itself. Finally, in addition to conforming with the logic of the system, and as the discussion above itself reveals, interpreting the requirement for an invention in the manner herein suggested explains the decisions of contemporary (EPO boards) and courts. For example, it explains the view of Lord Hoffmann on one hand that the threshold requirement for an invention will 'invariably be academic',[68] and his emphasis on the other on the importance of properly conceiving subject matter *qua* invention when determining issues of validity and scope.[69] Similarly, it explains the view of the Technical Board of Appeal (TBA) on one hand that the threshold requirement for an invention is a discrete requirement (for technical character) separate from the other patentability criteria, and its support on the other hand for restricting novelty and inventive step enquiries to the technical features of a claimed invention.[70] Further, it explains the decision in *Generics Ltd v Lundbeck A/S*,[71] and where the Courts fell into error in that case, even as they supported the analysis above by affirming the importance of not analogizing between different categories of product inventions on account of their essentially different nature;[72] a point again with a literary copyright counterpart.[73] Where the courts fell into error in *Lundbeck* was in conceiving escitalopram inconsistently *qua* invention, namely, as an *isolated product* for novelty purposes when showing that escitalopram was not anticipated by its racemate and as a *product as*

[64] ibid, [77].

[65] ibid.

[66] ibid, [33].

[67] [2007] EPOR 38; see also T_939/92 (*AGREVO/Triazole Herbicides*) [1996] EPOR 171; T_641/00 (*COMVIK/Two Identities*) [2004] EPOR 10.

[68] *Biogen Inc v Medeva plc* [1997] RPC 1, HL, 41.

[69] See *Kirin-Amgen* [2004] UKHL 46, [77].

[70] See [2007] EPOR 38, [5], [10].

[71] [2009] UKHL 12, affirming *Lundbeck A/S v Generics Ltd* [2008] EWCA Civ 311, overruling [2007] EWHC 1040, Pat.

[72] See especially [2009] UKHL 12, [22] per Lord Walker.

[73] See *Baigent & Leigh v Random House Group* [2007] EWCA Civ 24, [142].

such for sufficient support purposes when explaining why the principles of *Biogen Inc v Medeva plc,* based on the *Mullard Radio* rule, were distinguishable in the case.

Among other things, the decision in *Lundbeck* underlines the shift that has occurred in legal conceptions of the invention since the introduction of the EPC into UK law, and the effect of that shift in detaching the UK system from its normative origins. For example, it confirms the redundancy of the concept of the invention as an industrial art, protected for the limited and specific purpose of facilitating industrial (and through it social) progress. Indeed, Article 52(2) and (3) contain a list of subject matter that appear more as exceptions to a presumptive statutory right than as a definition of the boundaries of the patent system. The opacity of the Article 52(2) list is consistent with its origins in the pursuit of European and international harmonization, and the inattention of the EPC authors when considering its form to its substantive policy justification in the context of the European Patent Convention. Hence the normative vacuum into which Article 52(2) and (3) were introduced; which vacuum is at least partly to blame for the legalism and casuistry that, with some important exceptions, has dominated the approach of the boards and courts. Manifestations of that approach include statutory literalism of the type seen in *Lundbeck*, the courts' reliance on fact-specificity to avoid the need to reconcile decisions in individual cases and thereby mask inconsistent decision-making, and the acceptance of harmonization as the *raison d'etre* of the post-EPC patent system, limiting the role for principled policy debate. A striking example of the latter is the decision in *Symbian v Comptroller-General,*[74] where the Court of Appeal implied regret for the earlier attempt by Jacob LJ in *Aerotel Ltd v Telco Holdings*[75] to initiate a frank discussion with the EPO regarding the requirement for an invention. By implication of Neuberger LJ, that attempt was contrary to the spirit of harmonization, and 'the strong desirability of the approaches and principles in the [UK and European Patent] offices marching together as far as possible'.[76] With respect, however, the idea that the development of substantive patent law ought to be driven by a vision of patent offices 'marching together' is unfortunate, and also somewhat misleading, for between the UK courts and the EPO it is the courts alone which are charged with responsibility for ensuring harmonious law, with the result that it is the courts alone which can be expected to conform in cases of conflict, as the *Aerotel–Duns Licensing Associates–Symbian* exchange itself amply demonstrates.

The allusion here to section 130(7) of the Patents Act 1977[77] raises the issue above of the constraints imposed on judicial conceptions of the invention by the

[74] [2008] EWCA Civ 1066.

[75] [2006] EWCA Civ 1371.

[76] [2008] EWCA Civ 1066, [61].

[77] Declaring ss 1(1) to (4), 2 to 6, 14(3), (5), and (6), 37(5), 54, 60, 69, 72(1) and (2), 74(4), 82, 83, 100, and 125 of the PA to have been 'so framed as to have, as nearly as practicable, the same effects in the United Kingdom as the corresponding provisions of the European Patent Convention, the

text and context of legislation. The primary constraints are Article 52(2) and (3) of the EPC themselves. However, rather than resolving to a requirement for subject matter having 'technical character', as the TBA has consistently maintained, those paragraphs are best read on their face, namely, as an inclusive list of categories of subject matter not constituting inventions, and as a requirement that individual subject matter not excluded from threshold protection not be conceived *qua* invention with reference to the same. Thus, if a subject matter depends for its claim to patentability on its conception as a computer program, it is not capable of being a 'patentable invention'. There then remains the nature of the Article 52(1) requirement for an invention in general. That requirement ought to be conceived in the positive and purposive manner above, namely, as a requirement for a subject matter of a type that is inherently suitable for a patent, having regard to the purpose of the European patent system and the nature of the objects which it seeks to regulate, and a restriction of protection to such subject matter when properly conceived *qua* invention.

In considering the first of these matters—the purpose of the system, and its implications for legal conceptions of the invention—it is necessary to consider the constraints imposed by other international patent law treaties. Of central importance in this regard are the EC Biotech Directive[78] and the TRIPS Agreement,[79] the significance of which might be argued to be two-fold, as follows. First, the Biotech Directive supports a conception of the invention as a subject matter that 'respect[s] the dignity and integrity of the person',[80] underlining the ethical dimension of the (patentable) invention. And second, Article 27.1 of the TRIPS Agreement supports a conception of the invention as a mechanism for anchoring the system to its purpose, and that purpose itself as a mechanism for ensuring the technological neutrality of the system; the point being that the requirement for technological neutrality is a requirement for such differential treatment of individual subject matter as is necessary to ensure the system's aims are consistently met.

This leaves the question of the purpose of the patent system. In the UK, that purpose was historically to facilitate industrial growth by rewarding the creation of

Community Patent Convention and the Patent Co-operation Treaty have in the territories to which those Conventions apply'.

78 Council Directive (EC) 98/44 of the European Parliament and of the Council of 6 July 1998 on the legal protection of biotechnological inventions (Biotech Directive) [1998] OJ L213/13.

79 Agreement on Trade-Related Aspects of Intellectual Property Rights (TRIPS Agreement), Marrakesh Agreement Establishing the World Trade Organization, Annex 1C, Legal Instruments—Results of the Uruguay Round, vol 31 (15 April 1994, 33 ILM 81); approved in Council Decision (of 22 December 1994) concerning the conclusion on behalf of the European Community, as regards matters within its competence, of the agreements reached in the Uruguay Round multilateral negotiations (1986–1994) [1994] OJ L336/1.

80 Group of Advisers on the Ethical Implications of Biotechnology to the European Commission, Opinion No 8 (Ethical Aspects of Patenting Inventions Involving Elements of Human Origin), 25 September 1996, available at <http://ec.europa.eu/european_group_ethics/docs/opinion8_en.pdf>. See Biotech Directive, Arts 15, 16, 19.

subject matter of mechanical or chemical utility directed to advancing the industrial arts. Within the context of this purpose, the invention was conceived by various approaches.[81] In the history of UK patent law prior to 1977, various approaches to conceiving the invention have been adopted.[82] They include *formal* approaches, focused on the existence of subject matter possessing (or otherwise) certain formal characteristics; *philosophical* approaches, focused on the relationship between people and the material world; and *historical* approaches, focused on that which has historically been regarded to be an invention. On the other hand, the history also reveals points of consistency. One already alluded to is a requirement for some degree of ingenuity, and concern with supporting the industrial arts. While not amounting to a 'technical advance' (*technischer Fortschritt*) in the sense of pre-1977 German law, this requirement has nonetheless made the invention in UK law a linear and progressive concept. Another is the (related) requirement for human action, and the tension between conceptions of the invention focused on that requirement and conceptions of the invention as a physical artefact. And a third is the asserted ethical neutrality of the invention; a point which warrants further comment.[83]

Historically, patents could not validly be granted for subject matter the establishment of which would cause social detriment. The reason was the 'general inconvenience' proviso of section 6 of the Statute of Monopolies, which was applied to support the refusal of patents for subject matter that would interfere with existing industrial practices, including by superseding manual labour. This was the case notwithstanding that such a subject matter might advance the industrial interests of society, thereby furthering the prima facie policy objectives of the system. Unclear, however, was the extent of the proviso, and whether it was confined to industry-related detriment, or whether it supported a morality or ethics-based exception.

Before 1977, this question arose in the context of methods of medical treatment. Prior to 1971, the UK courts held such methods to be unpatentable, but denied that ethics were a determining factor; relying instead on the nature of inventions themselves as methods of new *manufacture*.[84] Then in 1971, the High Court of Australia, applying the principles of UK law, recognized that if methods of medical treatment were to be denied protection, it could only be on the basis of

[81] [1997] RPC 1. HL.

[82] On these approaches, see generally T Misa, 'Theories of Technological Change: Parameters and Purposes' (1992) 17 Science, Technology & Human Values 3–12.

[83] On conceptions of technology as ethically neutral, and of the (non-legal) history of technology as progressive and linear, see C Mitcham, *Thinking Through Technology: The Path Between Engineering and Philosophy* (Chicago: The University of Chicago Press, 1994). On the tension in non-legal conceptions of *techne* between act and artefact, see R Laudan, 'Natural Alliance or Forced Marriage? Changing Relations Between the Histories of Science and Technology' (1995) 36 Technology and Culture 517–530.

[84] See, eg *C & W's Application* (1914) 31 RPC 23, SG; *London Rubber Industries Ltd's Patent* [1968] RPC 31, PAT.

general inconvenience.[85] That decision seems to have been accepted by the UK courts as well,[86] and is interesting in its support for the denial of patents on social morality or ethical grounds, and in its support for a conception of the invention as having a moral or ethical component.

An important early decision in this context is *A & H's Application*.[87] That case involved a patent application for a contraceptive device that was rejected by the Comptroller under a legislative provision enabling the refusal of protection for inventions 'of which the use would . . . be contrary to law or morality'.[88] On appeal the decision was upheld, though without mention of the relevant legislative provision. Instead, the Court denied patentability on the ground that 'it would not be fitting to sell [the device] as being protected by Royal Letters Patent'.[89] It said:[90]

> I express no opinion as to whether the use of these articles is consistent with morality, because I am not aware that the law has laid down what the exact standards of morality are. I am a Court of Law, and not a Court of Morality. All I say is I think these are not articles for which, whether the specification be amended or not, the Crown can be expected to exercise its discretion by way of granting a patent. I therefore dismiss the Appeal.

With the introduction of the Patents Act, the issue of patents for medical methods was resolved by their express exclusion from patentability under the EPC; initially on grounds of deemed insusceptibility of industrial application,[91] and after the 2000 EPC revision, on (unstated) grounds of public policy.[92] Consistent with this, consideration of the question whether a subject matter might be denied patentability on social detriment grounds has since been confined to consideration of the express legislative exclusion from patentability of inventions contrary to morality or *ordre public*.[93] Still, the argument can be made that properly conceived, the invention in law has an ethical component.[94]

Of interest again in this context is the distinction between patent and copyright law, and in particular, between inventions and copyright (expressive) works. The reason for that interest is that it is difficult to imagine morality or social benefit being expressed as a constitutive element of the literary work, even on a view of

[85] See *Joos v Commissioner of Patents* (1972) 126 CLR 611, HCA.

[86] See, eg *Eli Lilly & Company's Application* [1975] RPC 438, PAT; *The Upjohn Company (Robert's) Application* [1977] RPC 94, PAT.

[87] (1927) 44 RPC 298, SG, discussed in the House of Lords during the passage of the PA (see Hansard HL (series 5) vol 379, cols 290–291 (24 January 1977)).

[88] See Patents, Designs, and Trade Marks Act 1883, s 86. See also Patents and Designs Act 1907, s 75, amended in 1932 (see Patents and Designs Act 1932, s 11).

[89] (1927) 44 RPC 298, 298.

[90] ibid.

[91] See EPC 1973, Art 52(d); PA 1977, s 4(2).

[92] See EPC, Art 53(c); PA, s 4A(1).

[93] See EPC, Art 53(a); PA, s 1(3).

[94] The argument made in chapter 5 draws on the view of the Group of Advisers on the Ethical Implications of Biotechnology to the European Commission that the requirement for an invention must be maintained in order to safeguard the dignity and integrity of the person, which view arguably receives support in the provisions of the Biotech Directive.

'literary work' as a term of legal art defined to further the objectives of the copyright system. But in the patent context the position is less clear. Indeed, the fact that there is a parallel debate in the philosophy of science regarding whether technology is properly conceived as having an ethical component suggests that even if 'invention' is not regarded as a term of legal art, it might be plausibly conceived as having an ethical component. Among other things, such a conception would be consistent with the historical (seventeenth century) connection between the ideas of the practical arts and of social progress.[95] It would also suggest that the copyright work and patentable invention are fundamentally different in their constitutive aspects.

In other respects, however, I submit that they are (or ought to be regarded as) the same. Perhaps most importantly, the history of a subject matter's individual production ought to be regarded as part of what constitutes the subject matter *qua* invention, as I have elsewhere argued for the copyright work.[96] This also leads to my conclusion, that of the four approaches described above, the most appropriate is a combination of the philosophical and historical ones.[97] Thus conceived, an invention may be defined as *a purposive human method of working upon the physical world to produce an objectively discernible (material) result directed to advancing the industrial arts*. In the focus on purposive human methods directed to advancing the industrial arts, this definition rejects formalistic conceptions of the invention, and the view of Lord Hoffmann in *Kirin-Amgen* that inventions are not constituted (even) in part by the history of their individual creation.[98] On the other hand, by emphasizing the situatedness of inventions in a particular human inventive tradition, as well as in the material and social worlds, it is consistent with the understanding of the relationship between sections 1 and 125 above of the Patents Act, the *Kirin-Amgen* approach to claim construction, and the possibility which I have also suggested that the invention in law has an ethical component. As my discussion and the excerpt from *Kirin-Amgen* make clear, this understanding raises complex sociological and methodological issues regarding the identification and conception of the relevant tradition, and the proper perception of a subject matter *qua* invention in light of, among other things, the same.[99] Further study of those issues should prove rewarding.

[95] See at page 8.

[96] See J Pila, 'An Intentional View of the Copyright Work' (2008) 71 MLR 535–558; J Pila, 'Copyright and its Categories of Original Works' (2010) OJLS (forthcoming).

[97] Cf the historical conception of the invention rejected by the German Bundesgerichtshof (Federal Supreme Court) in *Ex parte Schreiner* (Red Dove/*Rote Taube*) [1969] Neue Juristische Wochenschrift 1713, 1713, reported in English in (1970) 1 IIC 136, 137.

[98] See [2004] UKHL 46, [88] per Lord Hoffmann, stating that 'the history of how a product was made is not an attribute which it carries around . . . It [is] still the same product, even if made in a different way'.

[99] See, eg, *Ex parte Schreiner* (Red Dove/*Rote Taube*) [1969] Neue Juristische Wochenschrift 1713, 1713; (1970) 1 IIC 136, 138 ('In this connection it is again of little importance what the legislature in 1877 considered to be "technology," but rather how the biological phenomena and forces are to be understood and classified within the present state of science.')

Index

abstract subject matter
immaterial goods, producing 197–8, 208
improvements of insubstantial nature 72
inherent patentability, concept of 43
physical and vendible product, requirement for 76
practical subject matter, distinguished from 72
principles 44–5, 72
technical character 219–20
utility 72
accurate and complete representation of the invention requirement 41–3
adaptations 45, 270, 273
aesthetic works 206–7, 210, 212, 312–13
agricultural processes and products 90, 95, 183, 184, 330
algorithms 263–4, 267, 269–70, 274, 280
alleged inventions 71–90, 315–16, 328–9
analogous uses 45–6, 66–7, 72, 270–1
animals
animal varieties 118–19, 148–9, 163, 195–6
treatment, methods of 148, 155–6, 190
antibiotics 95–7, 330–1
apparatus 203–4, 218, 270, 316–17
Article 52(2) and (3) EPC, inherent patentability under
absolute exclusions 312
abstract methods producing an immaterial good 197–8, 208
aesthetic works 206–7, 208, 312–3
animal varieties 195–6
'as such' proviso 151–3, 160–1, 163, 206, 233–4, 248–9
automated business methods 235–46, 247–8
automated transactional schemes 233–46
biological processes 195–6
Biotech Directive 224–8
business methods 197–9, 208, 235–6, 247–8, 312–13
computer programs 196, 199–217, 228–35, 246–8, 312–22, 334–5
contribution theory 211–17, 246–7, 316–17
definition of the invention 332–3
discoveries 312–13
drafting 195–6
European Patent Convention 146, 191–2
European Patent Office, interpretation by 195–250
games 312
harmonization 195–6, 224–49, 334
industrial application 196, 208, 313, 323, 324
interpretation 195–250, 312, 314–18
inventiveness 209, 324
liberalization 246
limited exclusions 312
Luxembourg and Munich Conferences 147–56, 159–63
mathematical methods 313
methods of medical treatment 195–6
novelty 209, 324
Patent Cooperation Treaty 195, 196, 249
Patents Act 1977 332–5
plant varieties 195–6
presentations of information 234–5, 312–13
public policy 196, 313, 324, 333–4
purposive interpretation 198
revision 231–33, 240–1
soft exclusions 312–13
Strasbourg Convention 195, 196
subject matter, different approaches to different categories of 195–209
substance and form of claims 208
Technical Board of Appeal 196–224, 246–10, 312
technical character 196, 209, 210–24, 232–3, 246–10
technical effect 313–14
technical information 231–46
technical means 208–9, 246
technical progress 196
teleological interpretation 312, 314–15
threshold requirement 209, 247–8
travaux préparatoires 248–9
TRIPS 196, 228–33, 248–50
utility 209
Vienna Convention on the Law of Treaties 315
whole contents theory 211–15
artistic content, arrangements of 74
'as such' provisions 151–3, 160–1, 163, 206, 233–4, 254, 257–60

automated methods 235–48, 271–74, 281–82, 314–15, 331
automated transactional schemes 214, 218–20, 233–46

Banks Committee 180, 182–6, 187–90
biological subject matter 79–81, 90–1, 119, 183, 191, 195–6
Biotech Directive 224–8, 335
biotechnology
agricultural products 95
antibiotics 95–7, 330–31
Biotech Directive 224–8, 335
definition of the invention 298–9
descriptions of 295–6, 301
disclosure 205, 301–3, 330–31
discoveries 294, 297, 300
DNA 221–24, 294–6, 301–302
equiparating claims 97–8, 105–6, 295
exclusions 95
fair basis 295, 296–7, 301
horticultural products 95, 330
immorality 222–4
inventiveness 297
legalistic conceptions of the invention requirement 220–24
living and natural phenomena 220–22
natural conditions of living phenomena, modification of 95
new technologies, modern era of 90, 95–8, 330
non-discriminatory approach 96–7
novelty 297
Patents Act 1977 254–8, 262–3, 294–303, 333
policy 96–7, 223–4, 330
proportionate claims 97–8
purposive definition of the invention 96–7
social policy 96–7, 330
specifications 294–303
speculative claims 295, 296–8
subject matter, identification of 295–6
sufficient support in specifications 293–4, 298–303
technical character 302
technical contribution 300, 301
threshold requirement 303, 333
business methods
'as such' provisions 254, 257
automated methods 235–46, 331
cerebral activities 199
computer programs 254–6, 276–7
economic advantages, conferring 198
economic transactions, facilitating 198
EPO Guidelines for Examiners 254–5
formalistic reasoning 247–8
function, characterised by 198
industrial application 257
inherent patentability, concept of 197–9, 208, 235–6, 247–8, 312–13
inventiveness 254, 256–7
new technologies, modern era of 104
novelty 256–7
prior art 254
schemes or plans 104
technical character 218, 219, 257–8
technical means, not described as requiring 198–9

cases, list of
1600 to 1851 30–40
1852 to 1882 48–65
contribution theory 216–17
examples of subject matter 30–40
inherent patentability,
2005, main cases after 309–11
2005, main cases before 288–93
précis of reasoning or comments of relevance 48–65
certainty 161–63, 181, 183, 186, 242, 253, 271–73
chemical products
absolute protection 176, 177
additions 326
agricultural products 95
antibiotics 95–7, 330–1
by process 164
comparative law 174–8
compulsory licensing 70, 82, 84, 88, 166–7
definition of the invention 25
disclosure 175, 305
discoveries 326
discrimination 96–7, 171, 172, 174–9
equiparating claims 97–8, 174
European Committee of Experts on Patents 131, 164–8, 175
European Patent Convention 164–75
exclusive rights 87
fair basis requirement 88, 97–8, 172–3, 329
German Association of Comparative Law Mannheim Conference 174–7
harmonization 85, 164–71
historical context 25–69, 82–8, 326–8
horticultural methods 90–1, 95, 330
improvements to 326
industrial application 165
inherent patentability, concept of 25, 82–3, 87–9, 171–2, 174
intermediary protection 174–5
inventiveness 85–7, 304–6
Luxembourg and Munich Conferences 168–71, 174
manner of new manufacture 82–4
Mannheim Conference 174–7

manufacture 25–69, 82–4, 326–8
monopoly rights 306–7
natural conditions of living phenomena, modification of 95
new technologies, modern era of 82–8, 95–8
non-discrimination 96–7, 174–9
novelty 85–6, 304, 307, 326, 327
obstructive claims 82, 84–7, 97–8, 329
Patents Act 1977 303–8
physical and vendible product, requirement for 25
policy 96–7, 164, 172, 330
prior art 304
process, products produced by 85–8, 105
proportionate claims 97–8, 172–3, 304, 327
purposive definition of the invention 96–7
social policy 96–7, 330
Statute of Monopolies section 6 25
statutory exclusion 84–7
Strasbourg Convention 167–70
sufficient support in the specification 304–7
technical contribution 304–6
technical progress 175
TRIPS 176–8
utility 85–7, 174–5, 326, 327
classical notion of the invention 244–6
combinations of two or more objects 45–7, 327
common law 18–22, 24, 26–7, 251–53, 323, 325
Community patent 121, 124–6, 186
complete and accurate representation of the invention requirement 41–3
compulsory licensing 82, 84, 88, 134–7, 166–7
computer programs
abstract schemes 214–15, 284
adaptations 270, 273
algorithms 263–4, 267, 269–70, 274, 280
alleged contributions 315–16
analogous use 270–1
apparatus, implementing 270, 316–17
Article 52(2) and (3) 196, 199–217, 228–35, 246–8, 312–22, 334–5
'as such' provisions 233–4
automated methods 233, 235–6, 241–44, 271–74, 281–82, 314–15
business methods 254–6, 276–7
collocations 270–1, 273–4
conservative approach of UK courts 272
contribution theory 211–12, 216–17, 241–42, 316–19
copyright 332
data handling systems 155
dedicated and non-dedicated products 273–4
definition of the invention 316
descriptions 276
directive, proposal for 241–44
discoveries 260–1, 268–9
European Patent Convention 191, 196, 199–217, 220, 228–35, 241–48, 263–4
European Patent Office 275–6, 282–3, 312–22
fact-specificity 271–87
hardware 200–201, 273–4, 316
harmonization 228–31, 308, 313–22
improvements 275
industrial application 200–202, 233–4, 263, 268, 280, 321
information-related subjects 101–4
instructions 266, 269–71
inventiveness 211, 267, 270
legal certainty 271–93
Luxembourg and Munich Conferences 148–9, 153–6, 160–1, 163
mathematical algorithms 263–4, 267, 269–70
mathematical methods 183, 199–202, 267, 270, 283–4
novelty 185, 211, 267–8, 270–73, 276–7, 286–7, 314, 319
Patent Cooperation Treaty 119
patentable inventions 334–5
Patents Act 1977 261–93, 308–22
physical form 199–202, 247–8, 266–9, 272, 275, 278–81, 284
policy 282–3, 287
practical and technical impact 280–81
prior art 203, 240, 262, 271, 276, 314
schemes and plans 104, 154, 160
specifications 203
substantial part 332
technical and non-technical means, mix of 208
technical apparatus, as part of 203–4
Technical Board of Appeal 199–206, 263–4, 275–6, 282–3, 317–18
technical character 211–15, 233, 240, 262–4, 269, 276–81, 287
technical contribution 283, 285–7, 320
technical effect 233, 314, 318–19
technical means 199–202, 208–9, 211, 220
technical problems, solutions to 205–6, 211, 220, 270
technical process 272–3
technical progress 267, 270–71, 277–8
threshold status 247
transactional schemes 214–15
TRIPS 228–33
United Kingdom law, inherent patentability in 252
utility 262, 273, 281, 284–5
whole contents theory 211–12

contract theory of patents 27, 41
contribution theory 211–13, 216–17, 228, 241, 288–93, 316–19
Cooper's Application 73–81
copyright 332–3, 337
custom 2, 13–21, 26

data processing 155, 242
definition of the invention 1–8
alleged inventions 71
Article 52(2) and (3) of EPC 332–3
automated business methods 239
biotechnology 298–9
common language meaning 2, 21–2
computer programs 316
custom 2, 13–14
EEC Patents Working Party 144
European Committee of Experts on Patents 130–1, 135–41
European Patent Convention 4, 5–7, 332–3
first definition of the invention in English statutory law 25
harmonization 5
historical and philosophical conception of the invention 2–4
inherent patentability, concept of 2–7
judicial law-making 3–4
list of non-inventions 4, 7
literal interpretations 72–3
negative definition 7
novelty 6, 71
ontology of the invention 7–8
origins of contemporary definition 5
Patent Cooperation Treaty 5, 118–20
patentable invention 24
Patents Act 1977 2, 4, 5, 180–4, 189–90, 316, 332–3
Patents, Designs and Trade Marks Act 1883 71, 328–9
physical product, requirement for 3
purposive definition of the invention 96–7
specifications 104–5
Statute of Monopolies section 6 1–3, 21–2, 47, 184, 324, 327–9
statutory interpretation 3–4, 72–3
Strasbourg Convention 141, 183, 84, 190
subject matter 7–8
Technical Board of Appeal 248–9
technical character 141
travaux préparatoires 119
uncertainty and controversy surrounding definition 5–7
descriptions 185–6, 276, 295–6, 301
disclosure 27, 29, 68, 175, 295, 302–5, 330–31

discoveries
actual and useful discoveries 42–3
analogous use 72
Article 52(2) and (3) of EPC 312–13
'as such' provisions 259
biotechnology 259–60
computer programs 260–61, 268–9
inventiveness 259, 260–61
Luxembourg and Munich Conferences 150, 151, 152–3
mechanical and chemical manufacture 326
novelty 258–9, 261
Patents Act 1977 258–61, 262–3
post-specification invention as idea or discovery 26–9
discrimination 84–5, 96–7, 124, 174–9
DNA 221–25, 294–6, 301–302
double uses 40–1

EEC Patents Working Party 126, 141–5 , 153, 159, 162–3
Elizabethan policy of monopoly grants 16–21, 26
abuse of policy by Crown 18–21
common law courts, devolution of jurisdiction to 19–21, 24
custom 17–21
established trades, at expense of 17–19, 23–4, 335–7
importation of foreign manufactures 17
odious monopolies 18–21
industrial growth, objective of 17
novelty 17, 20–1
procedure for obtaining patents 17–18
public concerns 17–19, 23–4, 335–7
renewals or extensions 18–19
Statute of Monopolies section 6 20–1, 323, 325
entitlement to a patent, establishing 1
EPC *see* **European Patent Convention**
ephemeral subject matter 79–81
equiparating claims 97–8, 174, 295, 329
European and Community Patent Court, proposal for 125
European Certificate of Invention 129
European Committee of Experts on Patents 1951–1963 129–41
chemical products 131, 164–8, 175
common features of exclusions 132–3
comparative study 130–1
compulsory licensing 131, 166–7
conditions of patentability 134–7
definition of the invention 130–1, 135–41
European Certificate of Invention 129
European Patent Convention 126, 129–41, 142–3

European system for granting patents, creation of 134
foodstuffs 131, 167–8
harmonization 129, 133, 140
horticultural products 167–8
industrial application 165
industrial character, meaning of 132, 134–6, 140–1
law, *ordre public* or morality, patents contrary to 131
list of non-inventions 137
Longchambon proposal 129–32
Paris Convention 132
pharmaceuticals 131, 167–8
plants 131
questionnaires 130–1, 132
Strasbourg Convention 126
technical character 163
technical progress 137, 140
technology, meaning of 133–4
threshold requirement for an invention 140–1
timeline 142–3
unification 129–36, 140

European Patent Convention *see also* **Article 52(2) and (3) of EPC, inherent patentability under**
'all fields of technology' 123
Article 9(2) 147–59
Article 10(b) 148–9
biological materials 191
Boards of Appeal 122
chemical products 164–75
computer programs 191, 199–211, 213–15, 220, 247–8, 263–4
context and background 109–28
criteria of patentability 109–11
definition of the invention 4, 5–7
diplomatic committees 126
EEC Patents Working Party 126, 141–5
equality, doctrine of 122–3
European Committee of Experts on Patents 126, 129–41, 142–3, 163
European context 112–26
European Patent Office (EPO), interpretation by 110, 192
European Patent Organisation (EPO) 122
Examining Division 122
expert committees 126
flexibility 192
grant, conditions for 122
harmonization 5, 191
Implementing Regulations 210, 225
information-related subject matter 191
inherent patentability, concept of 1, 5–6, 109–11, 191–2, 251–322
intention of drafters 126–74
international context 112–26
interpretation 109–92
list of non-inventions 192
Luxembourg and Munich Conferences 126, 145–64
manner of new manufacture 106
medical treatment, methods of 191
Munich Conference 122–4
Patent Cooperation Treaty 5
Patents Act 1977 106, 123, 179–80, 191–2
precedent 122–3
pre-grant phase 123
purpose 122
revisions 123
supremacy of EPC 122–3
technical character 196, 209, 210–24, 232–3, 246–50
threshold requirement 191
travaux préparatoires 110, 119, 126–7, 129, 211, 248–9
United Kingdom law, inherent patentability in 251–322
Vienna Convention on the Law of Treaties 110–11

European Patent Office *see also* **Technical Board of Appeal**
Article 52(2) and (3) of EPC 195–250
Biotech Directive 224–8
computer programs 275–6, 282–3, 312–22
Enlarged Board of Appeal (EPO) 317–19
European Patent Convention (EPC), interpretation of 110, 192
Guidelines for Examiners 210, 254–5
inherent patentability, concept of 5–6
Patents Act 1977 251, 275–6, 282–3, 312–22
technology 9

European patent system, proposal for 109–10, 134

European Union
barriers to trade 114, 116
Biotech Directive 224–8, 335
Committee of Coordination 116
Community Patent Convention 1975 (CPC) 121, 124
computer programs, proposal for directive on 241–44
contribution theory 241
Council of European Committee of Experts 116
directives 114, 241–44
EEC Patents Working Party 126, 141–5, 153, 159, 162–3
European Court of Justice 125
European Economic Community (EPC) 115–16

European Union (*cont.*)
EEC Patents Working Party 126, 141–5
European Patent Office 224–8
Haertel Working Party 116
harmonization 114, 126
Strasbourg Convention 116–17
TRIPS 9, 124
exclusive rights 87
experts *see* **European Committee of Experts on Patents**

fact-specificity 271–87
fair basis 84–8, 97–8, 172–3, 182, 295–7, 301, 329
fine arts 94
flexibility 159–60, 179, 181, 184, 189, 192, 253
foodstuffs 131, 167–8
foreign workers, encouragement of skilled 13–15, 24
formalistic reasoning 235–6, 247–8

general inconvenience 47, 66–7, 73, 99–101, 327–8, 336
German Association of Comparative Law
Mannheim Conference 174–7
guilds 13

harmonization
Article 52(2) and (3) of EPC 195–6, 224–49, 334
chemical products 85, 167–71
computer programs 208, 228–31, 313–22
European Committee of Experts on Patents 129, 133, 140
European Patent Convention (EPC) 5, 195–6, 224–49, 334
European Union 114, 126
Luxembourg and Munich Conferences 161
Patent Cooperation Treaty (PCT) 117–21
Patents Act 1977 179, 251
Strasbourg Convention 115–16
TRIPS 228–31
historical context 321–2 *see also* **Statute of Monopolies section 6**
1623, before 13–24
1623 to 1882 24–69
1833 to 1977 71–106
custom 2, 13–21, 26
definition of the invention 2–4
Elizabethan policy of monopoly grants 16–21, 26, 323, 325
function of patent system 8–9, 333–4, 337
guilds 13
Italian practices, influence of 15–16
letters of protection 14–15, 24
modern era of new technologies 71–106
monopolies 8, 13–21, 24
new industries, encouragement of 13–16, 21–4
origins of patents 13–24
post-specification invention as idea or discovery 26–9
pre-modern era 13–69, 326–8
United Kingdom patent law before 1977 335
horticultural methods 80–1, 90–5, 167–8, 183–4, 330

immorality 88–90, 131, 222–4, 336
importation of foreign manufactures 17
improvements 27–9, 42–6, 68–9, 72, 275, 325–7
industrial application
Article 52(2) and (3) of EPC 196, 208, 313, 323, 324
business methods 257
chemical products 165
computer programs 200–202, 233–4, 263, 268, 280, 321
EEC Patents Working Party 144, 153
European Committee of Experts on Patents 165
inherent patentability, concept of 1, 162
Luxembourg and Munich Conferences 149, 150, 152–3, 162, 164
Paris Convention 1883 162
Statute of Monopolies section 6 323
Strasbourg Convention 115
technical character 219–20
utility 323–4
industrial and social progress 8–9, 13–14, 21–3, 333–4, 337
industrial designs 182
industrial or mechanical process 233
information-related subjects 90, 101–4, 183, 191, 214
ingenuity requirement 23, 45–7, 327, 335
inherent patentability, concept of *see also* **Article 52(2) and (3) of EPC, inherent patentability under**
abstract inventions 43
cases, list of 288–93, 309–11
chemical products 82–3, 87–9, 171–2, 174
combinations of two or more objects 47, 327
common law 252–3
compulsory licensing 82
conjunctions of known objects 43
definition of the invention 2–7
discrete requirement, as 71
EEC Patents Working Party 144 162
ephemeral and biological subject matter 80–1

EPO Boards, practice of 5–6
European Patent Convention (EPC) 1, 5–6, 109–11, 191–2, 251–322
fine arts 94
horticultural methods 94–5
immorality 88–90
industrial application 1, 162
information–related subject matter 101–2
inventive step 1
literal interpretation 72
Luxembourg and Munich Conferences 145–6, 152–3, 159, 160, 162–4
mechanical and chemical manufacture 25
methods 43
non–economic subject matter 94, 330
novelty 1
Patent Cooperation Treaty (PCT) 119–20
Patents Act 1977 179, 180–92
physical inventions 43, 77, 90
presentations 103
schemes or plans 104
Statute of Monopolies section 6 43
subject matter, clear and concise 1
surgical and therapeutic methods 99–101
technology 9–10
uncertainty and controversy 5–7
United Kingdom 5–6, 251–322
uses, distinguished from 47
utility 43, 94

instructions 267, 269–70

inventiveness
automated business methods 238, 240–41
biotechnology 297
business methods 254, 256–7
chemical products 85–7, 304–6
computer programs 211, 267, 270
discoveries 259, 260–61
inherent patentability, concept of 1
novelty 323–4
obstructive patents 329
Patents Act 1977 264
Strasbourg Convention 115
utility 71

Italian practices, influence of 15–16

known mechanisms, application of 41, 45–6, 91, 196

legal certainty 159–61, 179, 181, 184, 242, 253, 271–3

letters of protection 14–15, 24

letters patents 17–18, 69, 91–4, 330

licensing 82, 84, 88, 134–7, 166–7

list of non–inventions 4, 7, 106, 137, 145, 148, 161, 192

literary or artistic content, arrangements of 74

living and natural phenomena 80–1, 93–5, 220–22

Luxembourg and Munich Conferences 126, 145–64
animal varieties 148–9, 163
Article 9(2) EPC 147–59
Article 10(b) EPC 148–9
Article 52(2) and (3) of EPC 147–56, 159–63
chemical products 168–71, 174
computer programs 148, 149, 153–6, 160–1, 163
diagram of issues and their resolution 157–8
discoveries 150, 151, 152–3
equiparating claims 174
European Patent Convention 122–4, 126, 145–64
exclusions 147–56, 159–63
flexibility 159–61
harmonization 161
industrial application 149, 150, 152–3, 162, 164
inherent patentability, concept of 145–6, 152–3, 159, 160, 162–4
legal certainty 159–61
list of non–inventions 148, 161
mathematical methods 149–51, 153, 163
mathematical theories 148–50, 153
natural products and information 150–1, 153
Paris Convention 153, 162
Patent Cooperation Treaty 148, 149, 150, 159–61
Patents Act 1977 189–90
plant varieties 148–9, 163
policy 163–4
presentations of information 148, 150, 156
schemes and plans 149, 154, 160
scientific theories 149–50, 151, 163
Strasbourg Convention 159–60
technical character 162–3
technical progress 162, 164
therapeutic methods 149, 152–3, 155, 160–1, 163–4
veterinarian treatment, methods of 148, 155–6

manifestly non–inventive (non–ingenious–subject) matter 45–6, 327

manner of new manufacture 3, 72–3, 76–84, 323

Mannheim Conference 174–7

manufacture
Boulton v Bull 29, 40–1
chemical products 25–69, 82–4, 326–8
Cooper's Application 73–81

manufacture (*cont.*)
ephemeral and biological subject matter 79–81
European Patent Convention 106
historical context 25–69, 326–8
horticultural methods 90–1
importation of foreign manufacture 17
manner of new manufacture 3, 72–84, 90–1, 106, 323
mechanical and chemical manufacture between 1623 and 1882 25–69, 326–8
medical treatment, methods of 78–9
natural conditions of living phenomena, modification of 80–1
Patents Act 1977 106
physical and vendible product, requirement for 41, 76–81
post-specification invention 29, 40–1
principles 40–1
process element 83
reference to a manufacture, invention with 29, 40–1, 73–81, 329
Statute of Monopolies section 6 3, 22, 72–3, 323
threshold exclusions from patentability 40
materiality requirement 28–9
mathematical algorithms 263–4, 267, 269–70
mathematical methods 149–53, 163, 181–3, 199–201, 267–70, 313
mathematical theories 148–50, 153
mechanical and chemical manufacture between 1623 and 1882 25–69, 326–8
mechanically inventive subject matter 43–6, 327
medical treatment, methods of
animals 148, 155–6
antibiotics 95–7, 330–31
Article 52(2) and (3) of EPC 195–6
commercial significance 99–100
European Patent Convention 191, 195–6
general inconvenience 336
human response 98–9
immorality 90, 336
manner of new manufacture 78–9
Patents Act 1977 337
physical and vendible product requirement 78–9, 90, 93
Statute of Monopolies section 6 99–101
surgical and therapeutic methods 98–101, 330
United Kingdom law, inherent patentability 336
mental methods 199, 202, 272–3, 276, 278–80, 283–6
microbiological processes 119
minimum standards of treatment 113, 115, 124
modern era of new technologies 1883 to 1977 71–106, 328–30
agricultural, horticultural and biotechnological subject matter, exclusion of 90, 330
alleged inventions 71–90, 328–9
biotechnological products 95–8, 330
business methods 104
chemical products 82–8, 95–8
computer programs, exclusion of 101–3
exclusions of non-manufacturing methods 101–4
immoral subject matter, exclusion of 88–90
information-related subject matter, exclusion of 90, 101–3
letters patent, invention as proper subject of 90–104
literal interpretations of statutory definition of the invention 72–3
manufacture, requirement for invention with reference to a 73–81, 329
new manufacture, manner of 72–90, 329
Patents, Designs and Trade Marks Act 1883 71, 328–9
presentation, methods of 103–4
schemes or plans 73–4, 90, 104
subject matter, expansion of 71
surgical and therapeutic methods, exclusion of 90, 98–101
utility 90–5, 329
vendible product, definition of 329
monopoly rights 1, 8, 13–21, 24, 28, 253–4, 306–7, 323 *see also* **Statute of Monopolies section 6**
***Mullard Radio* solutions** 84–7, 97–8
multilateral treaties by organization, overview of 112–13
Munich Conference *see* **Luxembourg and Munich Conferences**

national treatment 113, 115, 124
natural conditions of living phenomena, modification of 80–1, 93–5, 220–22
natural products 150–1, 153
new industries, encouragement of 13–16, 21–4
non-discrimination *see* **discrimination**
non-manufactured methods 103–4
novelty
accurate and complete representation of the invention requirement 42–3
Articles 52(2) and (3) of EPC 209, 324
automated business methods 238, 240
biotechnology 297
business methods 256–7
chemical products 85–6, 304, 307

combinations of two or more objects 46
computer programs 185, 211, 267–73, 276–7, 286–7, 314, 319
definition of the invention 6, 71
derivative forms 68
discoveries 258–9, 261
Elizabethan policy of monopoly grants 17, 20–1
improvements 28–9, 45–6
inherent patentability, concept of 1
inventiveness 323–4
mechanical and chemical manufacture 326, 328
origins of patents 68
Patents Act 1977 264
post–specification invention as idea or discovery 27, 28–9
presentations 103
Statute of Monopolies section 6 323
Strasbourg Convention 115
technical character 214

obstructive patents 69, 82, 84–7, 329
odious monopolies 18–21, 28
ontology of the invention 7–8
origins of patents for inventions and conceptions of the invention before 1623 13–24, 68

packaging 103
Paris Convention 1883 113–15, 124, 132, 162–3
Parker Committee 85
Patent Cooperation Treaty 1970 5, 117–21
animal varieties 118–19
Article 52(2) and (3) of EPC 195, 196, 249
biological processes 119
categories of subject matter 118
computer programs 119
entrenchment 159–60
European Patent Convention 5, 118–20, 195, 196, 249
harmonization 117–21
inherent patentability, concept of 119–20
legal certainty 159–60
Luxembourg and Munich Conferences 148, 149, 150, 159–61
microbiological processes 119
plant varieties 118–19
pre–grant procedures 117
procedures, harmonization 117–18
searches 117, 118–19
standards, harmonization of 117
surgery or therapy, methods of 119
patentable invention requirement 24, 331–32, 334–5

Patents Act 1977
agricultural processes 183, 184
Article 52(2) and (3) of EPC 332–5
Banks Committee 180, 182–6, 187–90
biological processes 183
biotechnology 294–303
business methods 254–8, 262–3
chemical products 303–8
common law 251–52
Community Patent Convention 186
computer programs 179, 183–4, 261–93, 308–22
conservatism 252, 272
definition of the invention 2, 4, 5, 180–1, 183–4, 189–90, 316, 332–3
descriptions and claims, correspondence between 185–6
discoveries 258–61, 262–3
European Patent Convention 106, 123, 179–90, 191–2
European Patent Office 251, 275–6, 282–3, 312–22
fact–specificity 271–87
fair basis 182
flexibility 179, 181, 184, 189
harmonization 179, 251
horticultural processes 183, 184
industrial designs 182
information–related subject matter 183
inherent patentability, concept of 179, 180–90
intention 179–90
interpretation 252–3
inventiveness 264
legal certainty 179, 181, 184
Luxembourg and Munich Conferences 189–90
manner of new manufacture 106
mathematical methods 183
medical methods 337
non–therapeutic methods 181–2
novelty 264
Patent Law Reform White Paper 180, 186–90
Patents Liaison Group 179–80, 180–2, 184, 187–90
policy 190, 265
printed sheets and analogous cases 182
schemes 181–2
section 1(2) 179–90, 191–2
Statute of Monopolies section 6 106
Strasbourg Convention 180–1, 183, 190
subject matter, different approaches to different categories of 253–65
sufficient support in specifications 294–5, 298–303
technical character 262–3
veterinary treatment 190
White Paper 180, 186–90

Patents, Designs and Trade Marks Act 1883 69, 71, 328–9
Patents Liaison Group 179–80
Patents Liaison Group report 180–2, 184, 187–90
pharmaceuticals 131, 167–8
philosophical conception of the invention 3–4, 335
physical products
alterations 331
computer programs 199–202, 247–8, 266–9, 272, 275, 278–81, 284
definition of the invention 3
information-related subjects 101–2
inherent patentability, concept of 43
vendible product, requirement for 25, 41, 76–81, 90–3, 329
plans *see* **schemes or plans**
plants
European Committee of Experts on Patents 131
horticultural methods 80–1, 90–5, 167–8, 183–4, 330
plant varieties 118–19, 148–9, 163, 195–6
policy
Article 52(2) and (3) of EPC 196, 313, 324, 333–4
Biotech Directive 225–7
biotechnology 96–7, 223–4, 300, 330
chemical products 96–7, 223–4, 330
computer programs 282–3, 287
Elizabethan policy of monopoly grants 16–21
Luxembourg and Munich Conferences 163–4
Patents Act 1977 190, 265
surgical and therapeutic methods 99–100
TRIPS 226
post-specification inventions 26–9, 40–1, 68
practical arts, failure to advance the 101, 335
precedent 122–3
presentations
Article 52(2) and (3) of EPC 234–5, 312–13
inherent patentability 103
Luxembourg and Munich Conferences 148, 150, 156
new technologies, modern era of 103–4
novelty 103
packaging 104
technical character 214
technical information 232–3
utility 103
principles 43–5, 72, 327
prior art 203, 240, 254, 262, 271, 276, 304, 314
processes
agriculture 183, 184
biological processes 119, 183
computer programs 272–3
EEC Patents Working Party 141
horticultural processes 80–1, 90–5, 167–8, 183–4, 330
manner of new manufacture 83
non-chemical products 87–8, 105
non-physical subject matter 43
programs *see* **computer programs**
proportionate claims 41, 97–8, 172–3, 304, 327
public concerns 17–19, 23–4, 335–7
public policy *see* **policy**

Royal Custom 13–21, 24

Sargant Committee 85–6
schemes or plans
business methods 104
computer programs 104, 154, 160
information-related subjects 102
inherent patentability, concept of 104
Luxembourg and Munich Conferences 149, 154, 160
new technologies, modern era of 73–4, 90, 104
Patents Act 1977 181–2
reference to a manufacture, invention with 73–4
technical character 218
scientific theories 149–50, 151, 163
social detriment 17–19, 23–4, 335–7
social policy 105–6
social progress 8–9, 333–4, 337
soft exclusions 312–13
software *see* **computer programs**
specifications
1830 to 1855 43–4
accurate and complete representation of invention requirement 41–3
biotechnology 294–303
computer programs 203
definition of the invention 104–5
discoveries 41–3
entitlement to a patent 1
equiparating claims 105–6
methods of production 41
post-specification invention as idea or discovery 26–9
reference to a manufacture, invention with 29, 40–1
relaxation of attitudes 43–4
sufficient support 294–5, 298–307
utility 41–3
speculative claims 295, 296–8

Statute of Monopolies section 6
Boulton v Bull 29, 40–1
codification 13–14, 325–6
common language meaning of invention 21–2
common law 21–3, 323, 325
conception of invention 13–14, 21–3
custom 13–14
definition of the invention 2–3, 47, 103–4, 184, 324, 327
Elizabethan principles 20–1, 323, 325
general inconvenience 47, 66–7, 73, 99–101, 327–8, 336
horticultural methods 91–4, 330
immorality 88–9
improvements 325–6
industrial and social progress 8
industrial application 323
ingenuity requirement 23
inherent patentability, concept of 43, 94
invention, meaning of 2–3, 47, 103–4, 184, 324, 327
letters patent, proper subject of 91–4
manner of new manufacture 3, 73, 323
mechanical and chemical manufacture 25
monopoly grants, prohibition of 323
new manufactures 3, 21–3, 73, 323
novelty 323
Patents Act 1977 106
post-specification invention as idea or discovery 26
reference to a manufacture, invention with 29, 40–1
renewal and reaffirmation of policy, as being 24
social impact of subject matter 23
utility 23, 326
Strasbourg Convention 1963
animal varieties 149
Article 52(2) and (3) EPC 195, 196
chemical products 167–70
definition of the invention 183, 184, 190
diplomatic committees 126
European Committee of Experts on patents 126
European Union 116–17
harmonization of substantive patent law 115–16
industrial application 115
inventive step 115
Luxembourg and Munich Conferences 159–60
minimum standards of treatment 114
novelty 115
Patents Act 1977 180–1, 183, 190
plant varieties 149
substance not form, patents construed with reference to 41
sufficient support in the specification 294–5, 298–307
surgical and therapeutic methods 90, 98–101, 105, 119, 330
Swann Committee 87–8

Technical Board of Appeal
206–7
aesthetic works 206–7
Article 52(2) and (3) of EPC 196–224, 246–50, 312
automated business methods 238, 239, 242
computer programs 199–206, 262–4, 269, 276–7, 287
definition of the invention 248–9
technical character 215, 218–20
technical information 233–46
TRIPS 228–33, 250
technical character
abstract subject matter 219–20
aesthetic subject matter 212
Article 52(2) and (3) of EPC 196, 209, 210–24, 232–3, 246–50
automated transactional schemes 215, 218–20
Biotech Directive 226
biotechnology 226, 302
business methods 218, 219, 257–8
computer programs 199–202, 208–15, 220, 233, 240
contribution theory 211–16
EEC Patents Working Party 162–3
European Committee of Experts on Patents 163
European Patent Convention 196, 209, 210–24, 232–3, 246–50
Guidelines for Examiners 210
Implementing Regulations 210
inconsistent conceptions of 215, 218–20
industrial application 219–20
informational subject matter 212–13
inventiveness 214
long-standing legal practice 210
Luxembourg and Munich Conferences 162–3
novelty 214
Patents Act 1977 265
presentations of information 214
schemes 218
Strasbourg Convention 141
technical and non-technical means, mix of 219
technical apparatus 218
Technical Board of Appeals (TBA) (EPC) 210–24

technical character (*cont.*)
technical means 215, 219–20
technical problem, solutions to 215, 220
travaux préparatoires 211
TRIPS 228, 240, 249–50
whole contents theory 211–15
technical progress
Article 52(2) and (3) of EPC 196
chemical products 175
computer programs 267, 270–71, 277–8
EEC Patents Working Party 141, 162
European Committee of Experts on Patents 137, 140
Luxembourg and Munich Conferences 162, 164
technology *see also* **technical character; technical progress**
contribution 283, 285–7, 300–301, 304–6, 320
definition 133–4
European Committee of Experts on Patents 133–4
European Patent Convention (EPC) 9
European Patent Office 9
information 231–46
inherent patentability, concept of 9–10
neutrality, principle of technological 248, 331, 335
problems, solutions to 205–6, 211, 215, 220, 270
role played by concept 9–10
technical and non–technical means, mix of 208, 219, 237–8
technical effect 233, 313–14, 318–19
technical means 198–9, 208–9, 219, 237–8
TRIPS 9
therapeutic methods
commercial significance 99–100, 105, 330
EEC Patents Working Party 144–5
Luxembourg and Munich Conferences 149, 152–6, 160–1, 163–4
medical treatment, methods of 98–101, 330
modern era of new technologies 90, 98–101
Patent Cooperation Treaty 119
threshold requirement
aesthetic works 206–7
Article 52(2) and (3) of EPC 209, 247–8
biotechnology 303, 333
computer programs 247
European Committee of Experts on Patents 140–1
European Patent Convention 191, 209, 247–8
general inconvenience 100
reference to a manufacture, invention with 40
Trade–Related Aspects of Intellectual Property Rights *see* **TRIPS Agreement**
***travaux préparatoires* of EPC** 110, 119, 126–7, 129, 211, 248–9
treaties and conventions *see also* **European Patent Convention; Paris Convention 1883; Patent Cooperation Treaty; Strasbourg Convention 1963**
Community Patent Convention 1975 121, 124, 186
multilateral treaties by organization, overview of 112–13
Vienna Convention on the Law of Treaties 110–11, 315
World Intellectual Property Organization, administration of treaties by 112–13
TRIPS Agreement
'all fields of technology' 124
Article 52(2) and (3) of EPC 196, 228–33, 248–50
automated business methods 240
Biotech Directive 226–7
chemical products 176–8
computer programs 228–33, 240
contribution theory 227
discrimination 124, 176–8
dispute resolution 124
European Union 9, 124
GATT 114
harmonization 228–31
minimum treatment 114, 124
national treatment 124
Paris Convention 114, 124
policy 228–9
prior art 228
Technical Board of Appeal 228–33, 248
technical character 228, 240, 248–50
technological neutrality, principle of 248, 335
technology 9
World Trade Organization 114
'turned to account', invention as principle 43–5, 327

Unified Patent Litigation System, agreement creating 125
United Kingdom law, inherent patentability after EPC in 251–322 *see also* **Patents Act 1977**
common law traditions 252–3
computer programs 252
conservatism 252–3
discoveries 258–65
flexibility 253
formal approach 335
historical approaches 335

legal certainty 253
medical treatment 336
monopoly grants, prohibition on 253–4
philosophical approaches 335
utility
abstract subject matter 72
alleged mechanical utility 74–6
Article 52(2) and (3) of EPC 211–15
chemical products 85–7, 174–5
combinations of two or more objects 46, 327
computer programs 262, 273, 281, 284–5
horticultural methods 94
industrial application 323–4
inherent patentability, concept of 94
inventiveness 71
letters patent, proper subject of 94
mechanical and chemical manufacture 326, 328
new technologies, modern era of 90–5, 329
post–specification invention as idea or discovery 26, 28
presentations of information 103
reference to a manufacture, invention with 74–5
specifications 42
Statute of Monopolies section 6 23, 326

vendible and physical products, requirement for 25, 41, 76–81, 90–3, 329
veterinary treatment 148, 155–6, 190
Vienna Convention on the Law of Treaties 110–11, 215

whole contents theory 211–15
working directions 72
World Intellectual Property Organization 112–14, 162 *see also* **Paris Convention 1883; Patent Cooperation Treaty 1970**
World Trade Organization 114